Engineering Statics

Engineering Statics

M. Rashad Islam
Colorado State University – Pueblo

Md Abdullah Al Faruque
Rochester Institute of Technology

Bahar Zoghi
State University of New York – Farmingdale

Sylvester A. Kalevela
Colorado State University – Pueblo

CRC Press is an imprint of the
Taylor & Francis Group, an informa business

First edition published 2021
by CRC Press
6000 Broken Sound Parkway NW, Suite 300, Boca Raton, FL 33487-2742

and by CRC Press
2 Park Square, Milton Park, Abingdon, Oxon, OX14 4RN

Library of Congress Cataloging-in-Publication Data

Names: Islam, M. Rashad, author. | Al Faruque, Md Abdullah, author. | Zoghi, Bahar, author. | Kalevela, Sylvester A., author.
Title: Engineering statics/M. Rashad Islam, Colorado State University Pueblo, Md Abdullah Al Faruque, Rochester Institute of Technology, Bahar Zoghi, State University of New York--Farmingdale, Sylvester A. Kalevela, Colorado State University Pueblo.
Description: First edition. | Boca Raton: CRC Press, 2021. | Includes bibliographical references and index.
Identifiers: LCCN 2020020323 (print) | LCCN 2020020324 (ebook) | ISBN 9780367561062 (hardback) | ISBN 9781003098157 (ebook)
Subjects: LCSH: Statics.
Classification: LCC TA351 .I85 2021 (print) | LCC TA351 (ebook) | DDC 620.1/053--dc23
LC record available at https://lccn.loc.gov/2020020323
LC ebook record available at https://lccn.loc.gov/2020020324

ISBN: 978-0-367-56106-2 (hbk)
ISBN: 978-1-003-09815-7 (ebk)

Typeset in Computer Modern font
by KnowledgeWorks Global Ltd.

Contents

Preface

This textbook is intended for engineering and applied engineering undergraduate students of subjects such as civil engineering, mechanical engineering, civil engineering technology, mechanical engineering technology, architecture, construction engineering, construction management, etc. It is also suitable for applied science students who are pursuing a two-year degree with or without the intention to transfer to four-year-degree programs. The cutting-edge statics theory required for engineering has been described using plain text, appropriate equations, figures and worked out examples. Examples are mostly taken from the real world and are solved using step-by-step explanations. The authors believe that this text will be helpful for engineering, applied engineering and applied science students to understand and apply statics to succeed in their degree programs and careers.

The authors would like to express their sincere gratitude to Rafid Shams Huq from Stamford University, Arifur Rahman from Khulna University of Engineering & Technology, Jarrett Schirmer from Colorado State University-Pueblo, for reviewing the book. The authors also thank their good friends who helped by providing photos. Students of Colorado State University, Pueblo, who helped to identify some issues during the classroom experiment are acknowledged as well.

The book has been thoroughly inspected with the help of professional editors to fix typos, editorial issues, and poor sentence structure. Despite this, if there is any issue, please excuse us and report it to islamunm@gmail.com. Any suggestion to improve this book or any issue reported will be fixed in the next edition with proper acknowledgment. Thank you.

The Authors

About the Authors

Dr. M. Rashad Islam is an Assistant Professor at Colorado State University–Pueblo and a licensed professional engineer. He is also an ABET Inc. program evaluator of civil engineering, architecture, and construction management programs. Dr. Islam has more than 100 publications, including textbooks, reference books, over 60 scholarly journal articles, several major technical reports, and conference papers. Dr. Islam's other major textbooks are Pavement Design – Materials, Analysis and Highways published by McGraw Hill, and Civil Engineering Materials – Introduction and Laboratory Testing published by CRC Press.

Dr. Md Abdullah Al Faruque is an Associate Professor in the Department of Civil Engineering Technology, Environmental Management and Safety of the College of Engineering Technology at the Rochester Institute of Technology, New York.

Dr. Bahar Zoghi is an Assistant Dean and an Associate Professor of Architecture and Construction at the State University of New York at Farmingdale.

Dr. Sylvester A. Kalevela is a Professor at Colorado State University–Pueblo. He is a licensed professional engineer and a member of the American Society of Civil Engineers, the American Society for Engineering Education, and a fellow of the Institute of Transportation Engineers.

1 Introduction

1.1 MECHANICS

Mechanics is a branch of science dealing with the state of bodies - at rest, in motion or resulting in deformation when acted upon by a force system. Upon applying a force system to a body, at least one of the following outcomes may occur: the body may move; if there is not adequate restraint, or it may not move but noticeably deform. Generally, if the deformation is noticeable, then it is discussed under *deformable-body mechanics* (very often called strength of materials or mechanics of materials). If the deformation is negligible or no movement occurs, then it is discussed under *rigid-body mechanics. Both are* divided into two areas, *statics* and *dynamics,* as shown in Figure 1.1. Statics deals with bodies at rest, and dynamics deals with bodies in motion. The branch of mechanics that deals with liquids is called *fluid mechanics.*

1.2 PRINCIPLES OF MECHANICS

Engineering mechanics is formulated on the basis of Newton's three laws of motions which are stated as follows:

First Law. An object at rest remains at rest unless acted upon by an unbalanced force. An object in motion continues in motion with the same speed and in the same direction unless acted upon by an unbalanced force.

Second Law. If an unbalanced force acts on a body, the body will experience an acceleration proportional to the magnitude of the unbalanced force and the same direction of the unbalanced force.

Third Law. For every action, there is an equal and opposite reaction. This statement means that for every force on an object there is a reaction force that is equal in magnitude, but opposite in direction or sense.

An overview of Newton's three laws of motion indicates that the second law provides the basis for the study of dynamics. The concept of engineering statics is largely dependent on the first and third laws of motion. For example, the concrete beam shown in Figure 1.2 has its self-weight which acts downward. Two vertical supports carry (or resist) this self-weight. As the self-weight is downward, the supports apply equal-upward reaction to resist this self-weight. Here, the downward self-weight and the upward vertical reaction are numerically equal and opposite.

Another example is shown in Figure 1.3 where a truss is being lifted up to place it at the top of a residential building under construction. The truss has its self-weight which acts downward. The cable attached to the truss carries this self-weight. The cable tension is equal to this self-weight. Here, the downward self-weight and the upward cable tension are numerically equal and opposite.

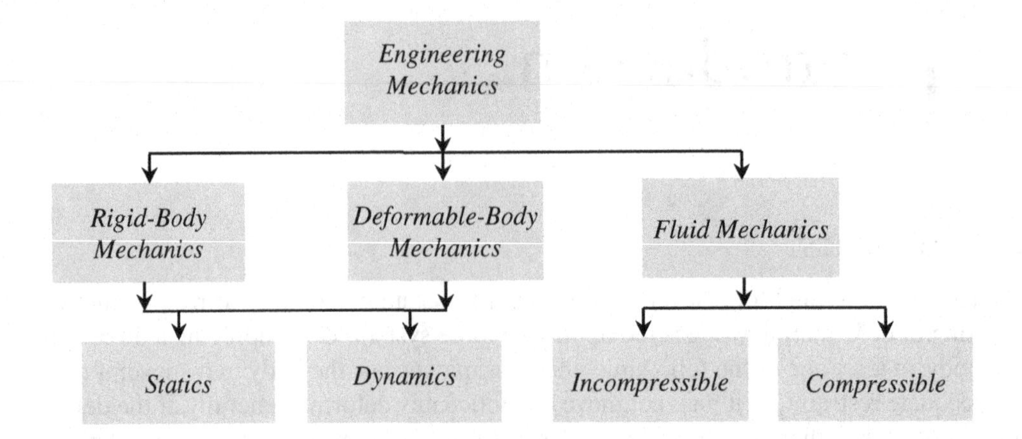

FIGURE 1.1 Branches of engineering mechanics.

FIGURE 1.2 An isolated beam with two supports. *(Photo by Armando Perez taken in Pueblo, Colorado)*

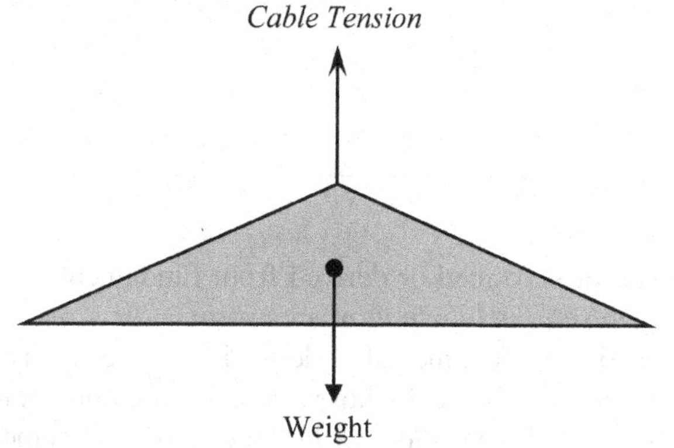

FIGURE 1.3 Lifting a truss during the construction of a building. *(Photo taken in Pueblo, Colorado)*

1.3 BASIC QUANTITIES

The three basic quantities used in mechanics are time, length, and mass.

Time means the duration spent on certain activities, action, etc. Time is naturally an absolute quantity and does not depend on any other outside object and proceeds uniformly at a fixed rate. The time interval between two events is equal for all observers.

Length is the linear measurement of an object. It is independent of observers, position, and time.

Mass is a measure of quantity of the matter in a body. It does not depend on the state of motion, place, and position.

1.4 BASICS OF UNITS

1.4.1 Types of Units

There are three types of units:

Fundamental Units. This unit does not depend on any other unit such as unit of length, mass or time.

TABLE 1.1
Prefixes

Multiple	Prefix	Symbol
10^{-18}	atto	a
10^{-15}	femto	f
10^{-12}	pico	p
10^{-9}	nano	n
10^{-6}	micro	μ
10^{-3}	milli	m
10^{-2}	centi	c
10^{-1}	deci	d
10^{1}	deka	da
10^{2}	hecto	h
10^{3}	kilo	k
10^{6}	mega	M
10^{9}	giga	G
10^{12}	tera	T
10^{15}	peta	P
10^{18}	exa	E

Derived Units. This unit is formed or derived from fundamental unit(s) such as the unit of force is kg.m/sec^2, and the unit of area is m^2.

Practical Units. Sometimes, fundamental or derived units are so large or so small that they are inconvenient for daily use. In those cases, sub-multiples or multiples are used as units such as km and micron. Table 1.1 lists the prefixes of some practical units.

1.4.2 Unit Systems

There are several types of unit systems. Two are very popular and discussed here:

US Customary System (USCS) Units or foot-pound-second *(FPS) units.* In the US, length, mass, and time are expressed in foot (ft), pound-mass (lbm) or slug, and second (sec), respectively. The unit of mass also uses slug (pound-mass) which is pound divided by the gravitational constant (g = 32.2 ft/sec^2). More clearly,

$$\text{Slug or lbm} = \frac{\text{lbf}}{\frac{\text{ft}}{\text{sec}^2}}$$

The unit of force is used as lbf or lb. Therefore, one must distinguish the pound-force (lbf) from the pound-mass (lbm). The pound-force is that force which accelerates one pound-mass at 32.2 ft/sec^2. Thus, 1 lbf = 32.2 lbm-ft/ sec^2. Again, to confirm, lbf is written as lb and lbm is written as slug. Another parameter to remember is the weight. It is the vertical downward attraction force on an object by gravity. Therefore, weight is a special form of force.

TABLE 1.2
Relationship Between SI Units and USCS Units

Quantity	SI units	Multiply by	To obtain USCS Units
Length	m	3.281	ft
Length	m	39.36	in.
Mass	kg	2.205	lb
Mass	kg	0.069	lbm or slug
Force	N or kg.m/sec^2	0.225	lb

System of International (SI) Units. SI units are very often used to maintain consistency worldwide. In this system, length, mass, and time are expressed in meter (m), kilogram (kg), and second (sec), respectively.

Table 1.2 can be used to convert SI to USCS units. Another confusing area is reading a value on the weighing scale or balance. It gives reading mostly in lb or kg. Remember that lb is the unit of force (or weight) and kg is the unit of mass (not weight).

Example 1.1:

Convert 25 kN.m into the appropriate USCS units.

SOLUTION

1 kN = 1,000 N
1N = 0.225 lb
1 m = 3.281 ft

$$25 \text{ kN.m}\left(1{,}000\ \frac{\text{N}}{\text{kN}}\right)\left(0.225\ \frac{\text{lb}}{\text{N}}\right)\left(3.281\ \frac{\text{ft}}{\text{m}}\right) = 18{,}456 \text{ lb.ft}$$

Answer: 18,456 lb.ft

Example 1.2:

Convert 2,500 lb.in. into the appropriate SI units.

SOLUTION

1 lb = 1/0.225 N = 4.44 N
1 m = 39.36 in.

$$2{,}500 \text{ lb.in.}\left(4.44\ \frac{\text{N}}{\text{lb}}\right)\left(\frac{1}{39.36}\ \frac{\text{m}}{\text{in.}}\right) = 282 \text{ N.m}$$

Answer: 282 N.m

Example 1.3:

Convert 750 lb.sec into the appropriate SI units.

SOLUTION

1 lb = 1/0.225 N = 4.44 N
750 lb.sec = 750 lb.sec (4.44 N/lb) = 3,330 N.sec = 3.33 kN.sec

Answer: 3.33 kN.sec

Note: to confirm again, the unit of force (or weight) in USCS units is lbf and is also written as lb. The unit of mass in USCS units is lbm and is also written as slug. The relationship between lbf and lbm is lbf = lbm x g. In SI unit, the units of force (or weight) and mass are N and kg, respectively.

1.5 ROUNDING OFF

Rounding off is a very confusing topic. Following basic customary rules for rounding off numbers is recommended to avoid confusion. Any one of the following three rules can be used.

- If the last digit (the digit to be removed) is greater than 5, round up
- If the last digit (the digit to be removed) is smaller than 5, round down
- If the last digit (the digit to be removed) is 5:
  - If the digit preceding the 5 is an even number, then the digit is not rounded up
  - If the digit preceding the 5 is an odd number, then the digit is rounded up

Now, how many digits are rounded? It depends on the number value, application, etc. To better understand this, consider the weight of gold versus the weight of a vehicle. We try to express the weight of gold in three to four digits after the decimal point if the unit is in gram or ounce (oz). For kilogram or lb, the decimal places may be even larger. For example, 1.4123 g of gold and 0.354267 lb of gold. However, the weight of a car is commonly expressed as 3,200 lb or 3,250 lb. Expressing the weight of a car as 3,237.25 lb looks non-practical. Another example is your age. If someone asks your age, you commonly answer as 21 or 23 years or so. You do not answer as 21 years, 5 months, 3 days, and 12 hours or so. However, to determine the seniority by age, if two persons are 21 years. Then, we may need to include the month or day or even to hour. Ideally, we should not round off in every single step of computation as it may accumulate errors. In mechanics, very often after the computation, the question of rounding off occurs. Students are advised to think logically. In this text, the authors round off using the first rule; if the last digit of a number is 5 or more, then the preceding digit can be increased by one.

In engineering analysis, where there are many steps of computation, rounding off is performed at the final stage of the calculation. Otherwise, rounding in every step of calculation may overestimates or underestimates the final results. For example, consider your test score in a class. Say, you scored 2.7% in all of the 10 short quizzes that your instructor gave you. Your total quiz score is $2.7 \times 10 = 27\%$. However, if your instructor rounded off your score in whole numbers in every quiz, then the total score is $3.0 \times 10 = 30\%$.

Example 1.4:

Round off the following numbers using two decimal places.

a. 2.365
b. 2.364
c. 2.367
d. 2.357

SOLUTION

a. 2.36 (If the digit preceding the 5 is an even number, then the digit is not rounded up. The digit, 6 will not change).
b. 2.36 (If the digit to be removed is smaller than 5, round down. The digit, 6 will not change)
c. 2.37 (If the last digit (the digit to be removed) is greater than 5, round up. The digit, 6 will round up to 7)
d. 2.36 (If the last digit (the digit to be removed) is greater than 5, round up. The digit, 5 will round up to 6)

1.6 SUPPORT TYPES

Knowledge of the support condition is required to understand and apply the equilibrium conditions to be discussed in the upcoming chapters. There are broadly five kinds of support as shown in Figure 1.4: roller, pin (or hinge), fixed, link and rocker. Roller support can resist translation/movement of a member in one direction. Pin support resists translation/movement in both orthogonal directions. Fixed support resists translation in both orthogonal directions and the rotation. Link support applies reaction force along the axis of the link. Examples of link support are cabling, a tie bar, a strut bar, etc. Rocker support applies a reaction force perpendicular to the surface at the point of contact. It is very similar to roller support.

Two beams supported by a column, shown in Figure 1.5, can be assumed to be roller support, as the beams can translate horizontally to some degree. Note that beams are

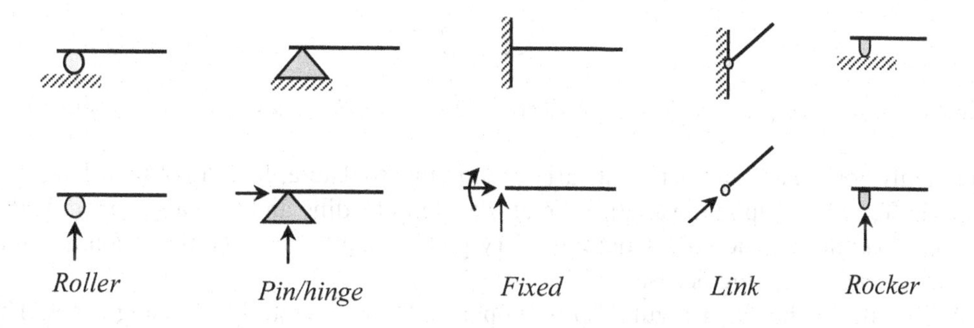

FIGURE 1.4 Support types.

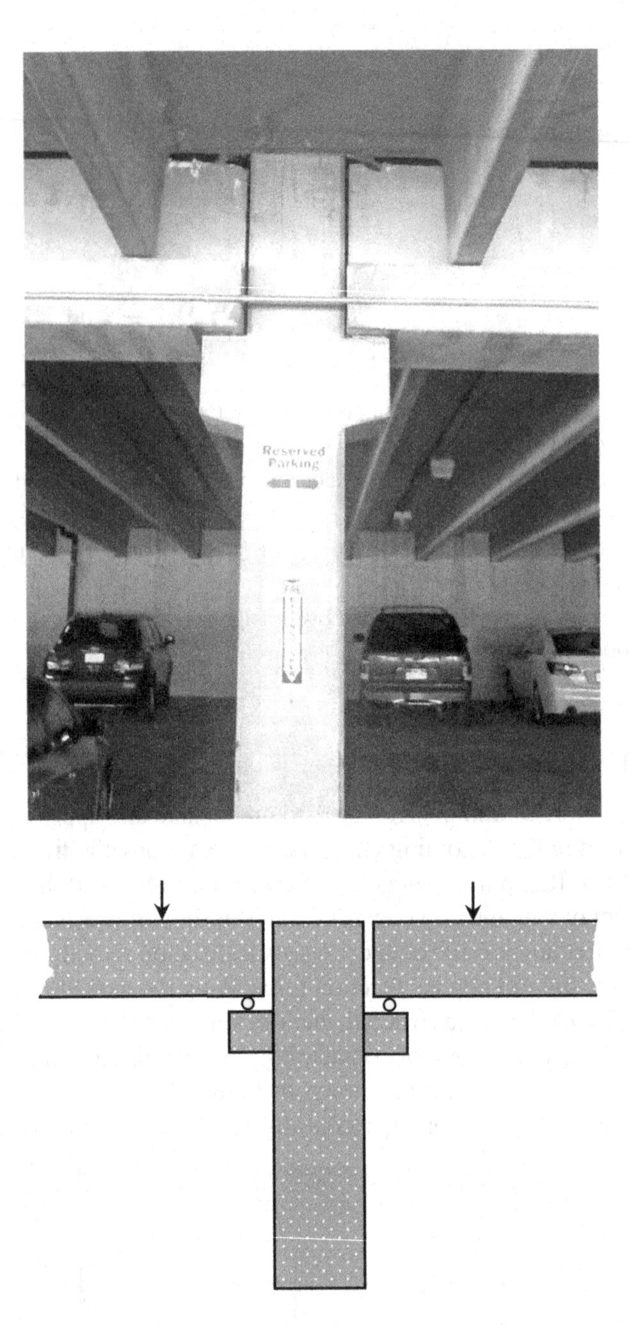

FIGURE 1.5 Roller support (beams can move horizontally a little). *(Photo taken in Colorado Springs)*

commonly horizontal members that carry predominantly lateral loading of their longitudinal axis. They bend upon the application of the lateral loading and may also carry axial or twisting loading occasionally. Columns carry predominantly axial compression and may also carry lateral loading occasionally.

A steel beam shown in Figure 1.6 is supported by a concrete block using a pin. This beam cannot move vertically or horizontally and can be considered a pin or hinge support.

The beam shown in Figure 1.7 is cast monolithically, the joint resists both horizontal and vertical movements and the rotation and thus, can be considered a fixed support.

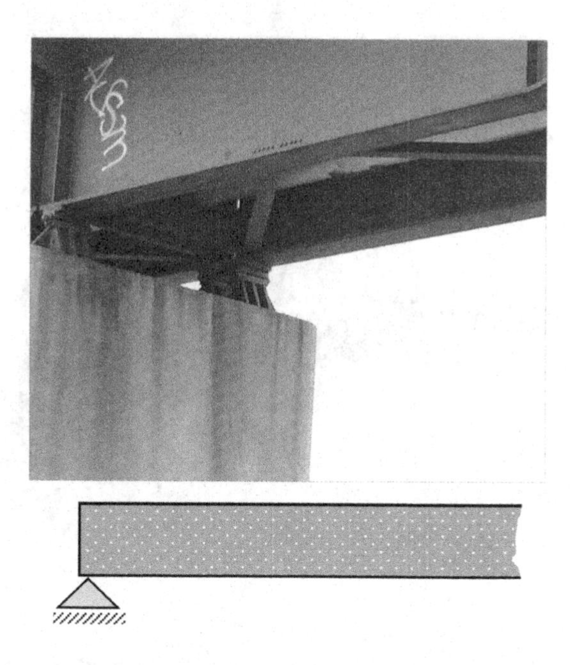

FIGURE 1.6 Pinned/hinged support (the joint resists both horizontal and vertical movements but cannot resist the rotation). *(Photo taken in Pueblo, Colorado)*

FIGURE 1.7 Fixed support (cast monolithically and the joint resists both horizontal and vertical movements and the rotation). *(Photo by Armando Perez taken in Pueblo, Colorado)*

FIGURE 1.8 Rocker support in an old structure (the rockers are transferring load to the underneath support with axial reaction). *(Photo taken in Pueblo, Colorado)*

In Figure 1.8, two beams are supported by a concrete block using steel devices which can resist the vertical movement but not the horizontal movement. Note that if the beams move horizontally, the steel device cannot resist it. This support can be considered a rocker.

1.7 LOADING TYPES

When analyzing a structure, the loads are idealized to some general shapes to simplify computations. A few common types of loading are discussed here.

Concentrated or Point Load. If the load application area is very small compared to the member, then the load is idealized as concentrated or point load. For example, a beam AB is being supported by two columns as shown in Figure 1.9. The T-beams are applying concentrated loads at C, D, E and F. The beam AB also applies a concentrated load on the columns.

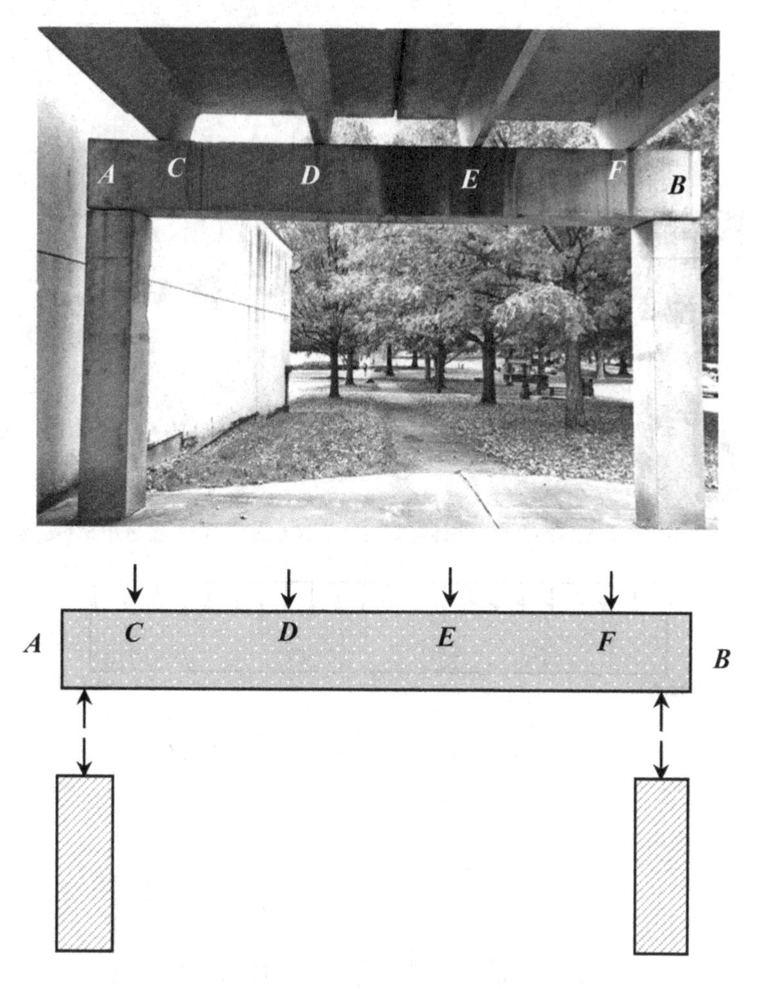

FIGURE 1.9 Concept of a concentrated load. *(Photo taken in Greenville Technical College, South Carolina)*

Uniformly Distributed Load. If the load acts on a larger area with a constant value all over, then it is termed as a uniformly distributed load (UDL). This loading is also known as rectangular loading (see Figure 1.10). The self-weight of a uniform member (such as a rod) is an example of a uniformly distributed load.

Uniformly Varying Load (UVL). If the load acts on a larger area with zero at one end and increases or decreases uniformly, then it is termed as a uniformly varying load (UVL). This loading is also known as triangular loading (see Figure 1.11). A common example of UVL is water pressure on vertical boundaries.

Concentrated or Point Moment. Sometimes, a moment can be applied to a point of a member. Say the beam *AB* in Figure 1.12 is supporting another beam *CD*. If *CD* experiences a twist, then the twisting transfers a concentrated moment at the beam *AB*.

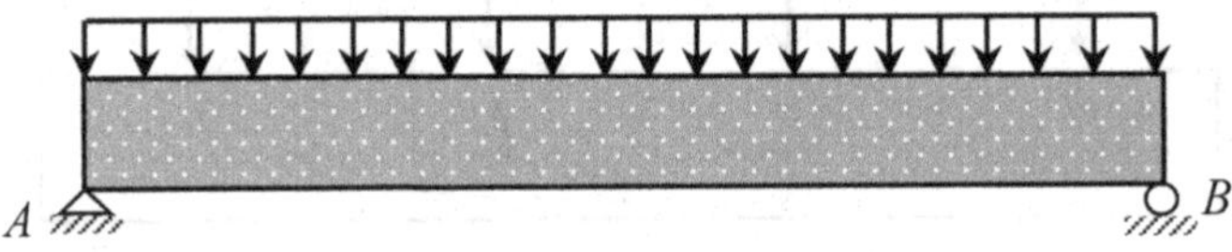

FIGURE 1.10 Uniformly distributed load. *(Photo taken in Pueblo, Colorado)*

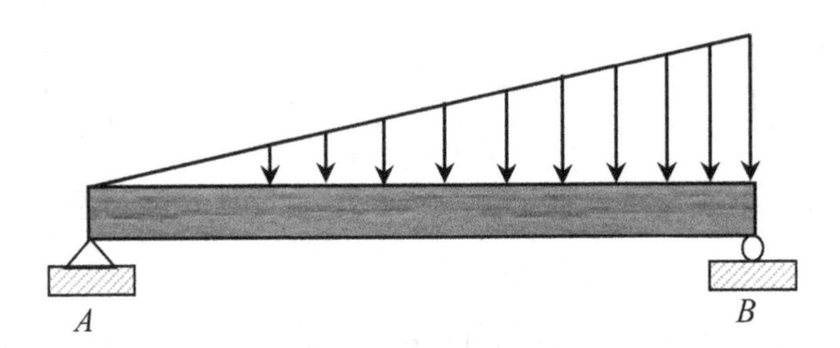

FIGURE 1.11 Uniformly varying load.

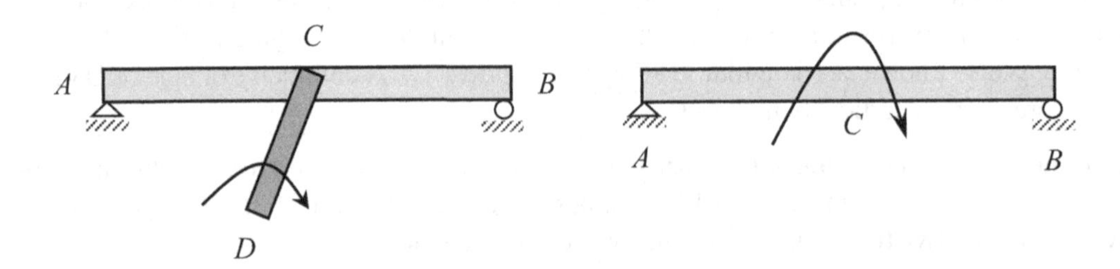

FIGURE 1.12 Concentrated or point moment.

Type 1 Simply supported beam:
One pin and one roller supports.

Type 2 Continuous beam:
One beam with three or more supports.

Type 3 Cantilever:
One end rigidly supported; the other end is free.

Type 4 Overhanging:
One or both ends are free; it may be continuous.

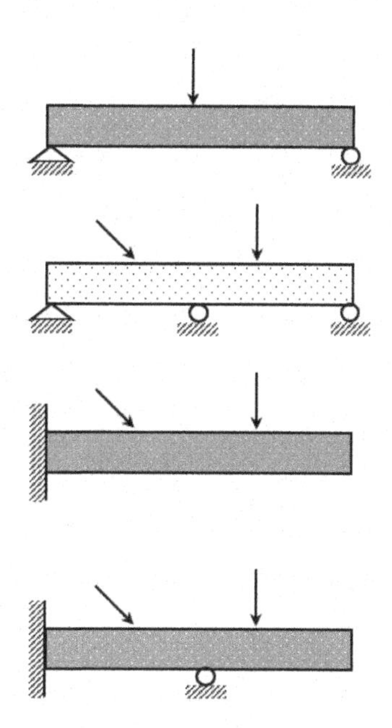

FIGURE 1.13 Different types of beams.

1.8 BEAM TYPES

A beam is a major structural member in engineering. It is defined as the essential bending member which can bend upon lateral loading. It can support axial force as well. As loads act perpendicularly to the longitudinal axis, the shear force also occurs in beams. Deflection is also a major design consideration. Depending on the support condition, a beam can be broadly divided into four types as shown in Figure 1.13:

1.9 ENGINEERING PRACTICE

In engineering, structural members are designed in such a way that the structure can resist the applied loading or the effect of loading with reasonable serviceability, economy and safety for its design life. Therefore, load calculation, load tracing, calculating their effects, etc. are important. Engineering statics discusses the analysis of forces and its actions-reactions in different types of structures. While designing a structure, the first step is to the calculate possible load that may act on it, and how this load will be distributed to different members of a structure. This subject starts the discussion of how the load is resolved into members, and so on.

2 Coplanar Force Systems

2.1 GENERAL

Most engineering structures are three-dimensional (3D). Many of the members in the structures are two-dimensional (2D) and loading acts on it are also two-dimensional. Even if the structures are 3D, a planar assumption (2D) is very often made to simplify analysis. This chapter deals with 2D force systems. Two types of coplanar force systems exist (Figure 2.1):

a. Concurrent forces – all forces act on a single point
b. Non-concurrent forces – forces act on different points

As shown in Figure 2.1a, three forces (F_1, F_2 and F_3) are acting at a point, O and can be considered a concurrent force system. Three other forces shown in Figure 2.1b, F_4, F_5 and F_6 are acting at different points on a beam and can be considered a non-concurrent force system.

Concurrent force systems have two options:

a. Two-force systems
b. More than two-force systems

Non-concurrent force systems can have two options:

a. All forces are parallel
b. All forces are not parallel

For all of the above cases, the main targets are to find out the resultant of the force systems and their orientation/direction. The resultant and its direction are required to determine the effect of the force system, the support reactions, or internal reactions. Knowledge of force resolution is required to study the rest of the chapter. Let us study force resolution now. Consider a force, F is acting at O along OA at an angle of ϕ with the line OB as shown in Figure 2.2. This force needs to be resolved into two perpendicular axes, OB and OC. Draw a right-angled triangle OAB.

$$\cos\phi = \frac{Adjacent}{Hypotenuse}$$

$$\cos\phi = \frac{OB}{OA}$$

$$OB = OA\cos\phi$$

Therefore, $OB = F\cos\phi$

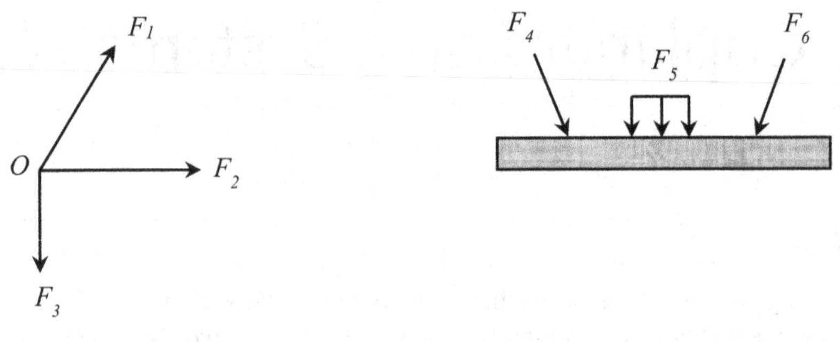

FIGURE 2.1 Concurrent versus non-concurrent forces.

Similarly, $\sin\phi = \dfrac{Opposite}{Hypotenuse}$

$$\sin\phi = \frac{AB}{OA}$$

$$AB = OA\sin\phi$$

Therefore, $AB = F\sin\phi$

Consider another force, R is acting at O along OA at an angle of α with the line OD as shown in Figure 2.3. This force needs to be resolved into two perpendicular axes, OD and OE. Draw a right-angled triangle OAD.

$$\cos\phi = \frac{Adjacent}{Hypotenuse}$$

$$\cos\alpha = \frac{OD}{OA}$$

$$OD = OA\cos\alpha$$

Therefore, $OD = R\cos\alpha$

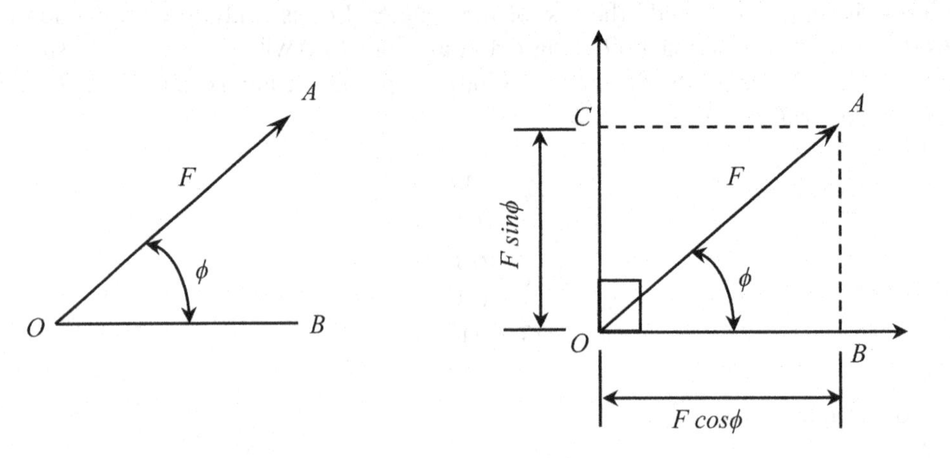

FIGURE 2.2 Resolution of a load.

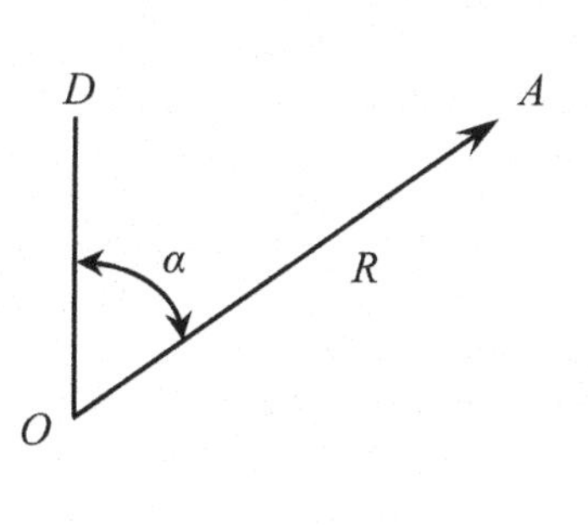

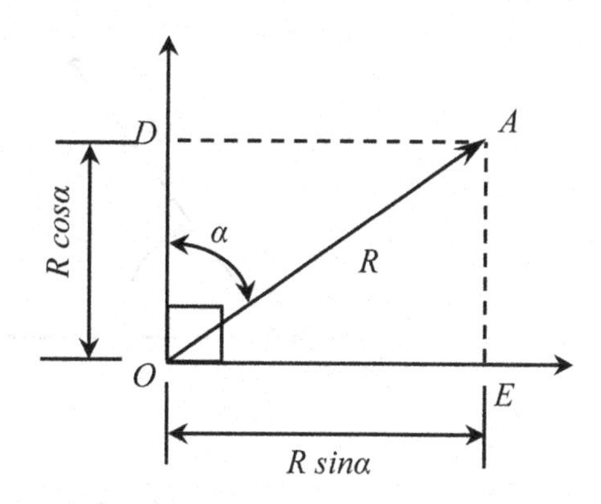

FIGURE 2.3 Resolution of another load.

Similarly, $\sin\phi = \dfrac{Opposite}{Hypotenuse}$

$$\sin\alpha = \frac{AD}{OA}$$

$$AD = OA\sin\alpha$$

Remember: the component of a force along a horizontal direction is not the cosine component; it depends on the angle. The cosine component is adjacent to the angle.

Therefore, $AD = R\ \sin\alpha$

2.2 CONCURRENT FORCES

A concurrent force system has two or more forces acting at a common point. For such a case, the resultant and its orientation are particularly important in engineering analysis and design. There are several methods to determine the resultant of a concurrent force system such as the parallelogram method, rectangular method, sine/cosine rules, vector analysis, and so on. The parallelogram method can be used more reasonably to determine the resultant of a two-force system. For a more than two-force systems, the rectangular method can be used for any number of forces. Vector analysis may be used for any number of force systems, although it is commonly used for a 3D force system and discussed in Chapter 8.

2.2.1 Two-Force Systems

Only two forces act on a point for two-force systems. The parallelogram method gives the magnitude and direction of the resultant in a very quick manner. Let us assume that two force vectors $\boldsymbol{P}$ and $\boldsymbol{Q}$ are acting at O in a plane along OA and OB, respectively, with an angle of α as shown in Figure 2.4a. According to the parallelogram rule, the resultant of $\boldsymbol{P}$ and $\boldsymbol{Q}$ can be represented by the diagonal of the parallelogram in terms of the magnitude of the resultant and its direction. Let us draw the parallelogram, $OBCA$ where AC is equal and parallel to OB and BC is equal and parallel to OA as shown in Figure 2.4b. Then, the diagonal OC is the resultant in terms of the magnitude of the resultant and its direction.

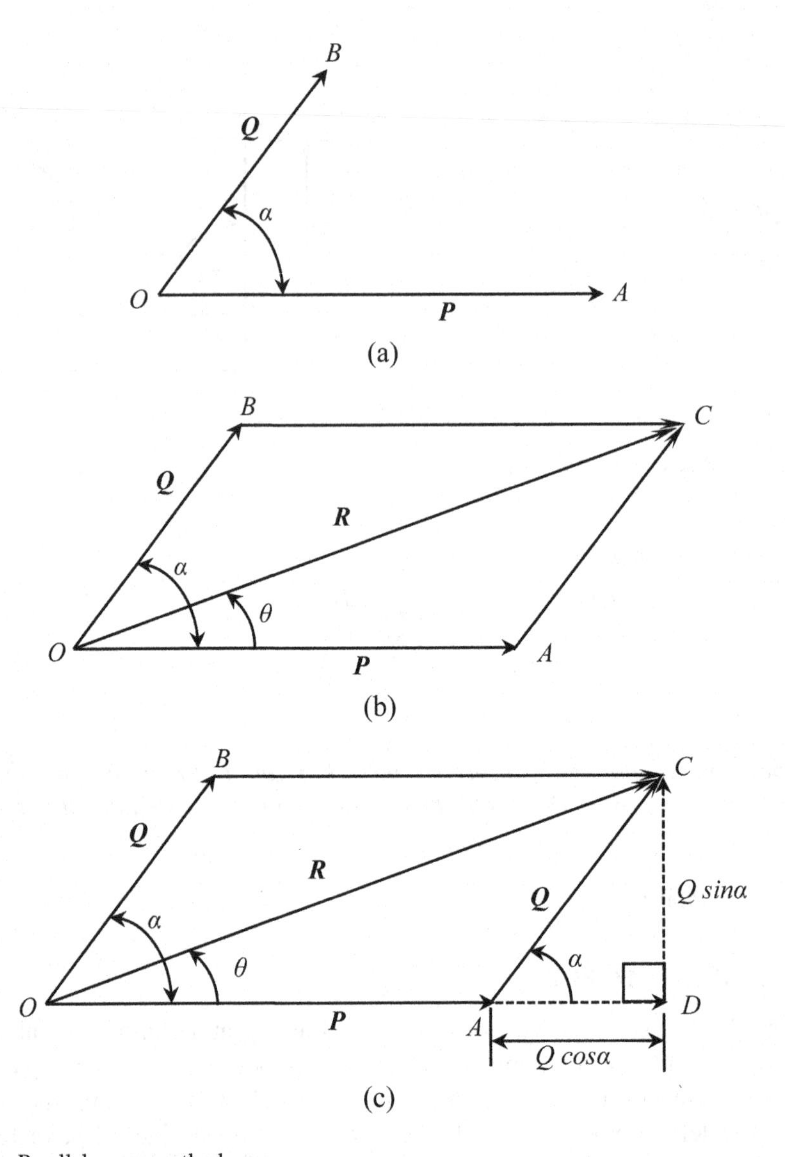

FIGURE 2.4 Parallelogram method.

More clearly, if a parallelogram *OACB* can be drawn such that *OC* is the diagonal, the diagonal *OC* can be represented as the resultant (***R***) of two forces ***P*** and ***Q***, that are acting at *O* in a plane along *OA* and *OB*, respectively.

Then, the resultant (***R***) can be expressed as:

$$R = \sqrt{P^2 + Q^2 + 2PQ\cos\alpha}$$

If the angle between the force (***P***) and the resultant (***R***) is θ, then it can be expressed as:

$$\theta = \tan^{-1}\left(\frac{Q\sin\alpha}{P + Q\cos\alpha}\right)$$

If the angle θ is positive, the orientation of the resultant is with respect to the force P toward Q as shown in Figure 2.4b,c. If the angle θ is negative, the orientation of the resultant is with respect to the opposite direction of the force P toward Q.

Now, let us see how these two handy equations can be derived. The force vector Q can be represented by the line AC also as per vector rule. Let us extend the OA member to OD and connect CD such that ACD is a right-angle triangle as shown in Figure 2.4c. Then, from ACD right-angled triangle:

$$\cos\alpha = \frac{Adjacent}{Hypotenuse}$$

$$\cos\alpha = \frac{AD}{AC}$$

Therefore, $AD = Q\cos\alpha$

Again, $\sin\alpha = \dfrac{Opposite}{Hypotenuse}$

$$\sin\alpha = \frac{CD}{AC}$$

Therefore, $CD = Q\sin\alpha$

From the OCD right-angle triangle, using the Pythagoras law:

$$OC^2 = OD^2 + CD^2$$
$$R^2 = (P + Q\cos\alpha)^2 + (Q\sin\alpha)^2$$
$$R^2 = (P + Q\cos\alpha)(P + Q\cos\alpha) + (Q\sin\alpha)^2$$
$$R^2 = (P^2 + PQ\cos\alpha + PQ\cos\alpha + Q^2\cos^2\alpha) + Q^2\sin^2\alpha$$
$$R^2 = (P^2 + 2PQ\cos\alpha) + (Q^2\cos^2\alpha + Q^2\sin^2\alpha)$$
$$R^2 = (P^2 + 2PQ\cos\alpha) + Q^2(\cos^2\alpha + \sin^2\alpha)$$
$$R^2 = (P^2 + 2PQ\cos\alpha) + Q^2(1.0)\ \text{[Remember: } \cos^2\alpha + \sin^2\alpha = 1.0]$$
$$R^2 = P^2 + Q^2 + 2PQ\cos\alpha$$

Therefore, $R = \sqrt{P^2 + Q^2 + 2PQ\cos\alpha}$

Again, from the OCD right-angled triangle:

$$\tan\theta = \frac{Opposite}{Adjacent}$$

$$\tan\theta = \frac{CD}{OD}$$

$$\tan\theta = \frac{Q\sin\alpha}{P + Q\cos\alpha}$$

Therefore, $\theta = \tan^{-1}\left(\dfrac{Q\sin\alpha}{P + Q\cos\alpha}\right)$

Note: if the angle between the two forces is 90°, then $cos90 = 0$ and thus, $2PQcos\alpha$ or $Qcos\alpha$ can be omitted.

Example 2.1:

The component of a 230-N force along the x-axis is 200 N. Determine the angle between the x-axis and the 230-N force.

SOLUTION

Let the angle between the x-axis and the 230-N force be θ –

Then, $F_x = F\cos\theta$

$$\theta = \cos^{-1}\left(\frac{F_x}{F}\right)$$

$$\theta = \cos^{-1}\left(\frac{200\text{ N}}{230\text{ N}}\right)$$

Therefore, $\theta = 30^{\circ}$

Answer: the angle between the x-axis and the 230-N force is 30°.

Example 2.2:

Two forces are acting on a point as shown in Figure 2.5. Determine the magnitude and the direction (with respect to 900-N force) of the resultant.

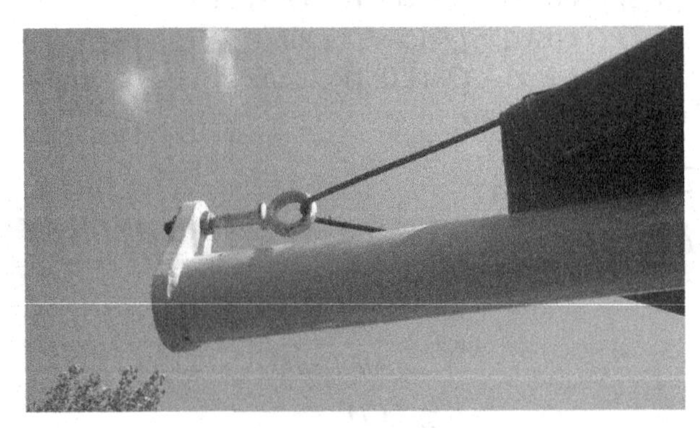

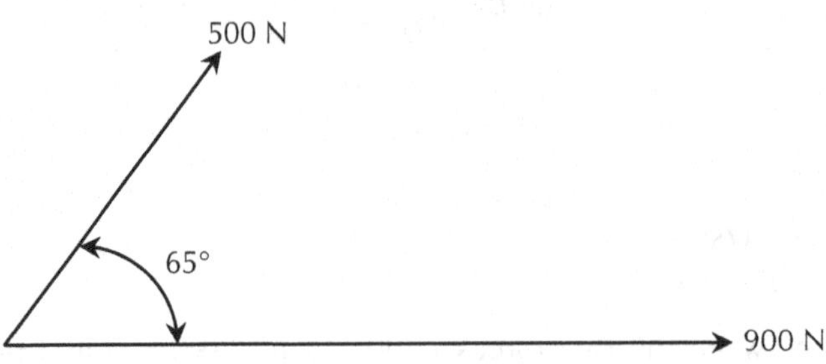

FIGURE 2.5 A concurrent two-force system for Example 2.2. *(Photo taken in Denver, Colorado)*

SOLUTION

There are only two concurrent forces acting in this example. Therefore, the parallelogram rule can be applied.

Using the parallelogram law, $R = \sqrt{P^2 + Q^2 + 2PQ\cos\alpha}$

Given,

First force, $P = 900$ N

Second force, $Q = 500$ N

Angle between P and Q, $\alpha = 65^\circ$

Resultant, $R = \sqrt{P^2 + Q^2 + 2PQ\cos\alpha}$

$$R = \sqrt{(900\text{ N})^2 + (500\text{ N})^2 + 2(900\text{ N})(500\text{ N})\cos 65}$$

Therefore, $R = 1{,}200$ N

Orientation of the Resultant, $\theta = \tan^{-1}\left(\dfrac{Q\sin\alpha}{P + Q\cos\alpha}\right)$

$$\theta = \tan^{-1}\left(\frac{500\text{ N}\sin 65}{900\text{ N} + 500\text{ N}\cos 65}\right)$$

Therefore, $\theta = 22.2^\circ$

Answer: the magnitude of the resultant is 1,200 N and it acts at an angle of 22.2° (counter-clockwise) with respect to the 900-N force (see Figure 2.6).

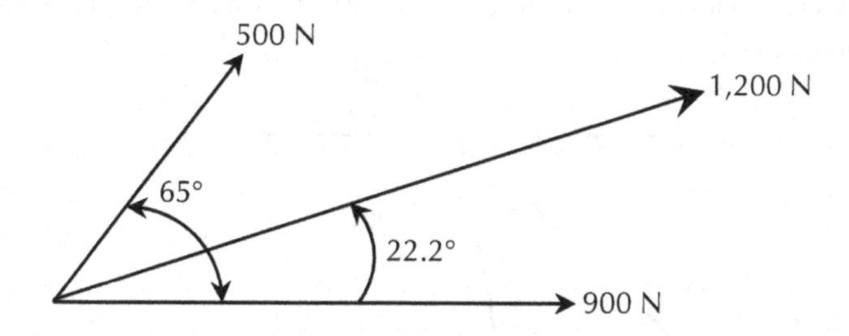

FIGURE 2.6 Results of Example 2.2.

Example 2.3:

Two forces are acting on a point as shown in Figure 2.7. Determine the magnitude and the direction (with respect to 230-lb force) of the resultant of this force system.

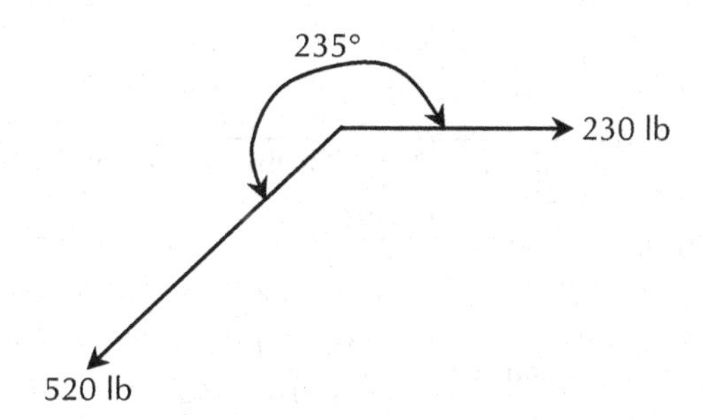

FIGURE 2.7 A concurrent two-force system for Example 2.3.

SOLUTION

There are only two concurrent forces acting here. Therefore, the parallelogram rule can be applied.

Given,

First force, $P = 230$ lb
Second force, $Q = 520$ lb
Angle between P and Q, $\alpha = 360 - 235 = 125^{\circ}$
(The angle of 235° is the exterior angle. The interior or smaller angle is always considered)

Resultant, $R = \sqrt{P^2 + Q^2 + 2PQ\cos\alpha}$

$$R = \sqrt{(230\text{ lb})^2 + (520\text{ lb})^2 + 2(230\text{ lb})(520\text{ lb})\cos 125}$$

Therefore, $R = 431.4$ lb

Orientation of the resultant, $\theta = \tan^{-1}\left(\dfrac{Q\sin\alpha}{P + Q\cos\alpha}\right)$

$$\theta = \tan^{-1}\left(\frac{520\text{ lb}\sin 125}{230\text{ lb} + 520\text{ lb}\cos 125}\right)$$

Therefore, $\theta = -80.9^{\circ}$

The angle θ represents the angle with respect to the direction of 230-lb force toward 520-lb force. The negative sign means it is 80.9° with respect to the opposite of the 230-lb force toward 520-lb force. Therefore, with respect to the direction of 230-lb force, the direction of resultant is −80.9 + 180 = 99.1°, as shown in Figure 2.8.

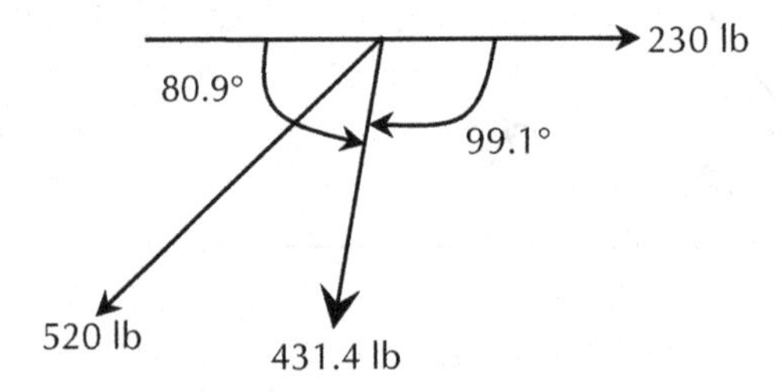

FIGURE 2.8 Resultant of the two-force system for Example 2.3.

Alternative Solution:

Consider,
$P = 520$ lb
$Q = 230$ lb

Resultant, $R = \sqrt{(520\text{ lb})^2 + (230\text{ lb})^2 + 2(520\text{ lb})(230\text{ lb})\cos 125} = 431.4\text{ lb}$

Orientation of the resultant, $\theta = \tan^{-1}\left(\dfrac{Q\sin\alpha}{P + Q\cos\alpha}\right)$

$$\theta = \tan^{-1}\left(\frac{230\text{ lb}\sin 125}{520\text{ lb} + 230\text{ lb}\cos 125}\right)$$

Therefore, $\theta = 25.9^{\circ}$

This angle of 25.9° is with respect to 520 lb. Then, with respect to the 230-lb force, the orientation of the resultant is 125 – 25.9 = 99.1°, as shown in Figure 2.9.

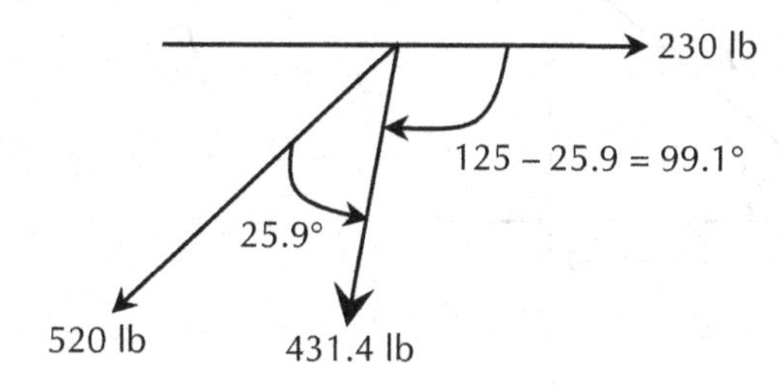

FIGURE 2.9 Alternative solution for the resultant of a two-force system for Example 2.3.

Answer: the magnitude of the resultant is 431.4 lb and it acts at an angle of 99.1° (clockwise) with respect to the 230-lb force.

Note: for the two-force system, any force can be assumed as P and the other one can be Q. The answer must be the same. Just remember that the equation, $\theta = \tan^{-1}\left(\frac{Q\sin\alpha}{P+Q\cos\alpha}\right)$ gives the orientation of the resultant with respect to the force P toward Q. If the angle θ is negative, the orientation of the resultant is with respect to the opposite direction of the force P toward Q.

2.2.2 More Than Two-Force Systems

The rectangular component method works well for two-dimensional force systems for any number of forces. However, if the number of force is two, then the parallelogram method is easier. The resolution of a single force, F, is shown in Figure 2.10 where θ is the angle between the x-axis and the force, F. The component of force, F, along the axis adjacent to the angle is $F\cos\theta$.

If there are a number of forces, then each force is resolved along two perpendicular axes (x and y). Then, the force components along two perpendicular axes are summed up as follows:

$$\sum_{i=1}^{n} F_{x,i} = F_{x,1} + F_{x,2} + F_{x,3} + \ldots\ldots$$

$$\sum_{i=1}^{n} F_{y,i} = F_{y,1} + F_{y,2} + F_{y,3} + \ldots\ldots$$

Once all the forces are resolved along two perpendicular axes, the resultant R, of n forces with components $F_{x,i}$ and $F_{y,i}$ can be calculated as:

$$R = \sqrt{\left(\sum_{i=1}^{n} F_{x,i}\right)^2 + \left(\sum_{i=1}^{n} F_{y,i}\right)^2}$$

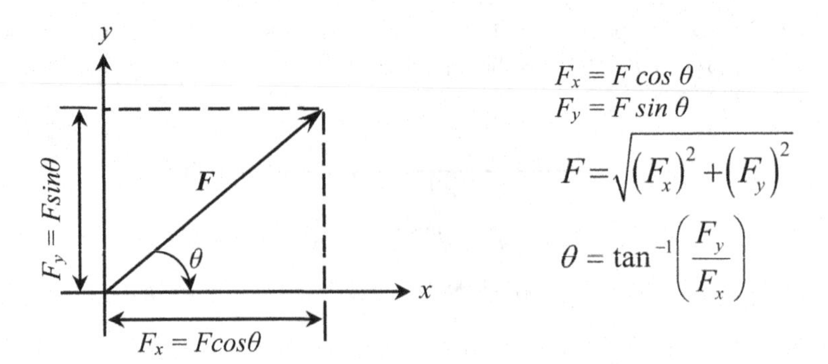

FIGURE 2.10 Rectangular component method.

The direction of the resultant with respect to the x-axis toward the y-axis is –

$$\theta = \tan^{-1}\left(\frac{\left(\sum_{i=1}^{n} F_{y,i}\right)}{\left(\sum_{i=1}^{n} F_{x,i}\right)}\right)$$

Note: cosine or sine is not related to the x-axis or y-axis. The component of the force adjacent to the angle is the cosine component.

Example 2.4:

Three forces are acting on a point as shown in Figure 2.11. The 900-N force is acting horizontally toward the right; the 600-N force is acting vertically downward. Determine the magnitude of the resultant and its direction (with respect to 900-N force) of it.

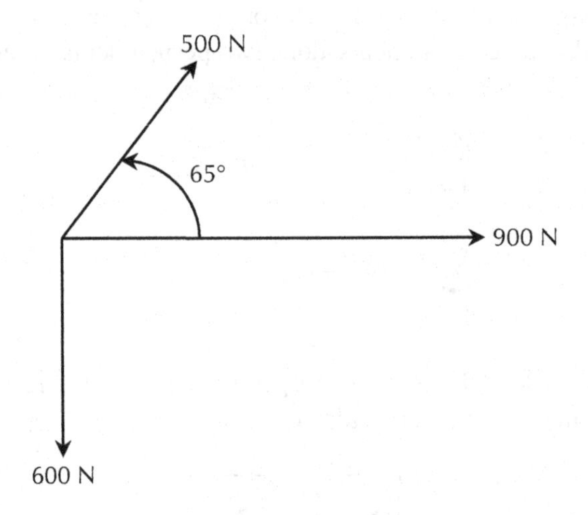

FIGURE 2.11 A concurrent three-force system for Example 2.4.

SOLUTION

Three forces are concurrently acting at a single point. The parallelogram method is good for two-force system. Therefore, we will be using the rectangular component method here.

Let us assume the following sign convention as shown in Figure 2.12.

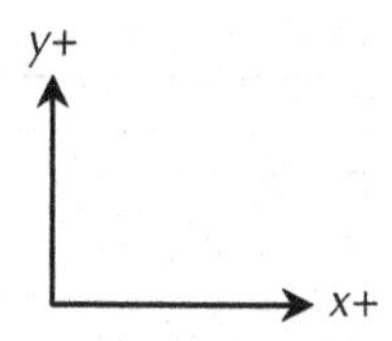

FIGURE 2.12 Cartesian coordinate system.

The *x-axis is positive horizontally right*
The *y-axis is positive vertically upward*

Given,

First force, $F_1 = 900$ N
First angle, $\theta_1 = 0°$
Second force, $F_2 = 500$ N
Second angle, $\theta_2 = 65°$
Third force, $F_3 = 600$ N
Third angle, $\theta_3 = 90°$

$$\sum F_x = F_1 \cos\theta_1 + F_2 \cos\theta_2 + F_3 \cos\theta_3$$
$$= 900\text{ N} + 500\text{ N}\cos 65 + 600\text{ N}\cos 90$$
$$= +1{,}111\text{ N}$$

$$\sum F_y = F_1 \sin\theta_1 + F_2 \sin\theta_2 + F_3 \sin\theta_3$$
$$= 900\text{ N}\sin 0 + 500\text{ N}\sin 65 - 600\text{ N}$$
$$= -146.8\text{ N}$$

Resultant, $R = \sqrt{\left(\sum F_x\right)^2 + \left(\sum F_y\right)^2}$

$$R = \sqrt{(1{,}111\text{ N})^2 + (-146.8\text{ N})^2}$$

Therefore, $R = 1{,}120$ N

Direction of resultant with respect to 900-N force, $\theta = \tan^{-1}\left|\dfrac{\sum F_y}{\sum F_x}\right|$

$$\theta = \tan^{-1}\left|\frac{-146.8\text{ N}}{1{,}111\text{ N}}\right|$$

Therefore, $\theta = 7.5°$

$\sum F_x$ is positive and $\sum F_y$ is negative; therefore, the resultant lies in the fourth quadrant. The quadrant is counted counter-clockwise.

Answer: the magnitude of the resultant is 1,120 N and it acts at an angle of 7.5° (clockwise) with respect to the 900-N force as shown in Figure 2.13.

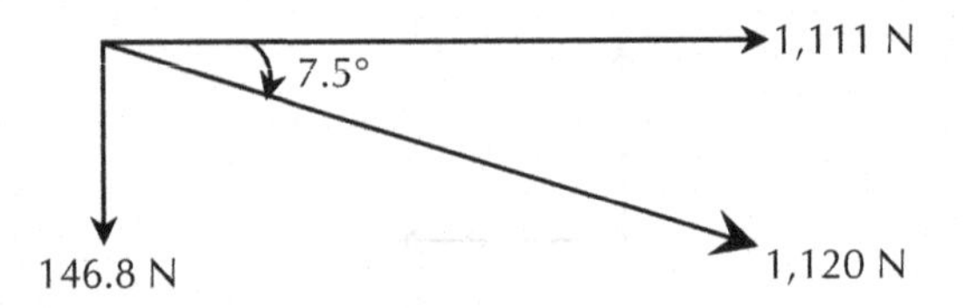

FIGURE 2.13 The resultant of the three-force system.

Example 2.5:

Three coplanar forces are acting on a point as shown in Figure 2.14. The 900-N force is acting horizontally toward the right. Determine the magnitude of the resultant and its direction (with respect to 900-N force).

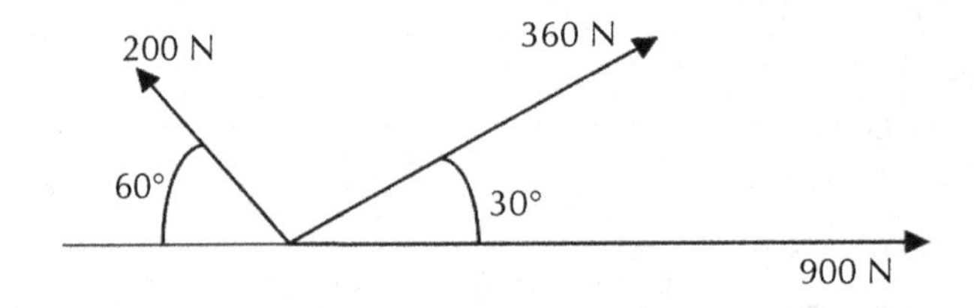

FIGURE 2.14 A concurrent three-force system for Example 2.5.

SOLUTION

We would use the parallelogram law, if the example had two forces. For this case, we will use the rectangular component method.

Let us assume the following sign convention as shown in Figure 2.15.

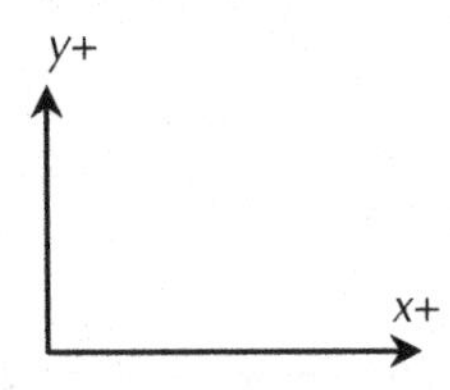

FIGURE 2.15 Cartesian coordinate system.

The *x-axis is positive horizontally right*
The *y-axis is positive vertically upward*

Given,

First force, $F_1 = 900$ N

First angle, $\theta_1 = 0°$
Second force, $F_2 = 360$ N
Second angle, $\theta_2 = 30°$
Third force, $F_3 = 200$ N
Third angle, $\theta_3 = 60°$

$$\sum F_x = F_1\cos\theta_1 + F_2\cos\theta_2 + F_3\cos\theta_3$$
$$= 900\ \text{N}\cos 0 + 360\ \text{N}\cos 30 - 200\ \text{N}\cos 60$$
$$= +1{,}112\ \text{N}$$

$$\sum F_y = F_1\sin\theta_1 + F_2\sin\theta_2 + F_3\sin\theta_3$$
$$= 900\ \text{N}\sin 0 + 360\ \text{N}\sin 30 + 200\ \text{N}\sin 60$$
$$= +353\ \text{N}$$

Resultant, $R = \sqrt{\left(\sum F_x\right)^2 + \left(\sum F_y\right)^2}$

$$R = \sqrt{(1{,}112\ \text{N})^2 + (353\ \text{N})^2}$$

Therefore, $R = 1{,}167$ N

The direction of the resultant with respect to the 900-N force, $\theta = \tan^{-1}\left|\dfrac{\sum F_y}{\sum F_x}\right|$

$$\theta = \tan^{-1}\left|\frac{353\ \text{N}}{1{,}112\ \text{N}}\right|$$

Therefore, $\theta = 17.6°$

As both $\sum F_x$ and $\sum F_y$ are positive, the resultant lies in the first quadrant.

Answer: the magnitude of the resultant is 1,167 N and it acts at an angle of 17.6° (counter- clockwise) with respect to the 900 N force as shown in Figure 2.16.

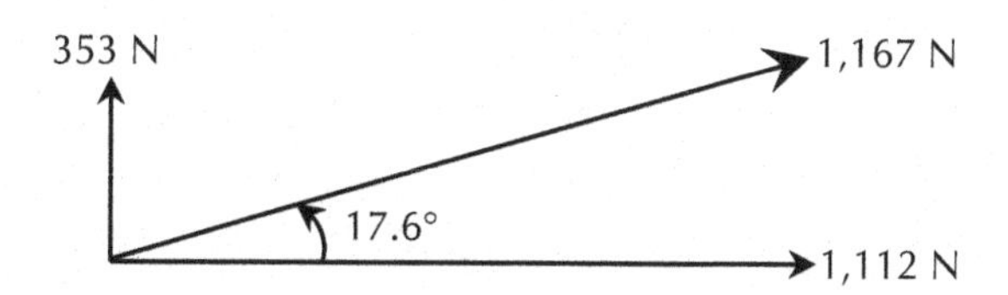

FIGURE 2.16 The resultant of the concurrent three-force system.

Example 2.6:

Three coplanar forces are acting on a point as shown in Figure 2.17. The 100-N force is acting horizontally toward the right. Determine the magnitude of the resultant and its direction (with respect to 100-N force).

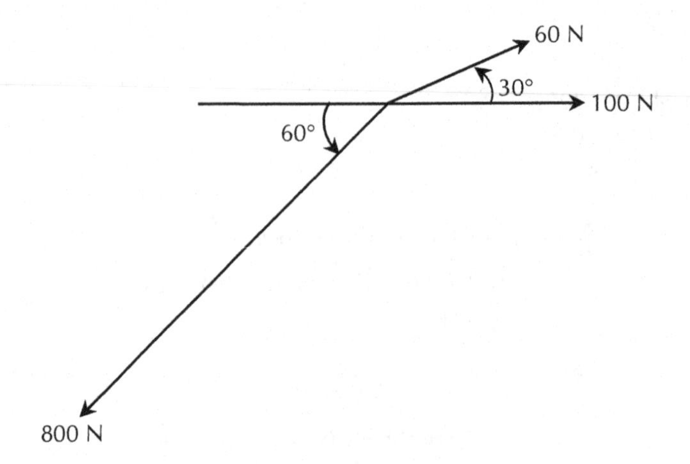

FIGURE 2.17 A concurrent three-force system for Example 2.6.

SOLUTION

We would use the parallelogram law, if the example had two forces. For the current case, we will use the rectangular component method.

Let us assume the following sign convention as shown in Figure 2.18.

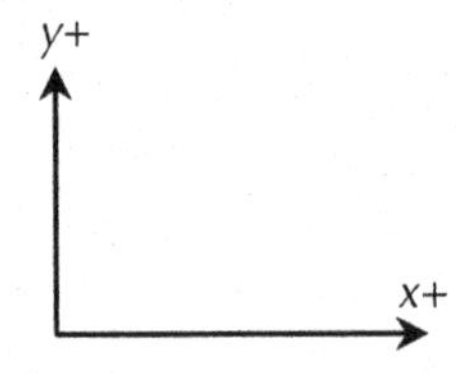

FIGURE 2.18 Cartesian coordinate system.

The *x-axis is positive horizontally right*
The *y-axis is positive vertically upward*

Given,

First force, $F_1 = 100$ N
First angle, $\theta_1 = 0^\circ$

Second force, $F_2 = 60$ N
Second angle, $\theta_2 = 30^\circ$

Third force, $F_3 = 800$ N
Third angle, $\theta_3 = 60^\circ$

$$\begin{aligned}\sum F_x &= F_1 \cos\theta_1 + F_2 \cos\theta_2 + F_3 \cos\theta_3 \\ &= 100 \text{ N} \cos 0 + 60 \text{ N} \cos 30 - 800 \text{ N} \cos 60 \\ &= -248 \text{ N}\end{aligned}$$

$$\sum F_y = F_1 \sin\theta_1 + F_2 \sin\theta_2 + F_3 \sin\theta_3$$
$$= 100\text{ N}\sin 0 + 60\text{ N}\sin 30 - 800\text{ N}\sin 60$$
$$= -663\text{ N}$$

Resultant, $R = \sqrt{\left(\sum F_x\right)^2 + \left(\sum F_y\right)^2}$

$$R = \sqrt{(-248\text{ N})^2 + (-663\text{ N})^2}$$

Therefore, $R = 708\text{N}$

The direction of the resultant with respect to the 100-N force, $\theta = \tan^{-1}\left|\frac{\sum F_y}{\sum F_x}\right|$

$$\theta = \tan^{-1}\left|\frac{-663\text{ N}}{-248\text{ N}}\right|$$

Therefore, $\theta = 69.5^\circ$

Answer: the magnitude of the resultant is 708 N as shown in Figure 2.19.

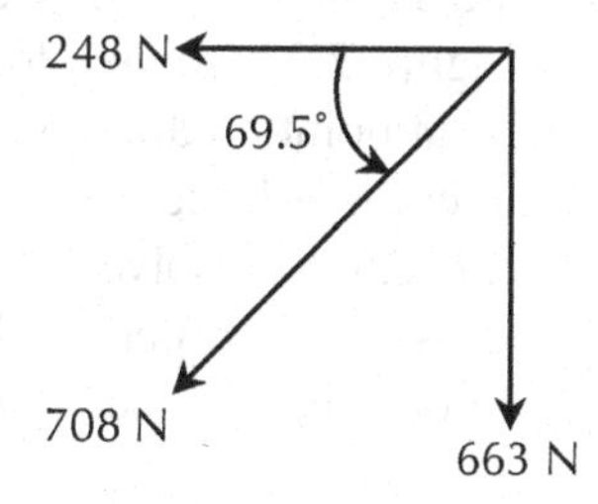

FIGURE 2.19 The resultant of the concurrent three-force system.

$\sum F_x$ is negative and $\sum F_y$ is positive; therefore, the resultant lies in the third quadrant.

2.3 NON-CONCURRENT FORCES

There are several types of non-concurrent force such as concentrated load, uniformly distributed load (UDL) or rectangular load, and uniformly varying load (UVL) or triangular. The resultant of these types of forces can be determined using simple arithmetic. Some background knowledge is required to study non-concurrent forces and is discussed here.

2.3.1 Centroid

'Centroid' means the center of gravity of an area. The centroids of two regular geometric areas (rectangle and triangle) are studied in this chapter; they are shown in Figure 2.20. If the sides of a rectangle are b and h, respectively, then the centroid is located at ($b/2$, $h/2$), as shown in Figure 2.20.

Chapter 9 is dedicated to determining the centroids of different shapes and composite shapes.

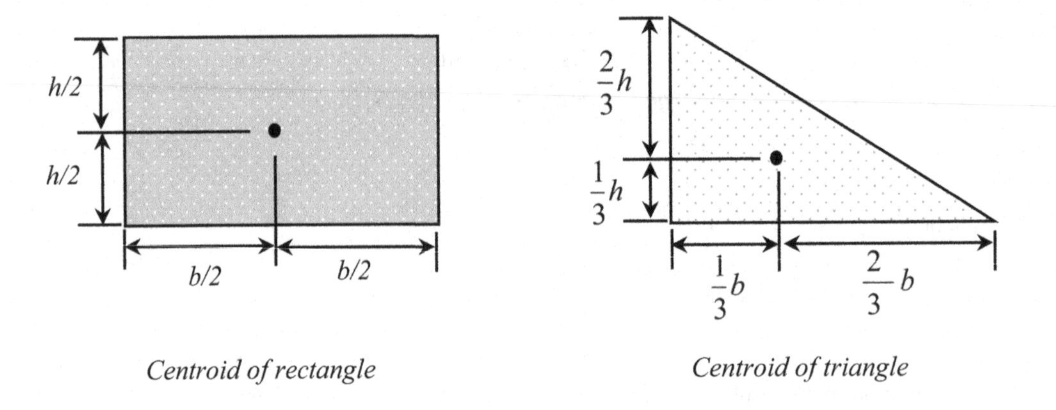

FIGURE 2.20 Centroids of a rectangle and a triangle.

2.3.2 Moment

Moment is defined as the force times the perpendicular distance. Let us consider a force, F, acts at d distance from a point, A, as shown in Figure 2.21 where the distance d is perpendicular to the direction of force. The moment at point A can be determined as $M = Fd$.

Now, consider the force system in Figure 2.22 where a force, F is applied with an angle of 60° with the distance d. The moment at point A cannot be determined as $M = Fd$, as the distance is not perpendicular to the direction of force.

To determine the moment, the force, F, can be resolved into two components as shown in Figure 2.23. The vertical component, F sin60 is perpendicular to the distance d. However, the horizontal component, F cos60 is along the distance d and the perpendicular distance

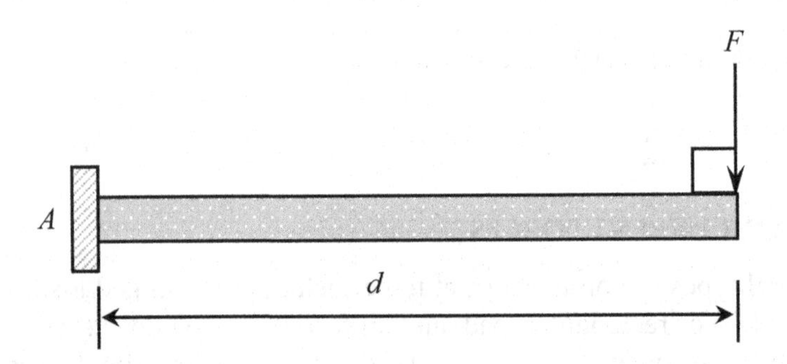

FIGURE 2.21 Concept of moment for perpendicular load.

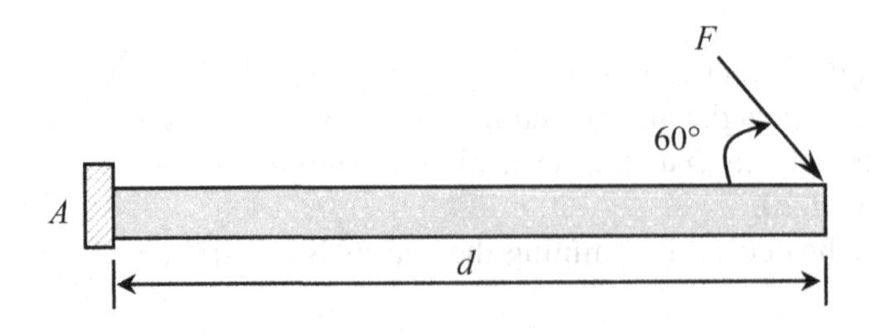

FIGURE 2.22 Concept of moment for non-perpendicular load.

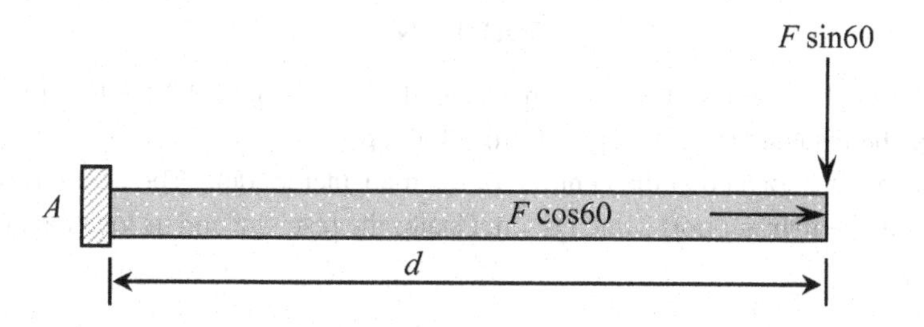

FIGURE 2.23 Analysis of moment for non-perpendicular load.

between point A and the component, $F\cos60$ is 0. The moment at A by the inclined force can be calculated as follows:

$$M = F\sin60(d) + F\cos60(0) = Fd\sin60$$

Remember: while calculating moment, the distance must be perpendicular distance.

2.3.3 Loads and Resultant

The actual loads that may occur in real structural members are quite difficult to be categorized. While analyzing the structure for designing, the loads are idealized into different regular shapes such as concentrated load, UDL, and UVL. Then, the resultant of these loads is determined by simple arithmetic calculation. The acting point of the resultant is determined using the principle that the summation of moment created by the load group with respect to a point is equal to the moment created by the resultant with respect to that point. This theory is also known as Varignon's theorem.

Remember: the summation of moment created by a load group with respect to a point is equal to the moment created by the resultant with respect to that point.

A few examples are given below for better understanding.

Example 2.7:

A simply supported structural beam has the loading configuration shown in Figure 2.24. Determine the resultant of the load and its acting point with respect to the left support.

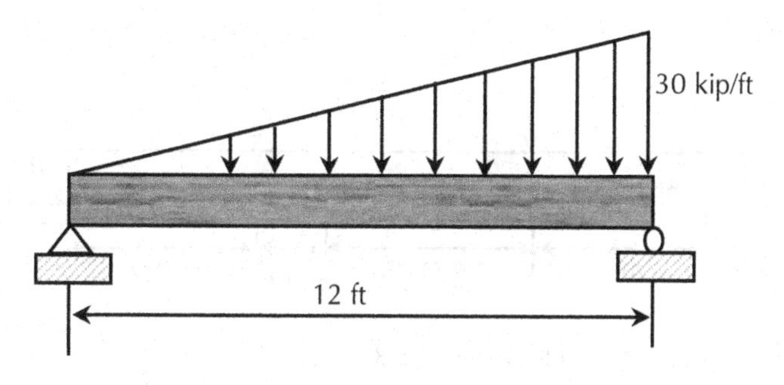

FIGURE 2.24 An arbitrary beam with a uniformly varying load for Example 2.7.

SOLUTION

The load acting on the beam is a UVL or triangular load. The resultant of UVL is the area of the load.

Therefore, the resultant is ½ (30 kip/ft)(12 ft) = 180 kip.

The acting point is located at the centroid of the triangular loading. The centroid is located at 1/3 of 12 ft from the right support, which is 4 ft. Finally, the resultant and its location are as shown in Figure 2.25:

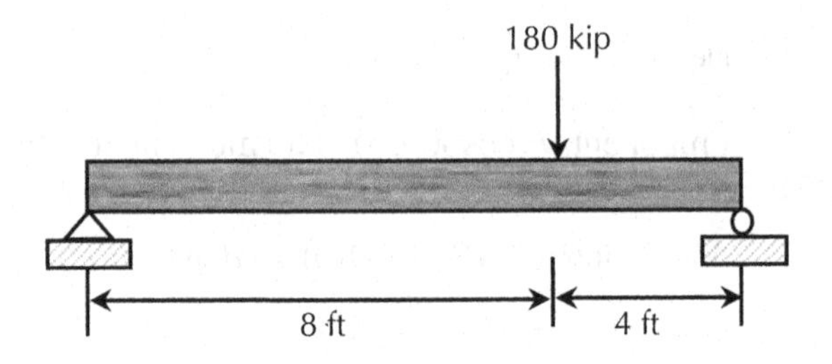

FIGURE 2.25 The resultant of the uniformly varying load.

Example 2.8:

A simply supported structural beam has the loading configuration shown in Figure 2.26. Determine the resultant of the load and its acting point with respect to the left support.

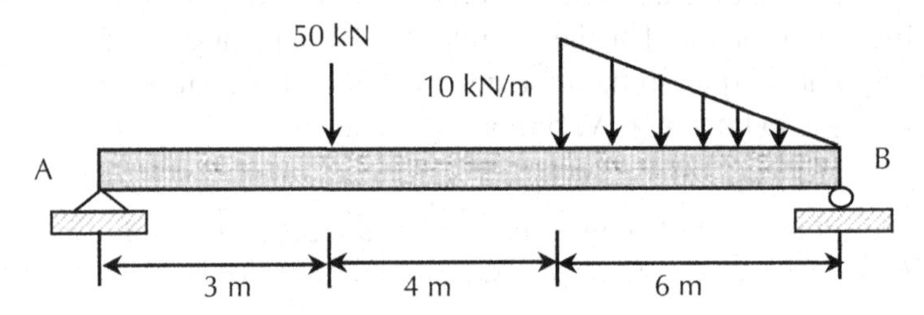

FIGURE 2.26 An arbitrary beam for Example 2.8.

SOLUTION

The equivalent load of the triangular loading is ½(10 kN/m)(6 m) = 30 kN. The loading configuration can be shown as shown in Figure 2.27:

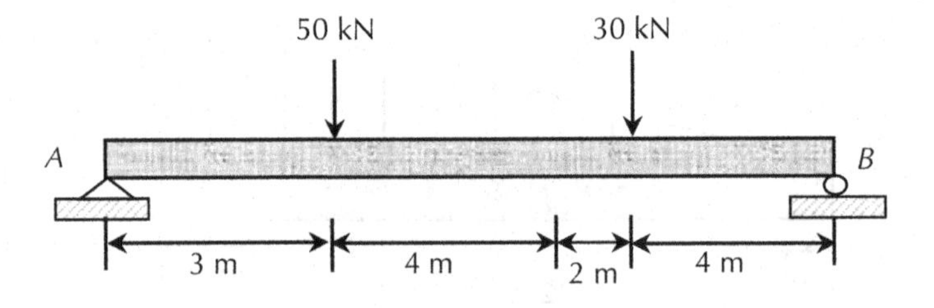

FIGURE 2.27 Analysis of the beam for Example 2.8.

Both the loads are parallel. Therefore, the resultant is 30 kN + 50 kN = 80 kN.

To determine the acting point of the resultant with respect to the left support, we need to take the moment about the left support. Again, to remind you, the moment created by the load group with respect to a point is equal to the moment created by the resultant with respect to that point.

Let us assume that the resultant acts at 'x' from the left support (A) as shown in Figure 2.28, then

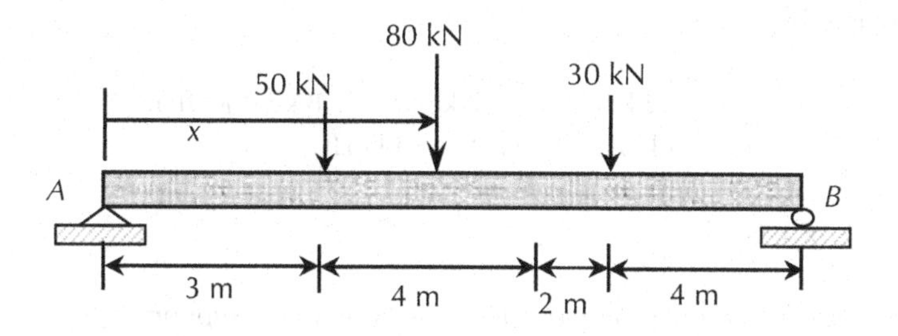

FIGURE 2.28 Determining the centroid of the loading for Example 2.8.

$$80 \text{ kN } (x) = 50 \text{ kN } (3 \text{ m}) + 30 \text{ kN } (9 \text{ m})$$
$$80x \text{ kN} = 150 \text{ kN.m} + 270 \text{ kN.m}$$
$$80x \text{ kN} = 420 \text{ kN.m}$$
$$x = 5.25 \text{ m}$$

Answer: the equivalent load is 80 kN and it acts at 5.3 m from the left support.

It is always better to check your answer by applying your common sense. The above figure shows the answer is reasonable.

Example 2.9:

A cantilever beam is supporting three types of loads as shown in Figure 2.29. Determine the equivalent concentrated load and its point of action with respect to A, which will cause the same reactions at the support.

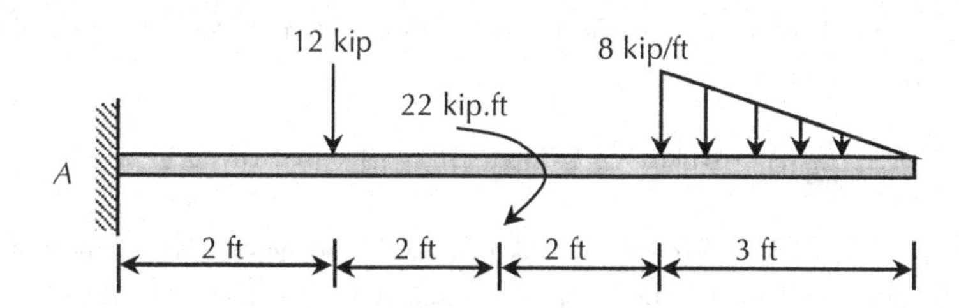

FIGURE 2.29 An arbitrary beam for Example 2.9.

SOLUTION

Assuming the downward force is positive, consider the summation of forces along the vertical direction to determine the resultant force.

$$\sum F_y = 12 \text{ kip} + \tfrac{1}{2}(8 \text{ kip/ft})(3 \text{ } ft) = 24 \text{ kip}$$

(Note that $\sum F_y$ means the summation of force along the y-direction (vertical direction). When we consider the summation of force, we do not consider the moment. This is why the moment of 22 kip.ft is not considered)

Let us assume that this resultant force of 24 kip is acting at x from the support. Moment at A caused by the resultant and by the load group will be equal. Let us assume that the clockwise moment is positive. Therefore:

$$24 \text{ kip } (x) = 12 \text{ kip } (2 \text{ ft}) + 22 \text{ kip.ft} + \tfrac{1}{2} (8 \text{ kip/ft})(3 \text{ ft}) (7 \text{ ft})$$
$$24x \text{ kip} = 24 \text{ kip.ft} + 22 \text{ kip.ft} + 84 \text{ kip.ft}$$
$$24x \text{ kip} = 130 \text{ kip.ft}$$
$$x = 5.42 \text{ ft}$$

Answer: the equivalent load is 24 kip and it acts at 5.4 ft from the support.

Example 2.10:

A cantilever beam is supporting three types of loads as shown in Figure 2.30. Determine the equivalent concentrated load and its point of action with respect to A, which will cause the same reactions at the support.

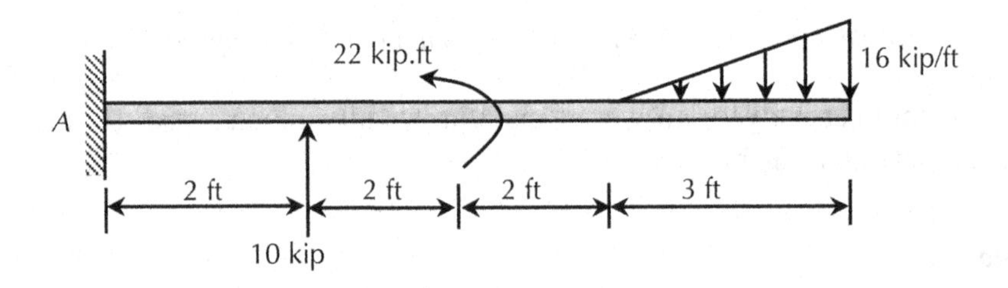

FIGURE 2.30 An arbitrary beam for Example 2.10.

SOLUTION

To determine the resultant force, consider the summation of forces along the vertical direction assuming downward force is positive.

$$\sum F_y = -10 \text{ kip} + \tfrac{1}{2}(16 \text{ kip/ft})(3 \text{ ft}) = 14 \text{ kip(downward)}$$

Let us assume that this resultant force of 14 kip is acting at x from the support. Moment at A caused by the resultant and by the load group is equal. Let us assume the clockwise moment to be positive. Therefore:

$$14 \text{ kip } (x) = -10 \text{ kip } (2 \text{ ft}) - 22 \text{ kip.ft} + \tfrac{1}{2} (16 \text{ kip/ft})(3 \text{ ft}) (8 \text{ ft})$$
$$14x \text{ kip} = -20 \text{ kip.ft} - 22 \text{ kip.ft} + 192 \text{ kip.ft}$$
$$14x \text{ kip} = 150 \text{ kip.ft}$$
$$x = 10.7 \text{ ft}$$

Answer: the equivalent load is 14 kip and it acts at 10.7 ft from the support.

Note: the resultant of this example is acting outside the beam.

PRACTICE PROBLEMS

PROBLEM 2.1

Two forces, 7 N and 8 N are acting at a point at an angle of 60°. Determine the magnitude of the resultant and its orientation.

PROBLEM 2.2

Two forces P and $2P$ are acting on a particle. If the first and second forces are increased by P and 6, respectively, the direction of the resultant is unaltered. Determine the magnitude of P.

PROBLEM 2.3

Two forces 7 N and 8 N are acting at a point at an angle of 260°. Determine the magnitude of the resultant and its orientation.

PROBLEM 2.4

Two parallel forces, P and Q have the resultant

I. $P + Q$ if they are parallel to each other and work along the same direction
II. Difference of P and Q if they are parallel to each other and work along the opposite direction
III. Acting at the midpoint of the line of action if they are equal, parallel to each other and working along the same direction

Which of the above statement(s) are true?

PROBLEM 2.5

Two equal forces of magnitude P acting at a point have the resultant

I. $\sqrt{3}P$ if acting at an angle of 60°
II. $\sqrt{2}P$ if they are at a right angle
III. 0 if they are in the opposite direction

Which of the above statement(s) are true?

PROBLEM 2.6

If two equal forces of magnitude Q are acting at a point at an angle of 60°, then

I. The resultant is $\sqrt{3}$ times of Q
II. The direction of the resultant bisects the included angle
III. The magnitude of Q is $5\sqrt{3}$ if they are in equilibrium by another force of magnitude 15

Which of the above statement(s) are true?

PROBLEM 2.7

Two forces P and Q are 4 N and 2 N, respectively. Their resultant is 2 N. Calculate the angle between them.

PROBLEM 2.8

The resultant of two forces acting at a point at an angle of 120° is perpendicular to the smaller force. If the greater force is 10 N, determine the smaller force.

PROBLEM 2.9

The resultant of two forces P and 12 N acting at a point is $3\sqrt{7}$ N, which makes an angle 90° with the direction of P. Calculate the magnitude of P.

PROBLEM 2.10

If two equal forces acting at a point at an angle of 60° are held in equilibrium by a force of 9 N. Calculate the magnitude of the equal force.

PROBLEM 2.11

The resultant of the forces $3P$ and $2P$ are R. If the first force is doubled, the resultant is also doubled. Calculate the angle between them.

PROBLEM 2.12

A simply supported beam has three concentrated forces as shown in Figure 2.31. Determine the resultant of the loading and its acting point with respect to the right support, B.

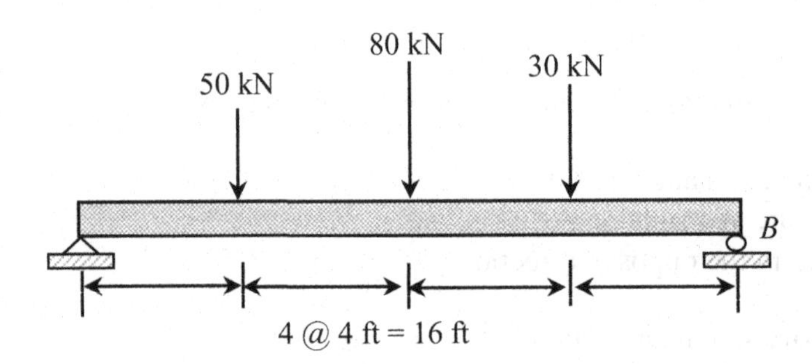

FIGURE 2.31 An arbitrary beam for Problem 2.12.

PROBLEM 2.13

Two planar-concurrent forces are acting on a hook as shown in Figure 2.32. Determine the magnitude of the resultant and its direction with respect to the 50-kN force. The 50-kN force is acting parallel to the ground.

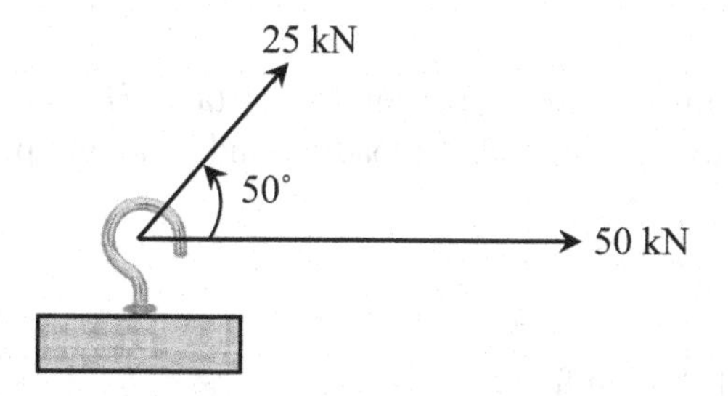

FIGURE 2.32 A concurrent two-force system for Problem 2.13.

PROBLEM 2.14

Two planar-concurrent forces are acting on a hook as shown in Figure 2.33. Determine the magnitude of the resultant and its direction with respect to the 50-kN force. The 50-kN force is acting parallel to the ground.

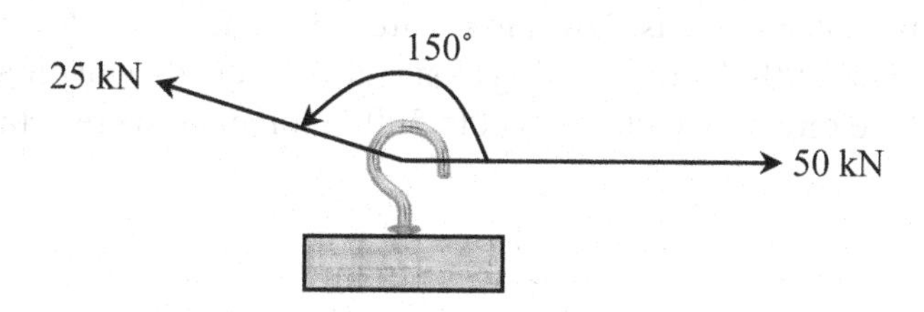

FIGURE 2.33 A concurrent two-force system for Problem 2.14.

PROBLEM 2.15

An over-hanging beam has three types of forces (a UVL, a concentrated, and a UDL) as shown in Figure 2.34. Determine the resultant of the loading and its acting point with respect to the right (roller) support.

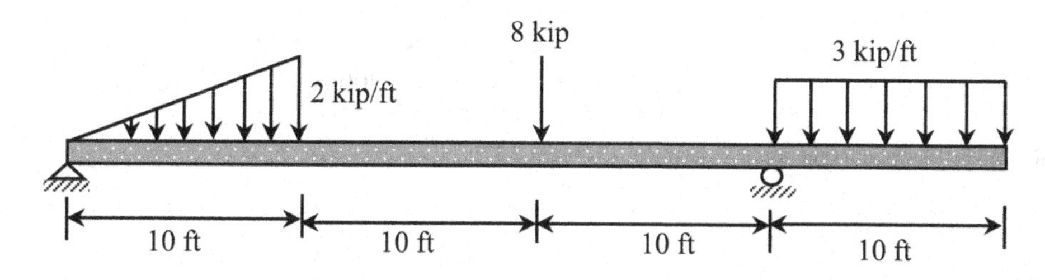

FIGURE 2.34 An arbitrary beam for Problem 2.15.

PROBLEM 2.16

A simply supported beam has two types of forces (a UVL and a UDL) as shown in Figure 2.35. Determine the resultant of the loading and its acting point with respect to the right (roller) support.

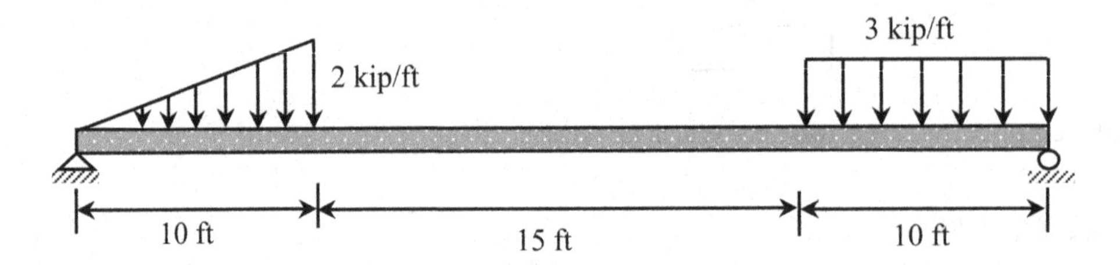

FIGURE 2.35 An arbitrary beam for Problem 2.16.

PROBLEM 2.17

Four forces are acting on a point as shown in Figure 2.36. The 200-N force is acting vertically downward and the 500-N force is acting along the horizontal-right direction. Calculate the magnitude and the direction (with respect to 500-N force) of the resultant.

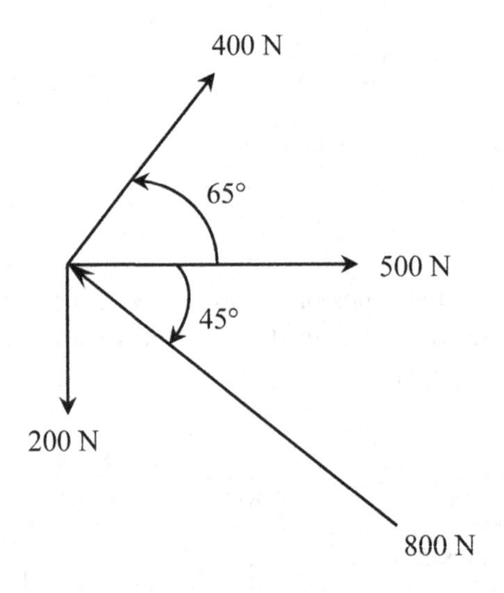

FIGURE 2.36 A concurrent force system for Problem 2.17.

PROBLEM 2.18

Four forces are acting on a point as shown in Figure 2.37. The 200-lb force is acting vertically downward and the 500-lb force is acting along the horizontal-left direction. Calculate the magnitude and direction of the resultant.

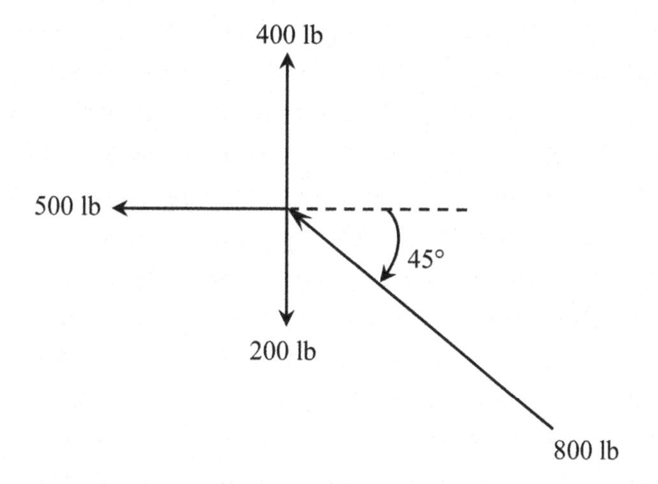

FIGURE 2.37 A concurrent force system for Problem 2.18.

PROBLEM 2.19

Four coplanar forces are acting on a point as shown in Figure 2.38. The 200-lb force is acting vertically downward and the 500-lb force is acting along the horizontal direction. Calculate the magnitude and the direction of the resultant.

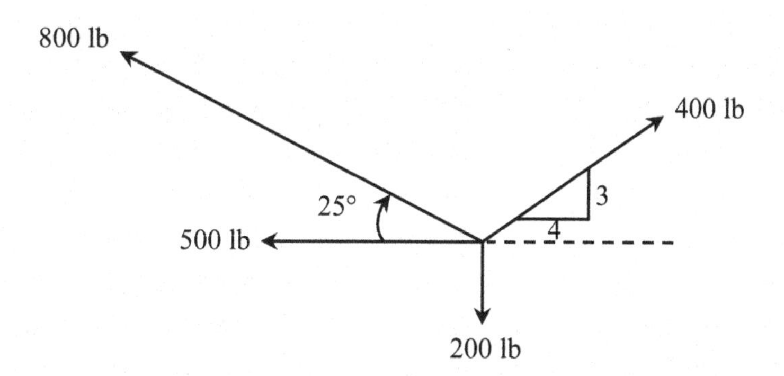

FIGURE 2.38 A concurrent force system for Problem 2.19.

3 Equilibrium of Particle and Rigid Body

3.1 CONCEPT OF EQUILIBRIUM

From Newton's third law of motion, we know that every action has its own and opposite reaction. That means, if you apply a force to a wall it will react with an equal amount of force on you, if the wall is not moving (rigid). Therefore, every action is equal to its reaction in any direction. This is called the equilibrium condition. If we sum up this action and reaction in a direction, it will be zero as these two are numerically equal but the signs are opposite. Thus, we can write the following three equations for two-dimensional conditions:

$\Sigma F_x = 0$ (the summation of forces along the x-direction (horizontal) is zero)
$\Sigma F_y = 0$ (the summation of forces along the y-direction (vertical) is zero)
$\Sigma M = 0$ (the summation of moment at any point about any axis perpendicular to the plane is zero)

Figure 3.1a shows a body is experiencing forces from different sides. The summation of forces in any direction must be zero to be in equilibrium. Figure 3.1b shows an inclined structure which is in equilibrium.

3.2 PARTICLE VERSUS RIGID BODY

A *particle* is a body of infinitely small volume and is considered to be concentrated at a point. Every point in a particle can be considered to have the same characteristics.

A rigid body is one which does not deform under the action of the loads or the external forces. In the case of a *rigid body*, the distance between any two points of the body remains constant, when this body is subjected to loads. Examples of a particle and a rigid body are shown in Figure 3.2.

3.3 IDEALIZATION OF STRUCTURES

Real structures are complicated and have different features in terms of loading application, loading distribution, support restrictions, etc. These conditions are simplified for analyzing the structures using paper and pencil or a computer. A simplified version of the structures is known as the *idealized structure*. Some general features of idealization are listed below:

- Linear members are represented by lines.
- Loading applied in a small area is considered a concentrated load.
- Support restrictions are simplified in three ways and is discussed in the next subsection.

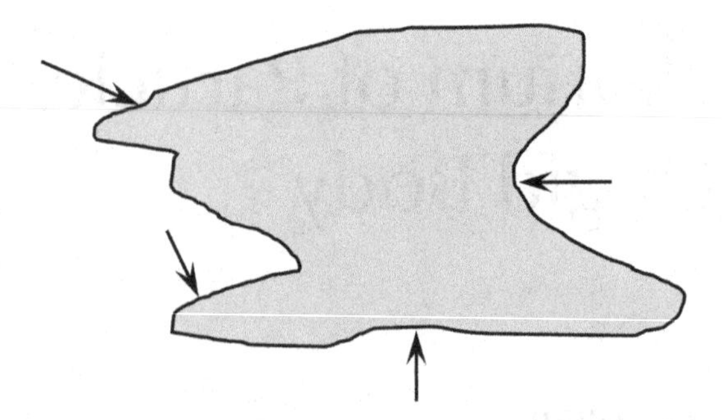

a) Concept of equilibrium

b) An inclined structure (Stade Olympique de Montréal) in equilibrium

FIGURE 3.1 Equilibrium of structure. (*Photo courtesy of Waymarking*)

The block to be lifted may be considered a particle; the whole body acts together.

(Photo by Quazi Shammas taken in Bangladesh)

A truss is considered a rigid body; different joints experience different loading.

(Photo taken in Durango, Colorado)

FIGURE 3.2 A particle versus a rigid body.

a) Actual structure

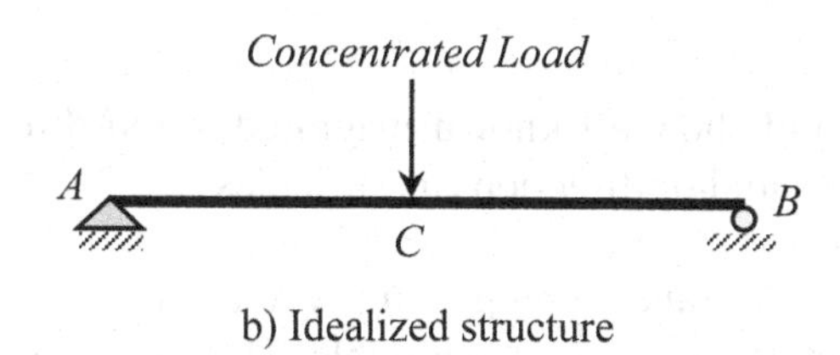

b) Idealized structure

FIGURE 3.3 An example of idealization. *(Photo taken in Pueblo, Colorado)*

For example, the beam AB, shown in Figure 3.3a can be idealized as shown in Figure 3.3b. Instead of showing the transverse beam, a concentrated load at C can be applied on the beam AB.

3.4 FREE-BODY DIAGRAMS

A free-body diagram is the state of a body without the support restrictions. Free-body diagram is essential to understand and apply the equilibrium condition. Free-body diagram is one of the most important concepts in the subject matter, specifically knowing how to draw and use it in statics, strength of materials, fluid mechanics and other similar subjects. This diagram shows all the external forces acting on the particle. We do not show the internal reactions. A free-body diagram is the key to writing the equations of equilibrium, which are used to solve for the unknowns (usually forces or angles).

The procedure to draw a free-body diagram can be summarized as follows:

a. Imagine the particle to be isolated or cut free from its surrounding support or restraints
b. Show all the forces (actions and reactions) that act on the particle

Active forces: They want to move the particle
Reactive forces: They tend to resist the motion

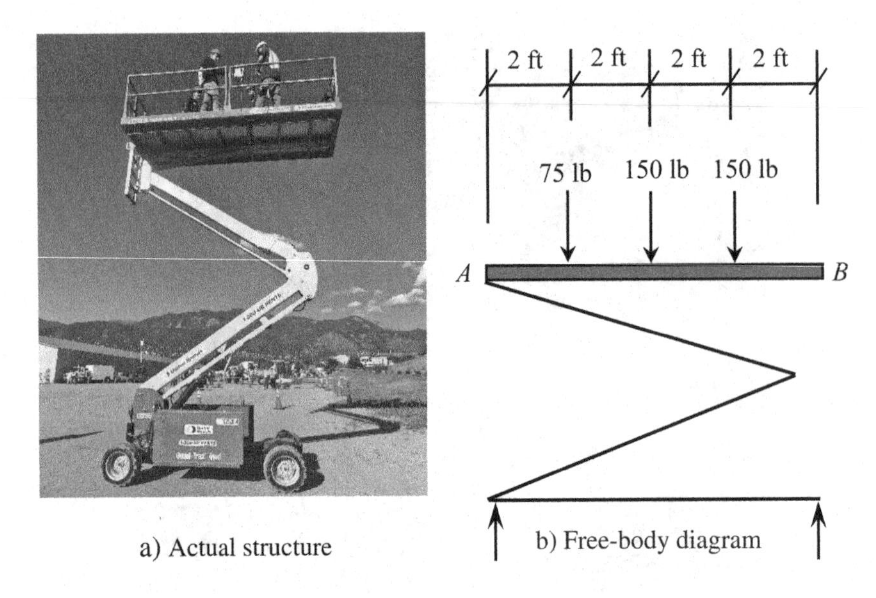

FIGURE 3.4 Drawing the free-body diagram. *(Photo taken in Colorado Springs)*

c. dentify each force and show all known magnitudes and directions and show all unknown magnitudes and/or directions as variables.

Let us see an example. A crane is supporting several crews as shown in Figure 3.4a. Let us consider the weight of the crane is negligible. The reactions of the tires are desired here. Then, the free-body diagram can be drawn as shown in Figure 3.4b where the reactions of the tires are shown by arrows. The weight and distances of the crews are assumed reasonable.

Let us see another example. The beam in Figure 3.5 is supported by two supports: the left support is a pin and the right support is a roller. A point load (concentrated load) is acting on the beam as shown.

To draw the free-body diagram of this beam, the restraints on the beam (the two supports) need to be removed and the all forces and reactions need to be shown. The free-body of the beam can be shown as in Figure 3.6:

Here, A_x and A_y are the reactions at A by the pin support along the x and the y directions, respectively. Remember that a pin support has two unknown reactions. B_y is the reaction along the y-direction at B. The support B has only one unknown reaction, as it is a roller support.

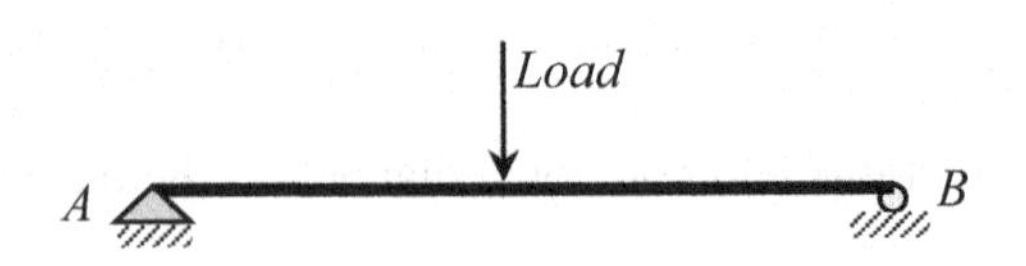

FIGURE 3.5 An idealized arbitrary beam.

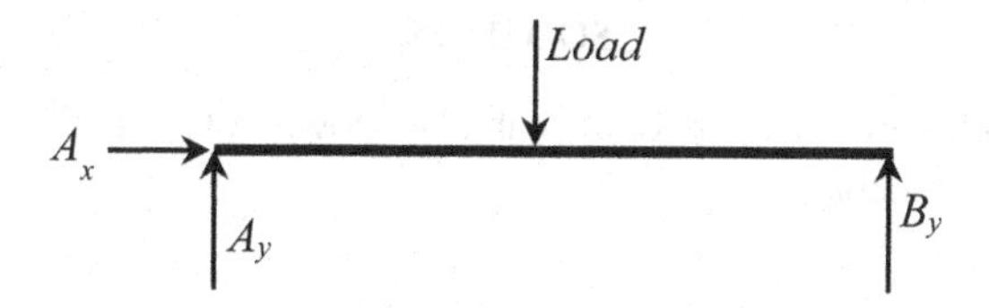

FIGURE 3.6 The free-body diagram of an arbitrary beam.

3.5 APPLICATION OF EQUILIBRIUM EQUATIONS TO PARTICLES

The concept of equilibrium, the equilibrium equations and the procedure to determine the free-body diagram are discussed above. Now, the application of these to analyze different types of structures will be discussed. In this section, the equilibrium equation will be applied to idealized particles and the unknown reactions/forces will be determined. Although the term 'particle' sounds like a tiny object, it more likely means an object dominated by external forces only without deformation. The summary of the analysis procedure can be listed as:

Step 1. Draw the free-body diagram of the concerned member(s)
Step 2. Apply the equilibrium equation(s) to the part of the system or in the whole system depending on the unknown forces.

The following examples will clarify how to apply the equilibrium equations on particles.

Example 3.1:

A 2-kip precast concrete post is being placed on a highway site using a crane. The actual structure and the idealized structure are shown in Figure 3.7. Calculate the tensile force developed in the cable of the crane.

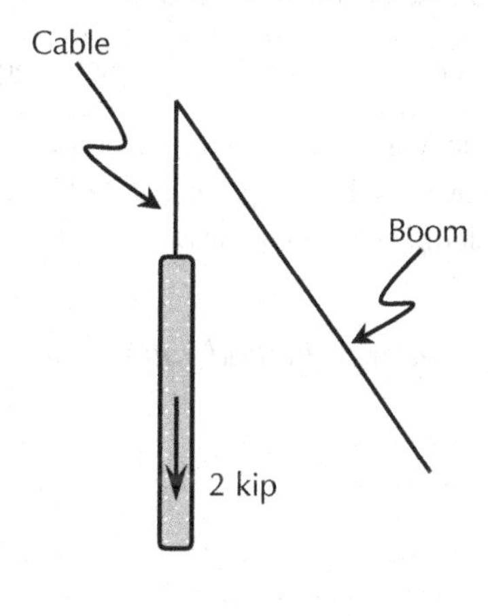

FIGURE 3.7 A crane holding a concrete pole. *(Photo by Armando Perez taken in Pueblo, Colorado)*

SOLUTION

Step 1. Draw the free-body diagram of the cable as shown in Figure 3.8.
Let the force in the cable be T.

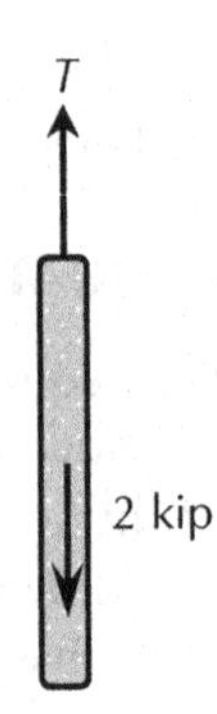

FIGURE 3.8 The free-body diagram of the cable of the crane.

Step 2. Apply the equilibrium equation

$$\sum F_y = 0\ (\uparrow +ve)$$

$$T - 2\text{ kip} = 0$$

Therefore, $T = 2$ kip

Answer: 2 kip.

Note that $\sum F_y = 0\ (\uparrow + ve)$ means the summation of force along the y-direction equals zero and the upward force along the y-direction is considered positive. The equilibrium equation to be used is dependent on the unknown force(s). For example, in Example 3.1, the application of $\sum F_x = 0$ or $\sum M = 0$ is of no benefit. This is why the equation $\sum F_y = 0$ is used to find out the cable force. This strategy is to be used in any problem throughout the textbook. Remember that there are different ways of solving a particular problem. Regardless of the equations or method used to solve the problem, the solution (answer) must be the same. Work on Example 3.2 to clarify this.

Remember: Whatever method you follow, the solutions must be the same.

Example 3.2:

A hydraulic crane lifting a pre-stressed concrete slab weighing 80-kip is shown in Figure 3.9a. The dimensions of the system are shown in the idealized structure in Figure 3.9b. The weight of the slab is 80 kip and is uniformly distributed. Both cables support the load equally.

(a) Actual structure

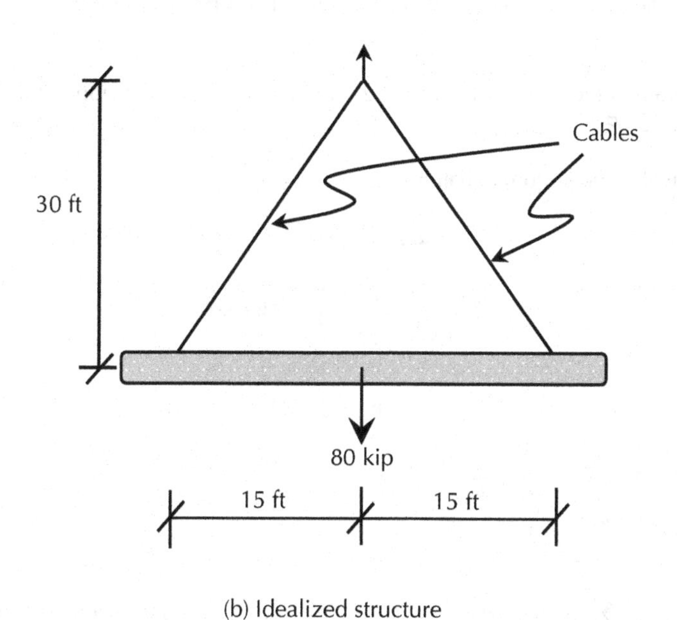

(b) Idealized structure

FIGURE 3.9 A hydraulic crane lifting a concrete slab for Example 3.2. *(Photo by Armando Perez taken in Pueblo, Colorado)*

Determine the tensile force developed in the cables.

SOLUTION

Step 1. Draw the free-body diagram as shown in Figure 3.10. The tension in each of the cables is T; A and B are the cable connection points.

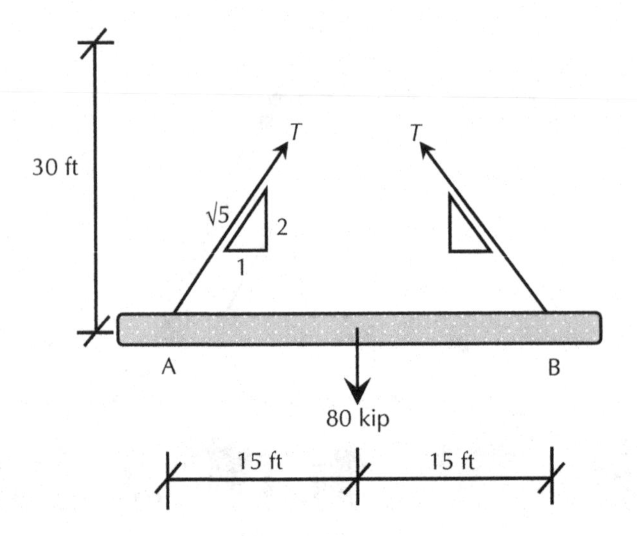

FIGURE 3.10 The free-body diagram of a hydraulic crane lifting a concrete slab.

The slope of the cable is 15:30 or 1:2. The hypotenuse of the slope of the triangle is $\sqrt{(1)^2+(2)^2}=\sqrt{5}$.

Step 2. Apply the equilibrium equations

$$\sum F_y = 0 \ (\uparrow +ve)$$

$$T\left(\frac{2}{\sqrt{5}}\right)+T\left(\frac{2}{\sqrt{5}}\right)-80\text{ kip}=0$$

$$1.79\ T = 80\text{ kip}$$

Therefore, $T = 44.7$ kip

Answer: 44.7 kip

Alternative:
Instead of applying $\sum F_y = 0\ (\uparrow +ve)$, we could use another method. Let us see another way of solving this problem.

$$\sum M \text{ at } A = 0\ (\circlearrowright + ve)$$

$$T\left(\frac{2}{\sqrt{5}}\right)(0\text{ ft})-T\left(\frac{2}{\sqrt{5}}\right)(30\text{ ft})+80\text{ kip}(15\text{ft})=0$$

$$0 - 26.83\ T\text{ ft} + 1{,}200\text{ kip.ft} = 0$$

$$26.83\ T\text{ ft} = 1{,}200\text{ kip.ft}$$

Therefore, $T = 44.7$ kip

Note that $\sum M$ at $A = 0$ (↲+ *ve*) means the summation of moment at A equals zero and the clockwise moment is considered positive.

Remember: there is no sequence of applying the equilibrium equations. You can apply them in any order to solve the problem.

Example 3.3:

A 90-kg man is hanging using two cables as shown in Figure 3.11. Determine the tensions at the cables AB and BC. Cable BC is horizontal and cable AB makes an angle of 20° with the surface.

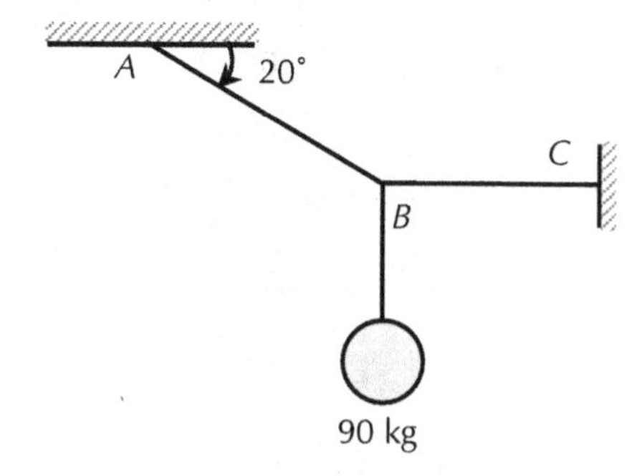

FIGURE 3.11 An idealized structure for Example 3.3.

SOLUTION

Step 1. Draw the free-body diagram as shown in Figure 3.12. The point B can be idealized as a particle and it will be in equilibrium (summation of forces in any direction = 0). Let T_{AB} and T_{BC} be the forces in cables AB and BC, respectively.

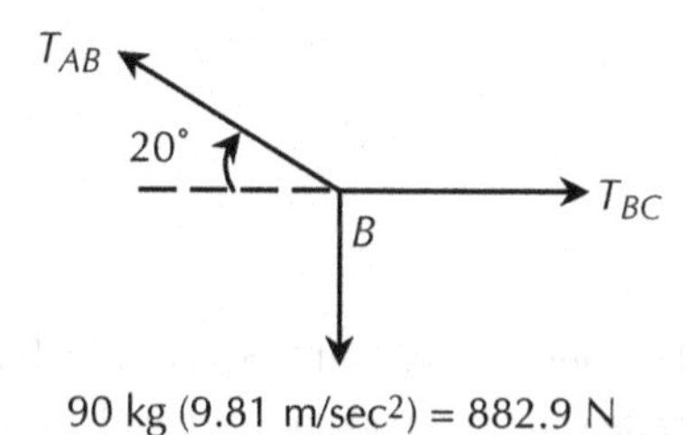

FIGURE 3.12 The free-body diagram of Point B for Example 3.3.

Step 2. Apply the equilibrium equations

If we apply $\sum F_x = 0$ there will be two unknown forces T_{AB} and T_{BC} and we will not get any answer. However, if we apply $\sum F_y = 0$ there will be only one unknown force, T_{AB}. Therefore, the application of $\sum F_y = 0$ is more beneficial for this problem.

$$\sum F_y = 0 \ (\uparrow +ve)$$

$$T_{AB}\,(\sin 20) = 882.9 \text{ N}$$

Therefore, $T_{AB} = 2{,}581.6$ N

Now, $\sum F_x = 0 \ (\rightarrow +ve)$

$$T_{BC} - T_{AB}\,(\cos 20) = 0$$

$$T_{BC} = 2{,}581.6 \text{ N}(\cos 20)$$

Therefore, $T_{BC} = 2{,}425.9$ N

Answers:

$$T_{AB} = 2{,}581.6 \ N$$

$$T_{BC} = 2{,}425.9 \ N$$

Note that $\sum F_x = 0 \ (\rightarrow +ve)$ means the summation of force along the x-direction equals zero and the rightward force along the x-direction is considered positive. Now, practice similar problems to get a better understanding.

Remember: whatever method you follow, the solutions must be the same.

Example 3.4:

A 10-kip concrete vault is to be lifted and placed in place by a bulldozer using four cables as shown in Figure 3.13. The concrete vault is square in shape with the dimensions of 8 ft × 8 ft. The slant length of each cable is 15 ft. The weight (10 kip) of the vault is considered to be acting at its centroid. Determine the tension in each cable.

(a) Actual structure

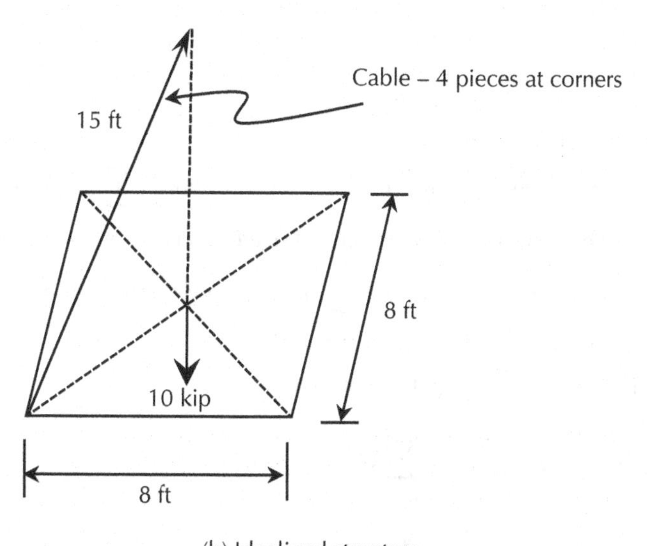

(b) Idealized structure

FIGURE 3.13 Lifting a concrete vault for Example 3.4. *(Photo by Armando Perez taken in Pueblo, Colorado)*

SOLUTION

Step 1. Draw the free-body diagram as shown in Figure 3.14. Only one cable is shown for simplicity.

Diagonal of the square = $\sqrt{(8\,\text{ft})^2 + (8\,\text{ft})^2} = 11.31$ ft

Half of the diagonal of the square = 11.31 ft/2 = 5.66 ft

Length of the cable = 15 ft (given)

Height of the cable, $h = \sqrt{(15\,\text{ft})^2 - (5.66\,\text{ft})^2} = 13.89$ ft

The slope of the cable is 5.66:13.89. The hypotenuse of the slope triangle is 15 ft.

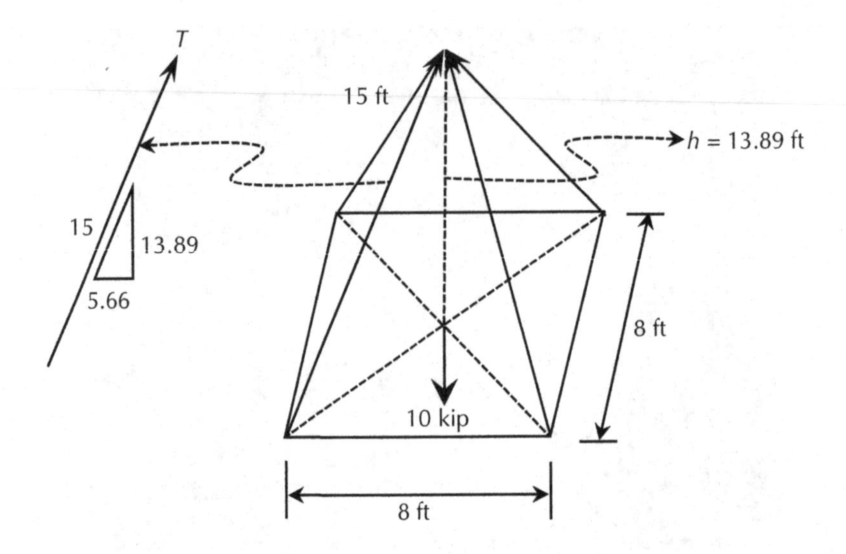

FIGURE 3.14 The free-body diagram of the lifting mechanism of the concrete vault.

Step 2. Apply the equilibrium equations

Again to remind you, we have three known equilibrium equations ($\sum F_x = 0$ or $\sum M = 0$ and $\sum F_y = 0$). Which one to be used first depends on the unknown force. The application of $\sum F_x = 0$ is of no use and the application of $\sum M = 0$ is very complex, as there are four forces at four corners and each one has both vertical and horizontal components. The application of $\sum F_y = 0$ would find out the unknown force quickly.

$$\sum F_y = 0 \ (\uparrow +ve)$$

Four cables will carry this combined load of the concrete vault weighing 10 kip.

Therefore, $4\left[T\left(\frac{13.89}{15}\right)\right] - 10 \text{ kip} = 0$

$$3.704\,T = 10 \text{ kip}$$

$$T = 2.699 \text{ kip} \approx 2.7 \text{ kip}$$

Answer: 2.7 kip.

Example 3.5:

A crane carries five crews with the loading as shown in Figure 3.15. The crane arms make an angle of 60° with the deck. Calculate the forces (F_1 and F_2) in the crane arms.

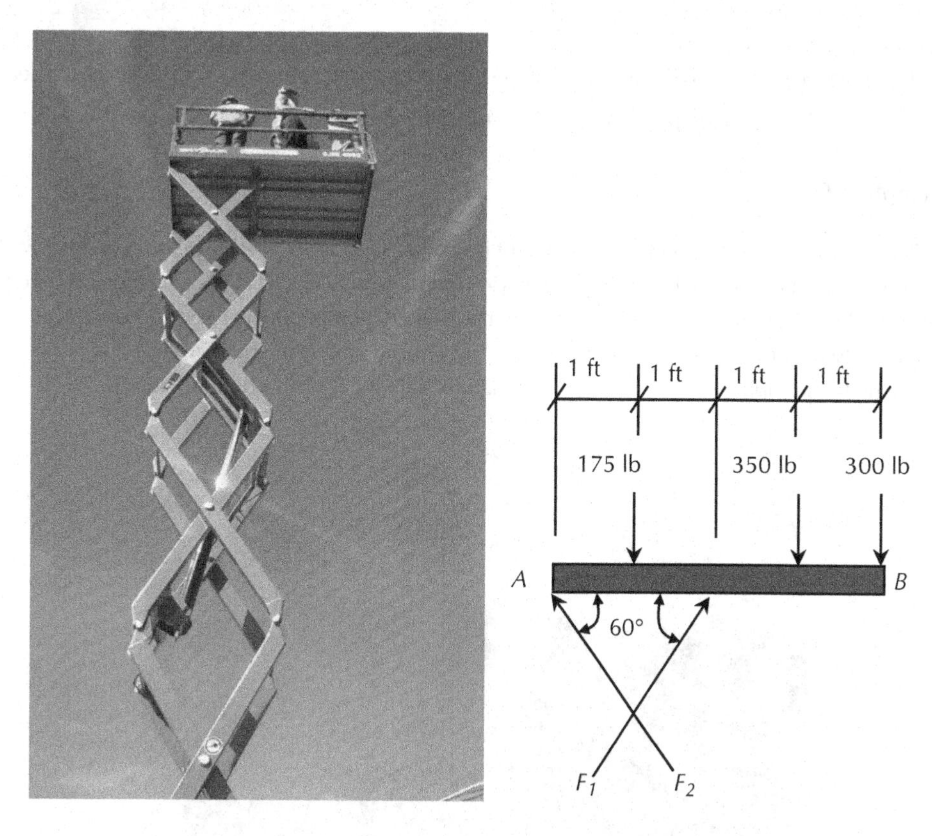

FIGURE 3.15 An actual and the free-body diagram for Example 3.6. *(Photo taken in Colorado Springs)*

SOLUTION

Step 1. Draw the free-body diagram of the support system (already provided).
Step 2. Apply the equilibrium equations.

In this problem, if we apply $\sum F_x = 0$ or $\sum F_y = 0$, there are two unknown forces F_1 and F_2. The application of $\sum M = 0$ would give us a quicker solution.To be in equilibrium, the summation of the moment about the point O must be zero.

$$\sum M \text{ at } A = 0 \ (\curvearrowright + ve)$$

$$(175 \text{ lb})(1.0 \text{ ft}) + (350 \text{ lb})(3.0 \text{ ft}) + (300 \text{ lb})(4.0 \text{ ft}) - (F_1 \sin 60)(2.0 \text{ ft}) = 0$$

$$175 \text{ lb.ft} + 1{,}050 \text{ lb.ft} + 1{,}200 \text{ lb.ft} - 1.732\ F_1 \text{ ft} = 0$$

$$2{,}425 \text{ lb.ft} - 1.732\ F_1 \text{ ft} = 0$$

$$1.732\ F_1 \text{ ft} = 2{,}425 \text{ lb.ft}$$

Therefore, $F_1 = 1{,}400$ lb

$$\sum F_y = 0 \ (\uparrow + ve)$$

$$F_1 \sin 60 + F_2 \sin 60 - 175 \text{ lb} - 350 \text{ lb} - 300 \text{ lb} = 0$$

$$1{,}400 \text{ lb} \sin 60 + F_2 \sin 60 - 825 \text{ lb} = 0$$

$$0.866\ F_2 - 387.44 = 0$$

Therefore, $F_2 = -447.4$ lb

(The negative sign means the direction of the force is opposite to the direction initially assumed)

Answers: $F_1 = 1{,}400$ lb (compression) and $F_2 = 447.4$ lb (tension)

Example 3.6:

While constructing a retaining wall, wood strut is used to support the freshly placed concrete wall as shown in Figure 3.16. At this stage, the soil exerts a force of 500 lb/ft at the height of 1.0 ft from the ground. The foundation of the wall can be considered a hinge support to be conservative.

Calculate the developed force in the wood strut if the system is in equilibrium.

Calculate the support reactions at the foundation wall, if the wall weighs 2.5 kip/ft.

(a) Actual structure

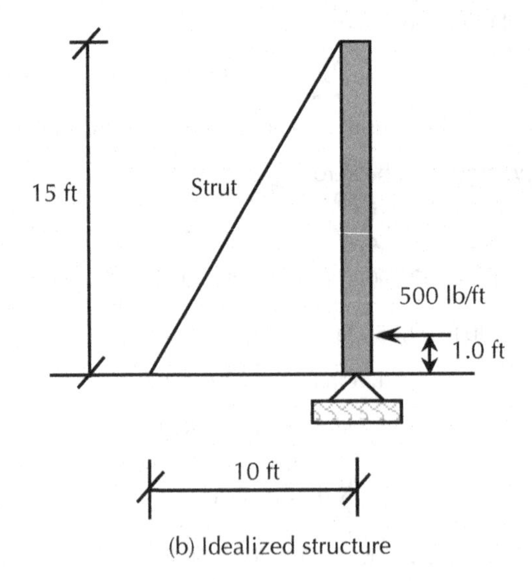

(b) Idealized structure

FIGURE 3.16 An actual and the idealized structure for Example 3.6. *(Photo by Armando Perez taken in Pueblo, Colorado)*

SOLUTION

Step 1. Draw the free-body diagram of the support system as shown in Figure 3.17. Let the force at the strut be P. The base foundation is pinned as the problem stated and the support reactions at the base foundation are O_x and O_y in directions x and y, respectively. The directions of the support reactions along the horizontal and the vertical axes can be assumed in any direction. After the analysis, if it yields a negative sign, it means the direction is opposite of the initial assumption. The weight of the wall (2.5 kip/ft) is also shown in the free-body diagram.

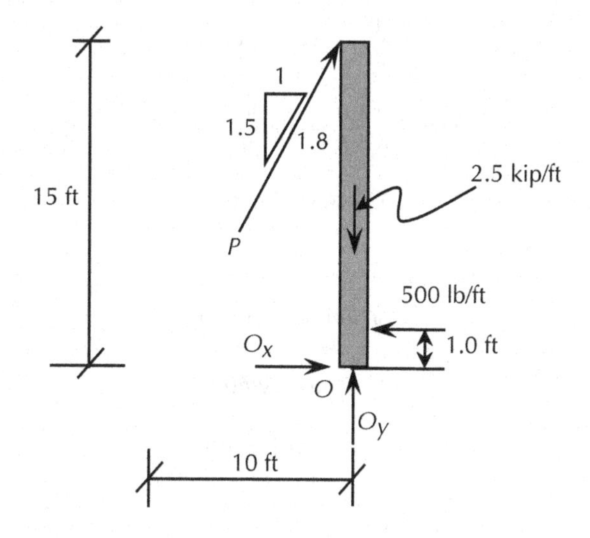

FIGURE 3.17 The free-body diagram of the support system for Example 3.6.

Step 2. Apply the equilibrium equations.

If we apply $\sum F_x = 0$ or $\sum F_y = 0$, there are still two unknown forces in each direction as the strut force P has both components. The application of $\sum M = 0$ would give us a quicker solution.

To be in equilibrium, the summation of the moment about the point O must be zero.

$$\sum M \text{ at } O = 0 \; (\circlearrowright + ve)$$

$$P\left(\frac{1}{1.8}\right)(15\,\text{ft}) - 500\,\frac{\text{lb}}{\text{ft}}(1.0\,\text{ft}) = 0$$

$$8.33P\ \text{ft} - 500\ \text{lb} = 0$$

$$P = 60\ \text{lb/ft}$$

Now, let us apply another equilibrium equation, $\sum F_y = 0\ (\uparrow +ve)$

$$P\left(\frac{1.5}{1.8}\right) - 2{,}500\ \text{lb/ft} + O_y = 0$$

$$O_y = 2{,}500 \text{ lb/ft} - P\,(1.5/1.8) \text{ lb/ft}$$

$$O_y = 2{,}500 \text{ lb/ft} - 60\,(1.5/1.8) \text{ lb/ft}$$

$$O_y = 2{,}500 \text{ lb/ft} - 50 \text{ lb/ft}$$

$$O_y = 2{,}450 \text{ lb/ft } (\uparrow)$$

Now, $\sum F_x = 0 \; (\rightarrow +ve)$

$$P\left(\frac{1.0}{1.8}\right) - 500\frac{\text{lb}}{\text{ft}} + O_x = 0$$

$$\left(60\frac{\text{lb}}{\text{ft}}\right)\left(\frac{1.0}{1.8}\right) - 500\frac{\text{lb}}{\text{ft}} + O_x = 0$$

$$O_x = 466.7 \text{ lb/ft } (\rightarrow)$$

Answers:

$$P = 60 \text{ lb/ft (compressive)}$$

$$O_x = 466.7 \text{ lb/ft } (\rightarrow)$$

$$O_y = 2{,}450 \text{ lb/ft } (\uparrow)$$

Example 3.7:

A 20-kip sphere is supported by two rigid planes as shown in Figure 3.18. The plane, *AB* is inclined with the horizon at an angle of 35° and the other plane *AC* is vertical. The joints of *AB* and *AC* are pin connected. If the contact surfaces are frictionless, determine the forces exerted by the spherical body on the planes *AB* and *AC*.

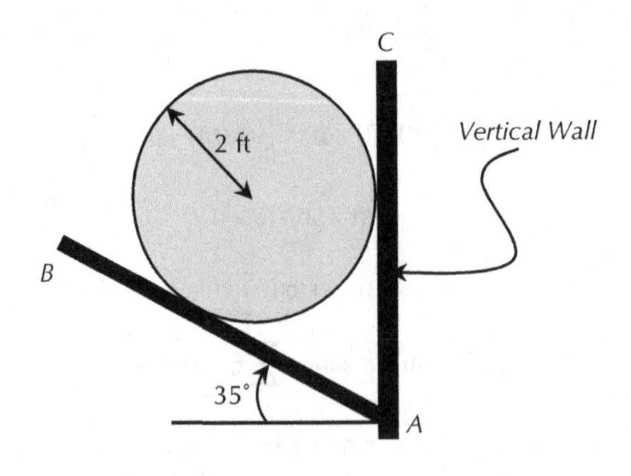

FIGURE 3.18 A spherical body for Example 3.7.

SOLUTION

Step 1. Draw the free-body diagram as shown in Figure 3.19. The free-body diagram of each individual part of the system is shown. Here, R_B is the reaction exerted by the *AB* wall on the spherical body. At the same time, R_B is the applied force by the spherical body to the *AB* wall. Remember Newton's third law: every action has its equal and opposite reaction. Similarly, R_C is the action-reaction between the spherical body and the wall *AC*.

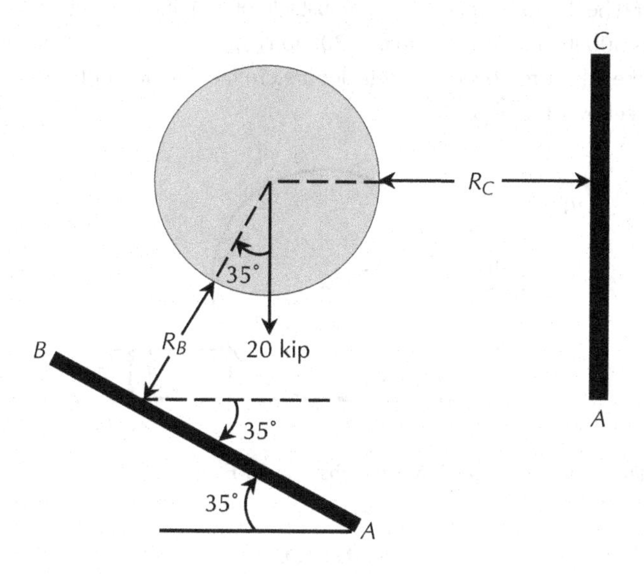

FIGURE 3.19 The free-body diagram of the spherical body for Example 3.7.

Step 2. Apply the equilibrium equations. Let us apply, $\sum F_y = 0$ in the free-body diagram of the spherical body only. The reason for choosing $\sum F_y = 0$ is that there is only one unknown force (R_B) along the *y*-direction. Along the *x*-direction, there are two unknown forces (R_C and the component of R_B).

$$\sum F_y = 0\ (\uparrow +ve)$$

$$R_B \cos 35 - 20 \text{ kip} = 0$$

$$R_B \cos 35 = 20 \text{ kip}$$

$$R_B = 20 \text{ kip}/\cos 35$$

$$R_B = 24.4 \text{ kip}$$

$$\sum F_x = 0\ (\rightarrow +ve)$$

$$R_B \sin 35 - R_C = 0$$

$$R_C = R_B \sin 35$$

$$R_C = 24.4 \text{ kip} \sin 35$$

$$R_C = 14 \text{ kip}$$

Answers:

$$R_B = 24.4\ kip$$

$$R_C = 14\ kip$$

Example 3.8:

A 30-in. diameter wheel is to be rolled over an obstacle of 7-in. in height as shown in Figure 3.20. The wheel load is 300 lb. If a pushing force of 200 lb is applied as shown at an angle of 40° with the ground, will the wheel roll over the obstacle? Determine the factor of safety against the overturning of the wheel over the obstacle.

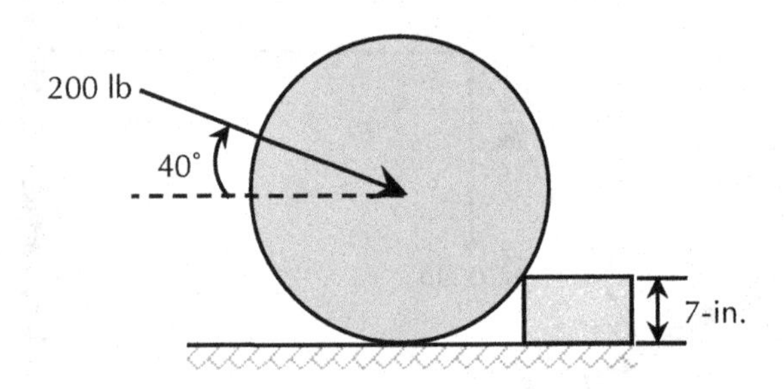

FIGURE 3.20 The wheel to be rolled over the obstacle for Example 3.8.

SOLUTION

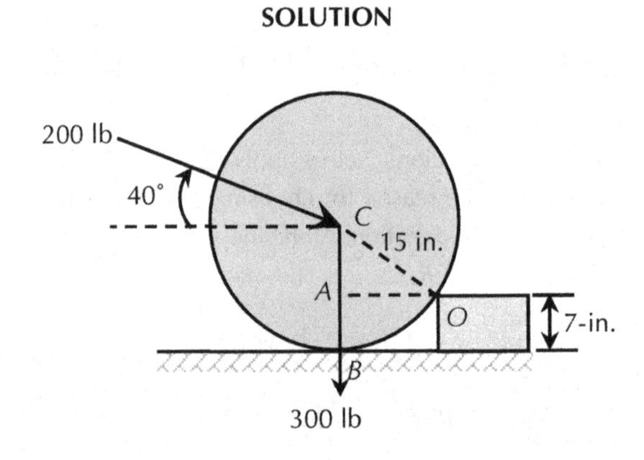

FIGURE 3.21 The free-body diagram of the wheel to be rolled over for Example 3.8.

Step 1. Draw the free-body diagram as shown in Figure 3.21.

Step 2. Apply the equilibrium equations. To intend to roll over the obstacle, the overturning moment with respect to the point O must be equal to the resisting moment (or the clockwise and the counter-clockwise moments must be equal).

Radius, $OC = 15$ in.

$$CA = BC - AB = 15\text{ in.} - 7\text{ in.} = 8\text{ in.}$$

$$OA = \sqrt{OC^2 - AC^2} = \sqrt{(15\text{ in.})^2 - (8\text{ in.})^2} = 12.69\text{ in.}$$

Overturning moment about O = 200 lb cos40 (8 in.) = 1,225.7 lb.in.

Resisting moment about O = 200 lb sin40 (12.69 in.) + 300 lb (12.69 in.) = 5,438.4 lb.in.

The resisting moment is much larger than the overturning moment. So, the body will not roll over the obstacle.

$$\text{Factor of Safety} = \frac{\text{Resisting Moment}}{\text{Overturning Moment}} = \frac{5,438.4\,\text{lb.in.}}{1,225.7\,\text{lb.in.}} = 4.4$$

Answer: the body will not roll over, and the factor of safety against overturning is 4.4.

3.6 APPLICATION OF EQUILIBRIUM EQUATIONS TO RIGID BODIES

In this section, the equilibrium equation will be applied to idealized rigid bodies such as beams or different members. Other types of engineering structures such as trusses, arches, cables, pulleys, frames etc. will be discussed in future chapters. The following examples will clarify how to apply the equilibrium equations to rigid bodies.

Example 3.9:

A 40-ft steel-beam bridge has a uniformly distributed load of 2 kip/ft as shown in Figure 3.22. The left support A can be considered as pinned and the right support B is a roller. What are the support reactions at both supports of the following beam?

(a) Actual beam

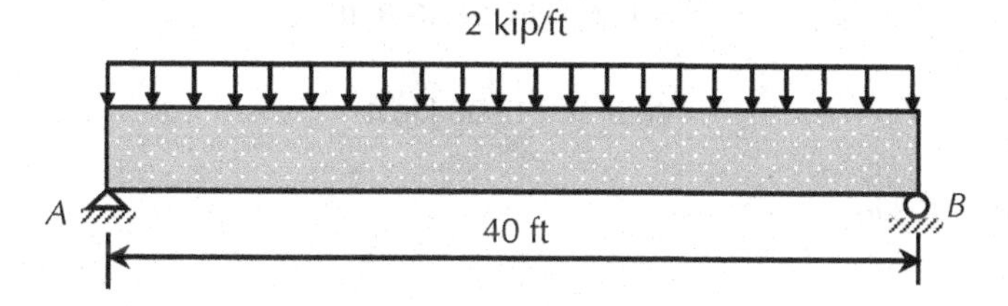

FIGURE 3.22 A beam with a uniformly distributed load for Example 3.9. *(Photo taken in Pueblo, Colorado)*

SOLUTION

Step 1. Draw the free-body diagram showing the unknown reactions and the applied force as shown in Figure 3.23. Let the support reactions at the support A be A_x and A_y along the horizontal and the vertical directions, respectively. The vertical reaction at B is B_y. Support A has two unknowns, as it is a pin support. Support B has one unknown, as it is a roller support. The uniformly distributed load (also called the rectangular load) of 2 kip/ft can be considered concentrated at the center of the rectangular loading area, i.e., in the middle of the beam. Thus, a uniformly distributed load of 2 kip/ft over the 40-ft span can be considered a concentrated load of 2 kip/ft(40 ft) = 80 kip acting at the center of the beam (40 ft/2 = 20 ft from the left support).

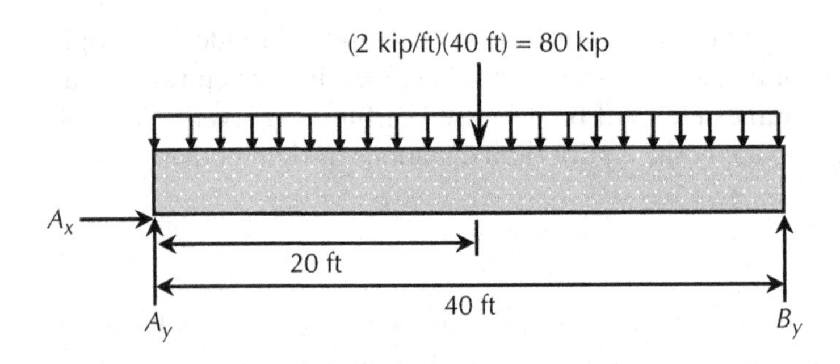

FIGURE 3.23 The free-body diagram of the beam for Example 3.9.

Step 2. Apply the equilibrium equations.
If we apply $\sum F_y = 0$, there will be two unknown forces: A_y and B_y. The application of $\sum M = 0$ would give us a quicker solution for A_y or B_y. The application of $\sum F_x = 0$ will find out A_x.

$$\sum F_x = 0\ (\rightarrow +ve)$$

$$A_x = 0$$

$$\sum M \text{ at } A = 0\ (\circlearrowright +ve)$$

$$A_x(0) + A_y(0) + (80\text{ kip})(20\text{ ft}) - B_y(40\text{ ft}) = 0$$

$$0 + 0 + 1{,}600\text{ kip.ft} - 40\ B_y\text{ ft} = 0$$

$$40\ B_y\text{ ft} = 1{,}600\text{ kip.ft}$$

Therefore, $B_y = 40$ kip
Now, $\sum F_y = 0\ (\uparrow +ve)$

$$A_y + B_y - 80\text{ kip} = 0$$

$$A_y + 40\text{ kip} - 80\text{ kip} = 0$$

$$A_y - 40 \text{ kip} = 0$$

$$A_y = 40 \text{ kip}$$

Answer: Reactions at both supports are 40 kip.

Remember this procedure of solving a simply supported beam. A good number of simply supported beam problems are required to be solved in your academic and professional life. Nonetheless, the simply supported beam with uniform loading is the most widely used beam in civil engineering. In real life, most beams are commonly considered as simply supported for an easy solution and to be conservative. The total load on the beam is divided by the span to determine the load per linear foot. This type of problem can be also be solved visually. The loading is symmetric about the centerline of the beam, and thus each support carries 50% of the total load (50% of 80 kip = 40 kip). Let us work on another similar problem.

Example 3.10:

A 30-ft I-shaped-steel beam has a uniformly distributed load of 750 lb/ft as shown in Figure 3.24. Calculate the support reactions at the supports of the beam.

(a) Actual I-beam

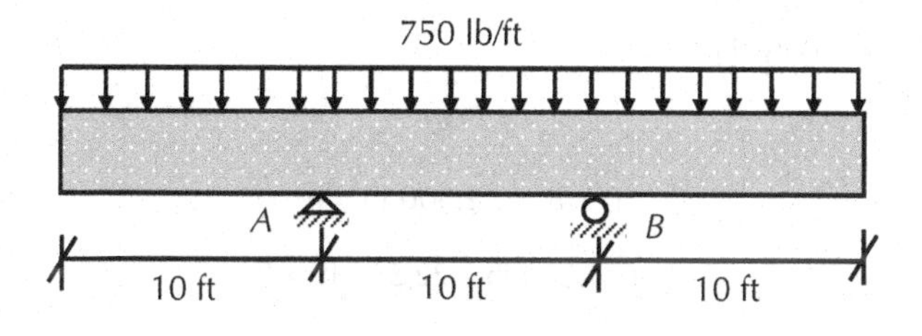

(b) Idealized beam

FIGURE 3.24 A Beam with a uniformly distributed load for Example 3.10. *(Photo taken in Canyon City, Colorado)*

SOLUTION

Step 1. Draw the free-body diagram showing the unknown reactions and the applied force as shown in Figure 3.25. Let the support reactions at the support A be A_x and A_y along the horizontal and the vertical directions, respectively. The vertical reaction at B is B_y. Support A has two unknowns, as it is a pin support. Support B has one unknown, as it is a roller support. The uniformly distributed load of 750 lb/ft over the 30-ft span can be considered a concentrated load of 750 lb/ft(30 ft) = 22,500 lb acting at the center of the beam (10 ft/2 = 5 ft from the left support).

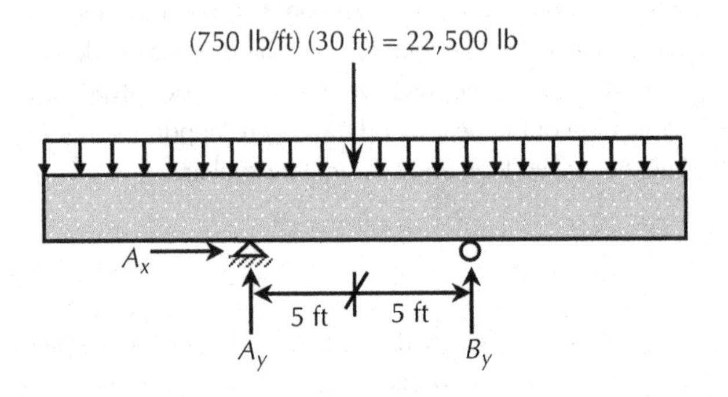

FIGURE 3.25 The free-body diagram of the beam for Example 3.10.

Step 2. Apply the equilibrium equations.

$$\sum F_x = 0 \ (\rightarrow +ve)$$

$$A_X = \mathbf{0}$$

$$\sum M \text{ at } A = 0 \ (\lrcorner + ve)$$

$$A_x(0) + A_y(0) + (22,500\,\text{lb})(5\,\text{ft}) - B_y(10\,\text{ft}) = 0$$

$$0 + 0 + 112,500 \text{ lb.ft} - 10\ B_y \text{ ft} = 0$$

$$112,500 \text{ lb.ft} - 10\ B_y \text{ ft} = 0$$

$$10\ B_y \text{ ft} = 112,500 \text{ lb.ft}$$

Therefore, $B_y = 11,250$ lb (↑)

Now, $\sum F_y = 0 \ (\uparrow +ve)$

$$A_y + B_y - 22,500 \text{ lb} = 0$$

$$A_y + 11,250 \text{ lb} - 22,500 \text{ lb} = 0$$

$$A_y - 11,250 \text{ lb} = 0$$

$$A_y = 11,250 \text{ lb } (\uparrow)$$

Answer: the reaction at both supports is 11,250 lb (↑).

Example 3.11

Determine the support reactions at the supports of the following beam shown in Figure 3.26.

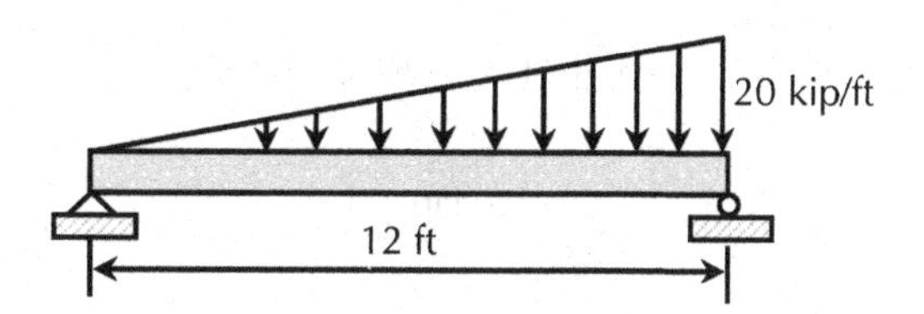

FIGURE 3.26 A beam with a uniformly varying load for Example 3.11.

SOLUTION

Step 1. Draw the free-body diagram showing the unknown reactions and the applied force as shown in Figure 3.27. Let the support reactions at the support A be A_x and A_y along the horizontal and the vertical directions, respectively. The vertical reaction at B is B_y. The support A has two unknowns, as it is a pin support. The support B has one unknown, as it is a roller support. The uniformly varying load (also called the triangular load) can be considered concentrated at the center of the triangular loading area, i.e., at one-third the distance from the longer side of the load. Thus, a uniformly varying load over the 30-ft span can be considered a concentrated load of ½ (20 kip/ft)(12 ft) = 120 kip acting at 1/3(12 ft) = 4 ft from the right support [2/3(12 ft) = 8 ft from the left support].

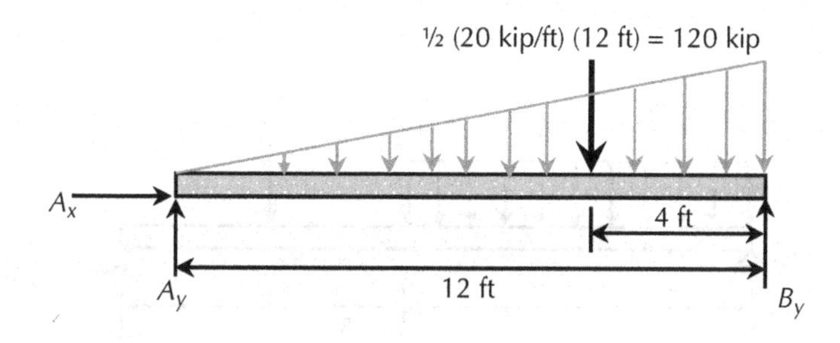

FIGURE 3.27 The free-body diagram of the beam for Example 3.11.

Step 2. Apply the equilibrium equations to determine the unknown reactions.

$$\sum F_x = 0\ (\rightarrow +ve)$$

Therefore, $A_x = 0$

$$\sum M \text{ at } A = 0\ (\circlearrowright + ve)$$

$$120 \text{ kip } (8 \text{ ft}) - B_y\ (12 \text{ ft}) = 0$$

$$960 \text{ kip.ft} - 12\ B_y \text{ ft} = 0$$

$$12\ B_y \text{ ft} = 960 \text{ kip.ft}$$

Therefore, $B_y = 80$ kip (↑)

$$\sum F_y = 0 \ (\uparrow +ve)$$

$$-120 \text{ kip} + A_y + B_y = 0$$

$$A_y = 120 \text{ kip} - B_y$$

$$A_y = 120 \text{ kip} - 80 \text{ kip}$$

Therefore, $A_y = 40$ kip (↑)

Answers:

$$A_x = 0$$

$$A_y = 40 \text{ kip } (\uparrow)$$

$$B_y = 80 \text{ kip } (\uparrow)$$

Example 3.12

Determine the support reactions at both supports of the beam shown in Figure 3.28.

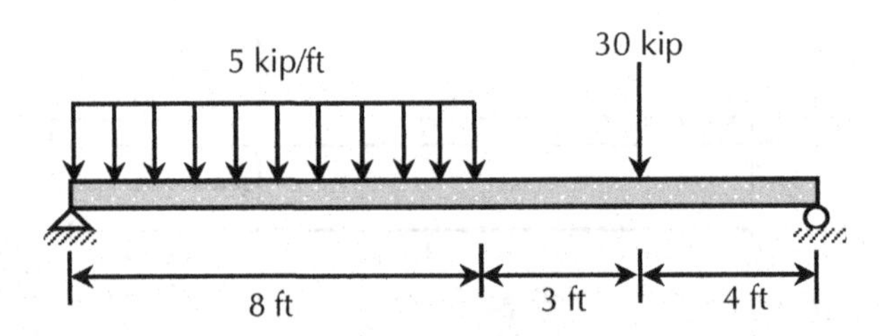

FIGURE 3.28 A beam with arbitrary loading for Example 3.12.

SOLUTION

Step 1. Draw the free-body diagram showing the unknown reactions and the applied force as shown in Figure 3.29. Let the support reactions at the support A be A_x and A_y along the horizontal and the vertical directions, respectively. The vertical reaction at B is B_y. The support A has two unknowns, as it is a pin support. The support B has one unknown, as it is a roller support. The resultant of the rectangular load [5 kip/ft (8 ft) = 40 kip] is applied at its centroid [½ (8 ft) = 4 ft from the left support]. The concentrated load of 30 kip remains as it is.

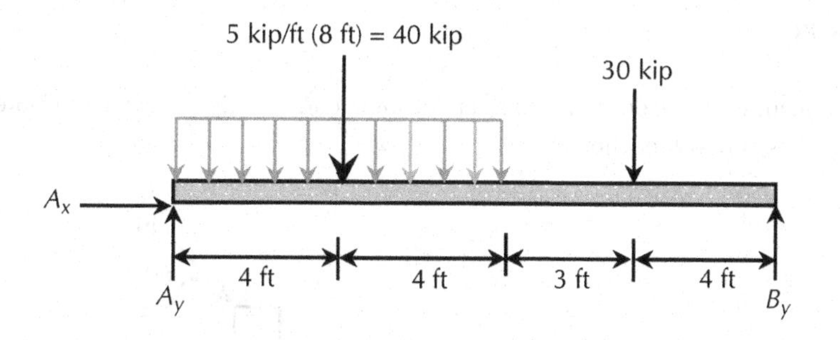

FIGURE 3.29 The free-body diagram of the beam for Example 3.12.

Step 2. Apply the equilibrium equations

$$\sum M \text{ at } A = 0 \; (\circlearrowright + ve)$$

$$40 \text{ kip } (4 \text{ ft}) + 30 \text{ kip } (11 \text{ ft}) - B_y \, (15 \text{ ft}) = 0$$

$$160 \text{ kip.ft} + 330 \text{ kip.ft} - 15 \, B_y \text{ ft} = 0$$

$$490 \text{ kip.ft} - 15 \, B_y \text{ ft} = 0$$

$$15 \, B_y \text{ ft} = 490 \text{ kip.ft}$$

$$B_y = 32.7 \text{ kip } (\uparrow)$$

$$\sum F_x = 0 \; (\rightarrow +ve)$$

Therefore, $A_x = 0$

$$\sum F_y = 0 \; (\uparrow +ve)$$

$$-40 \text{ kip} - 30 \text{ kip} + A_y + B_y = 0$$

$$A_y = 40 \text{ kip} + 30 \text{ kip} - B_y$$

$$A_y = 70 \text{ kip} - 32.7 \text{ kip}$$

Therefore, $A_y = 37.3$ kip (↑)

Answers:

$$A_x = 0$$

$$A_y = 37.3 \; kip \; (\uparrow)$$

$$B_y = 32.7 \; kip \; (\uparrow)$$

Example 3.13

A cantilever beam with a fixed support has an inclined load of 5 kip as shown in Figure 3.30. Determine the support reactions.

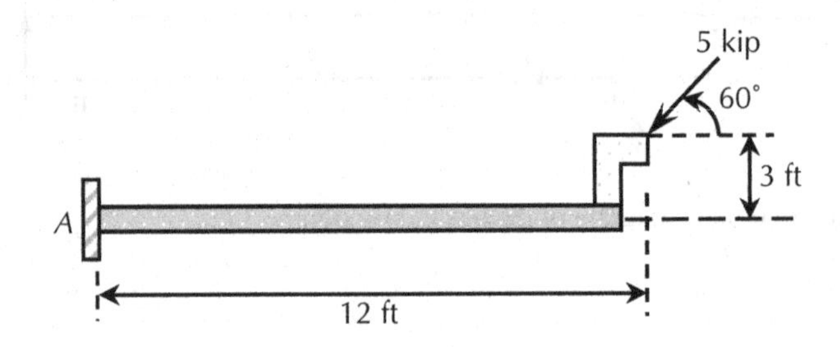

FIGURE 3.30 A beam with arbitrary loading for Example 3.13.

SOLUTION

Step 1. Draw the free-body diagram showing the unknown reactions and the applied force as shown in Figure 3.31. The support A is a fixed support; thus, it has three unknown forces of A_x, A_y and M. The directions of the support reactions along the horizontal and the vertical axes can be assumed in any direction. After the analysis, if it yields a negative sign, it means the direction is the opposite of the initial assumption. The inclined load of 5 kip acting at 60° with the horizon is resolved into horizontal and vertical components of 5cos60 and 5sin60, respectively.

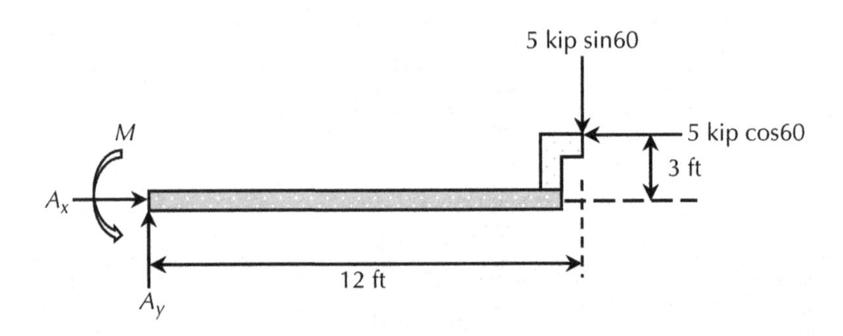

FIGURE 3.31 The free-body diagram of the beam for Example 3.13.

Step 2. Apply the equilibrium equations

$$\sum M \text{ at } A = 0 \; (\circlearrowright + ve)$$

$$-M + 5 \text{ kip } \sin 60 \text{ (12 ft)} - 5 \text{ kip } \cos 60 \text{ (3 ft)} = 0$$

$$-M + 51.96 \text{ kip.ft} - 7.5 \text{ kip.ft} = 0$$

$$-M + 44.5 \text{ kip.ft} = 0$$

$$M = 44.5 \text{ kip.ft (counter-clockwise)}$$

$$\sum F_x = 0 \; (\rightarrow +ve)$$

$$A_x - 5 \text{ kip} \cos 60 = 0$$

$$A_x - 2.5 \text{ kip} = 0$$

$$A_x = 2.5 \text{ kip} \; (\rightarrow)$$

$$\sum F_y = 0 \; (\uparrow +ve)$$

$$A_y - 5 \text{ kip} \sin 60 = 0$$

$$A_y = 4.3 \text{ kip} \; (\uparrow)$$

Answers:

$$A_x = 2.5 \; kip \; (\rightarrow)$$

$$A_y = 4.3 \; kip \; (\uparrow)$$

$$M = 44.5 \; kip.ft \; (counter\text{-}clockwise)$$

Example 3.14:

A simply supported beam has the loading shown in Figure 3.32. Determine the reactions developed at the supports. Support A is a pin, and support B is a roller.

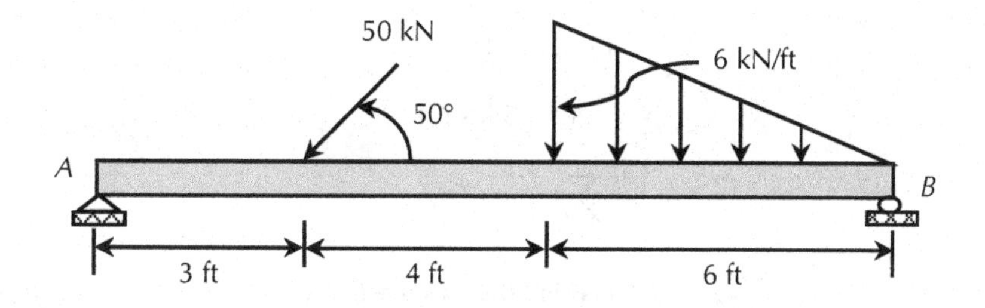

FIGURE 3.32 A beam with arbitrary loading for Example 3.14.

SOLUTION

Step 1. Draw the free-body diagram showing the unknown reactions and the applied force shown in Figure 3.33. Let the support reactions at the support A be A_x and A_y along the horizontal and the vertical directions, respectively. The vertical reaction at B is B_y. The resultant of the triangular load [½ (6 kN/ft)(6 ft) = 18 kN] is applied at its centroid (2/3 of 6 ft = 4 ft from the right support). The inclined load of 50 kN acting at 50° with the horizon is resolved into horizontal and vertical components of 50cos50 and 50sin50, respectively.

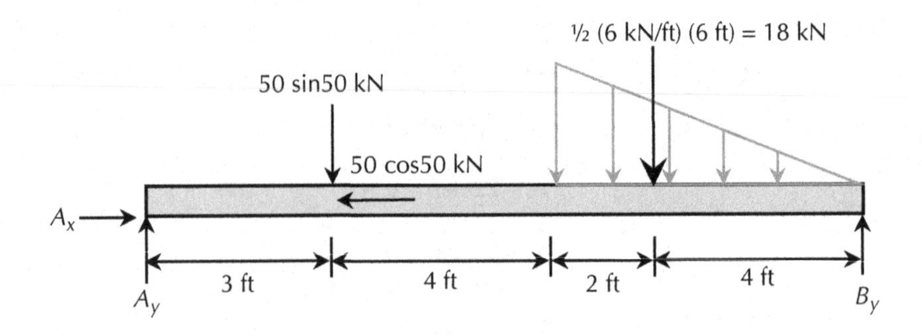

FIGURE 3.33 The free-body diagram of the beam for Example 3.14.

Step 2. Apply the equilibrium equations to determine the unknown reactions

$$\sum F_x = 0 \ (\rightarrow +ve)$$

$$A_x - 50 \text{ kN} \cos 50 = 0$$

$$A_x - 32.1 \text{ kN} = 0$$

$$A_x = 32.1 \text{ kN} \ (\rightarrow)$$

$$\sum M \text{ at } B = 0 \ (\lrcorner + ve)$$

$$A_y \,(13 \text{ ft}) - 50 \text{ kN} \sin 50 \,(10 \text{ ft}) - 18 \text{ kN} \,(4 \text{ ft}) = 0$$

$$A_y \,(13 \text{ ft}) - 383 \text{ kN.ft} - 72 \text{ kN.ft} = 0$$

$$13\, A_y \text{ ft} - 455 \text{ kN.ft} = 0$$

$$13\, A_y \text{ ft} = 455 \text{ kN.ft}$$

$$A_y = 35 \text{ kN} \ (\uparrow)$$

$$\sum F_y = 0 \ (\uparrow +ve)$$

$$A_y - 50 \text{ kN} \sin 50 - 18 \text{ kN} + B_y = 0$$

$$35 \text{ kN} - 50 \text{ kN} \sin 50 - 18 \text{ kN} + B_y = 0$$

$$B_y = 38.3 \text{ kN} + 18 \text{ kN} - 35 \text{ kN}$$

$$B_y = 21.3 \text{ kN} \ (\uparrow)$$

Answers:

$$A_x = 32.1 \ kN \ (\rightarrow)$$

$$A_y = 35 \ kN \ (\uparrow)$$

$$B_y = 21.3 \ kN \ (\uparrow)$$

Example 3.15:

Determine the supports reactions of the truss system shown in Figure 3.34. Support E is a pin, and support F is a roller.

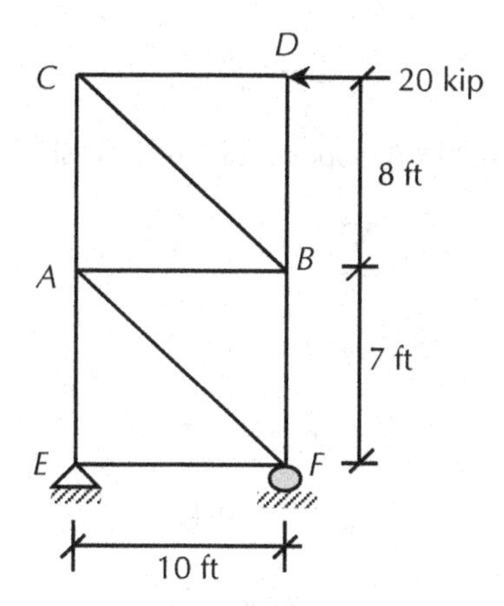

FIGURE 3.34 A truss with arbitrary loading for Example 3.15.

SOLUTION

Step 1. Draw the free-body diagram as shown in Figure 3.35. Support E has two unknowns for being a pin support, and support F has only one unknown for being a roller support.

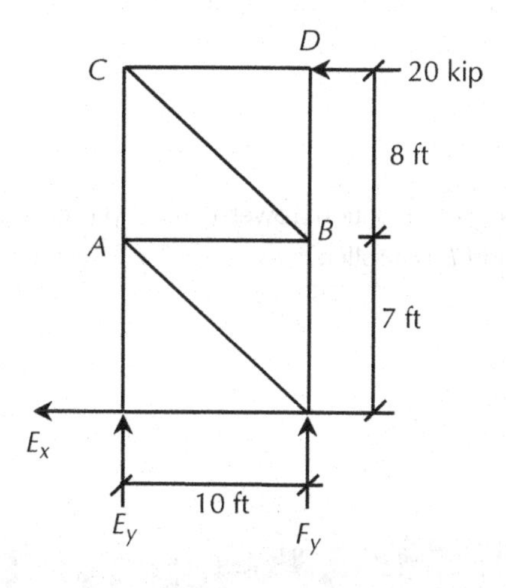

FIGURE 3.35 The free-body diagram of the truss for Example 3.15.

Step 2. Apply the equilibrium equations

$$\Sigma M_E = 0 \text{ (Assume, counter-clockwise moment is positive)}$$

$$20 \text{ kip } (15 \text{ ft}) + F_y (10 \text{ ft}) = 0$$

$$300 \text{ kip.ft} + 10F_y \text{ ft} = 0$$

$$10F_y \text{ ft} = -300 \text{ kip.ft}$$

$$F_y = -30 \text{ kip}$$

$$\sum F_x = 0 \; (\rightarrow +ve)$$

$$E_x + 20 \text{ kip} = 0$$

$$E_x = -20 \text{ kip (opposite of the arrow shown)}$$

$$\sum F_y = 0 \; (\uparrow +ve)$$

$$E_y + F_y = 0$$

$$E_y = -F_y$$

$$E_y = -(-30) \text{ kip}$$

$$E_y = 30 \text{ kip } (\uparrow)$$

Answers:

$$F_y = 30 \; kip \; (\downarrow)$$

$$E_y = 30 \; kip \; (\uparrow)$$

$$E_x = 20 \; kip \; (\rightarrow)$$

EXAMPLE 3.16:

The truss in the Niagara Falls Observation Tower can be presented as shown in Figure 3.36. Support A is a pin, and support B is a roller.

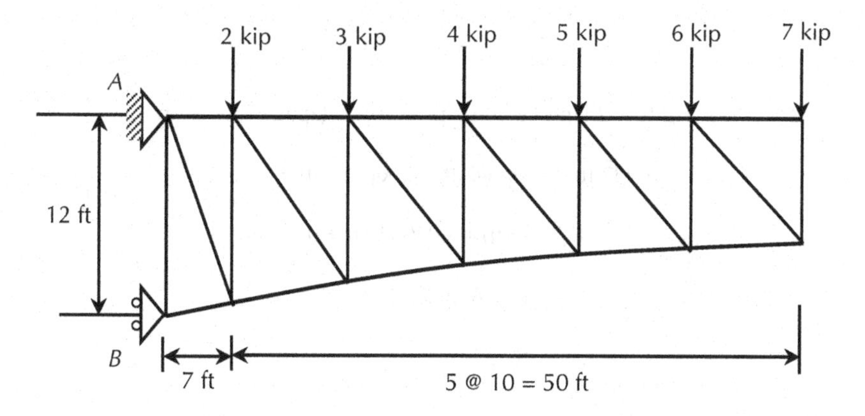

FIGURE 3.36 A truss with arbitrary loading for Example 3.16. *(Photo taken in Niagara Falls, New York)*

Determine the reactions at the supports A and B.

SOLUTION

Step 1. Draw the free-body diagram as shown in Figure 3.37. Three reactions, A_x, B_x and A_y are assumed.

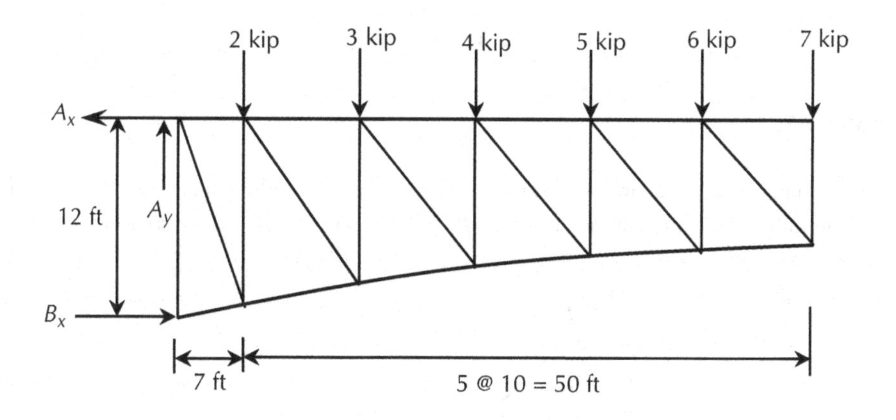

FIGURE 3.37 The free-body diagram of the truss for Example 3.16.

Step 2. Apply the equilibrium equations

$$\sum F_y = 0 \ (\uparrow +ve)$$

$$A_y - 2 \text{ kip} - 3 \text{ kip} - 4 \text{ kip} - 5 \text{ kip} - 6 \text{ kip} - 7 \text{ kip} = 0$$

$$A_y - 27 \text{ kip} = 0$$

$$A_y = 27 \text{ kip} \ (\uparrow)$$

$$\sum M \text{ at } A = 0 \; (\lrcorner + ve)$$

$$-B_x\,(12\text{ ft}) + 2\text{ kip}\,(7\text{ ft}) + 3\text{ kip}\,(17\text{ ft}) + 4\text{ kip}\,(27\text{ ft}) + 5\text{ kip}$$

$$(37\text{ ft}) + 6\text{ kip}\,(47\text{ ft}) + 7\text{ kip}\,(57\text{ ft}) = 0$$

$$-B_x\,(12\text{ ft}) + 1{,}039\text{ kip.ft} = 0$$

$$B_x = 86.6\text{ kip}\;(\rightarrow)$$

$$\sum F_x = 0\;(\rightarrow +ve)$$

$$B_x - A_x = 0$$

$$A_x = B_x$$

$$A_x = 86.6\text{ kip}\;(\leftarrow)$$

Answers:

$$A_y = 27\text{ kip}\;(\uparrow)$$

$$A_x = 86.6\text{ kip}\;(\leftarrow)$$

$$B_x = 86.6\text{ kip}\;(\rightarrow)$$

Example 3.17:

A 1,200-kg load is being sustained by a beam *AB* and a cable as shown in Figure 3.38. Determine the cable tension. The cable can be considered a link support and *A* is a pin support.

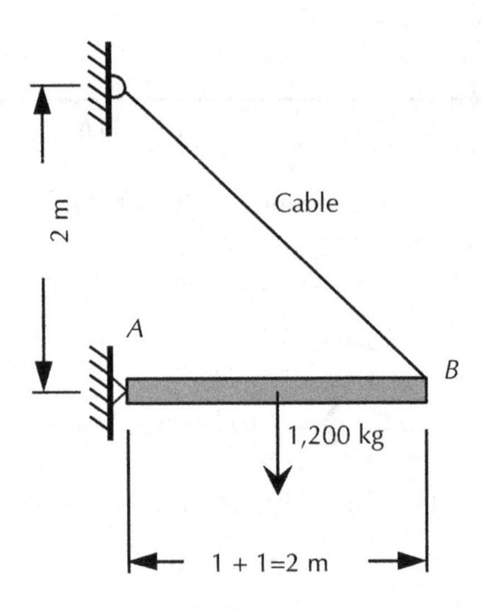

FIGURE 3.38 A structure with arbitrary loading for Example 3.17.

SOLUTION

Step 1. Draw the free-body diagram as shown in Figure 3.39.

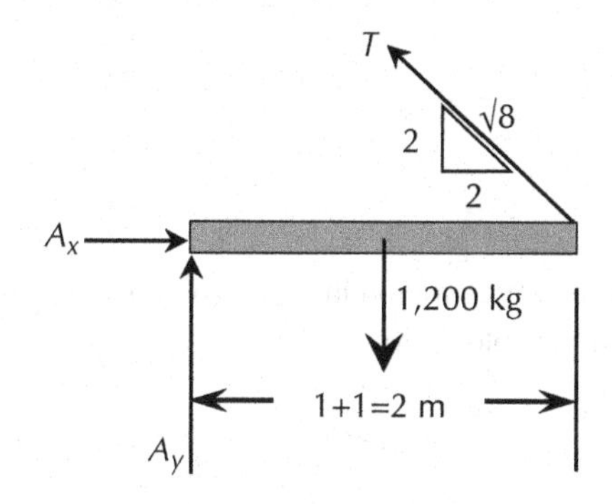

FIGURE 3.39 The free-body diagram of the structure for Example 3.17.

Step 2. Apply the equilibrium equations

$$\sum M \text{ at } A = 0 \; (\circlearrowleft + ve)$$

$$-T\left(\frac{2}{\sqrt{8}}\right) \times 2\text{m} + \left(1{,}200 \text{ kg} \times 9.81 \frac{\text{m}}{\text{sec}^2}\right) \times 1\text{m} = 0$$

$$T = 8{,}324 \text{N}$$

Note that kg is the unit of mass. To convert mass (kg) into force (N), the gravitation acceleration (9.81 m/sec^2 or 32.2 ft/sec^2) is to be multiplied as shown in the above computation.

$$\sum F_x = 0 \; (\rightarrow +ve)$$

$$A_x - T\left(\frac{2}{\sqrt{8}}\right) = 0$$

$$A_x - 8{,}324 \text{N}\left(\frac{2}{\sqrt{8}}\right) = 0$$

$$A_x - 5{,}886 \text{ N} = 0$$

$$A_x = 5{,}886 \text{ N } (\rightarrow)$$

$$\sum F_y = 0 \; (\uparrow +ve)$$

$$A_y + T\left(\frac{2}{\sqrt{8}}\right) - \left(1{,}200 \text{ kg} \times 9.81 \frac{\text{m}}{\text{sec}^2}\right) = 0$$

$$A_y + 8{,}324 \text{ N}\left(\frac{2}{\sqrt{8}}\right) - 11{,}772 \text{N} = 0$$

$$A_y - 5{,}886 \text{ N} = 0$$

$$A_y = 5{,}886 \text{ N } (\uparrow)$$

Answers:

$$T = 8{,}324\ N$$

$$A_y = 5{,}886\ \text{N}\ (\uparrow)$$

$$A_x = 5{,}886\ \text{N}\ (\rightarrow)$$

Example 3.18:

A 2-m cubic block is in equilibrium condition in an inclined plane as shown in Figure 3.40. If the inclination is raised, when will the block overturn?

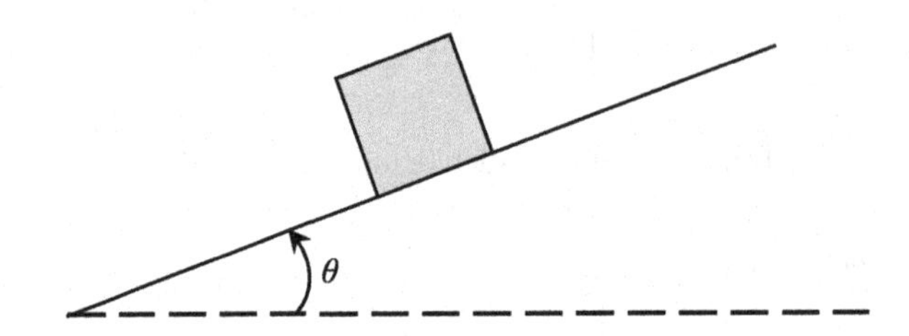

FIGURE 3.40 A block in an inclined plane for Example 3.18.

SOLUTION

Let the weight of the block be W. Then, its component of weight along the inclined plane be $W\sin\theta$ and normal to the plane be $W\cos\theta$ as shown in Figure 3.41.

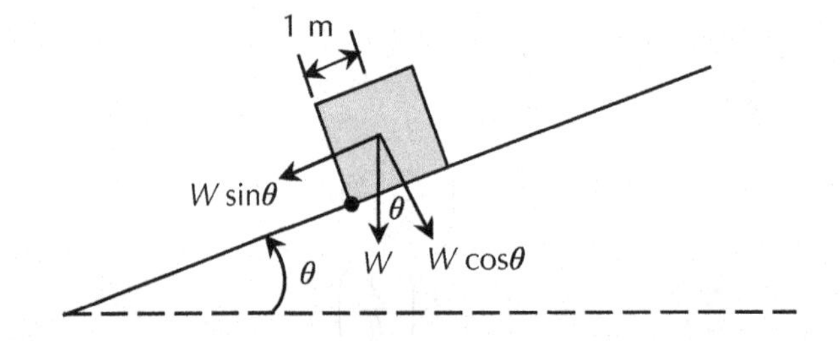

FIGURE 3.41 The free-body diagram of the block in an inclined plane for Example 3.18.

Taking the moment about the tipping point which is the lower bottom point of the block:
Resistive moment: $W(\cos\theta)(1.0)$

Overturning moment: $W(\sin\theta)(1.0)$

$$W\sin\theta(1.0) = W\cos\theta(1.0)$$

$$\tan\theta = 1$$

$$\theta = 45^\circ$$

Answer: just after 45°.

Example 3.19:

A 10-kip tractor is used to lift 1,800-lb of soil as shown in Figures 3.42 and 3.43. Determine the reactions at the tires (A_y and B_y)? Each axle is a single-axle single-tire each side.

FIGURE 3.42 An actual loader for Example 3.19.

SOLUTION

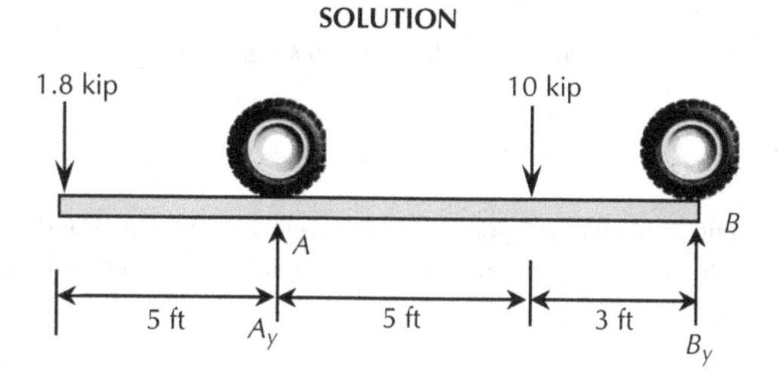

FIGURE 3.43 The free-body diagram of the loader for Example 3.19.

Step 1. Draw free-body diagram (already provided)
Step 2. Apply the equilibrium equations

The application of $\sum F_x = 0$ will not be of any use, as there is no horizontal load.If we apply $\sum F_y = 0$, there are two unknown forces (A_y and B_y). The application of $\sum M = 0$ would give us the solution quicker.

$$\sum M \text{ at } A = 0 \ (\lrcorner + ve)$$

$$10 \text{ kip } (5 \text{ ft}) - 1.8 \text{ kip } (5 \text{ ft}) - B_y \ (8 \text{ ft}) = 0$$

$$50 \text{ kip.ft} - 9.0 \text{ kip.ft} - 8B_y \text{ ft} = 0$$

$$41 \text{ kip.ft} - 8B_y \text{ ft} = 0$$

$$8B_y \text{ ft} = 41 \text{ kip.ft}$$

$$B_y = 5.125 \text{ kip } (\uparrow)$$

$$B_y = 5.125 \text{ kip/2 tires} = 2.56 \text{ kip per tire } (\uparrow)$$

$$\sum F_y = 0 \ (\uparrow +ve)$$

$$A_y + B_y - 1.8 \text{ kip} - 10 \text{ kip} = 0$$

$$A_y + B_y - 11.8 \text{ kip} = 0$$

$$A_y + 5.125 \text{ kip} - 11.8 \text{ kip} = 0$$

$$A_y = 11.8 \text{ kip} - 5.125 \text{ kip}$$

$$A_y = 6.675 \text{ kip } (\uparrow)$$

$$A_y = 6.675 \text{ kip/2 tires} = 3.34 \text{ kip per tire } (\uparrow)$$

Answers:

Reaction at the left tire = 3.3 kip per tire (↑)

Reaction at the right tire = 2.6 kip per tire (↑)

Example 3.20:

Two blocks are resting as shown in Figure 3.44. The weights of blocks *A* and *B* are 400 N and 600 N, respectively. If the surfaces are frictionless, determine the amount of force (*P*) required to maintain equilibrium.

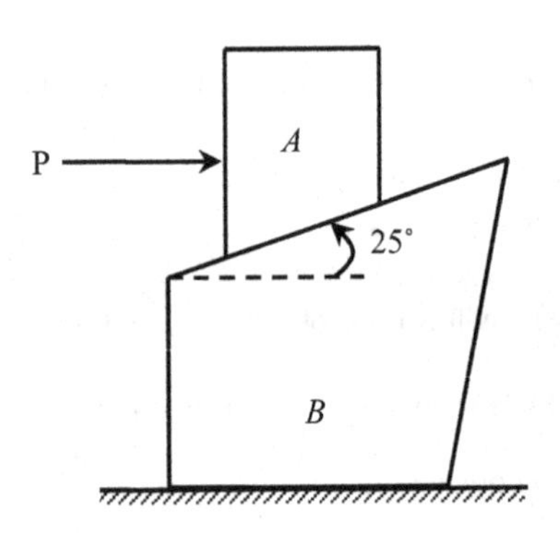

FIGURE 3.44 A wedge system for Example 3.20.

SOLUTION

Step 1. Draw the free-body diagram as shown in Figure 3.45. Let the normal force perpendicular to the contact surface be R_B. There is no friction force along the surfaces of the wedges, as the surfaces are frictionless. The sign conventions are as shown by the $x+$ and $y+$ axes.

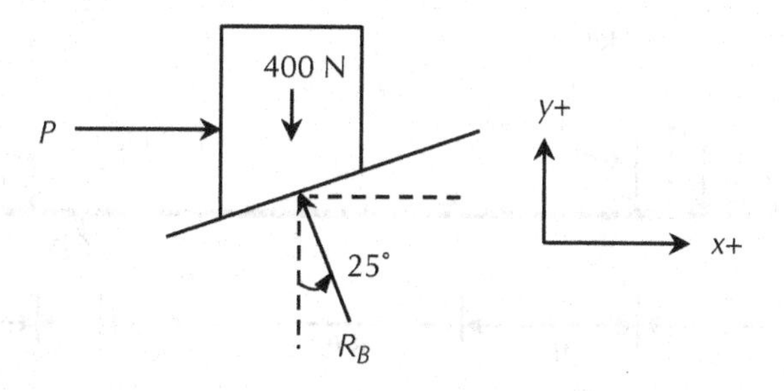

FIGURE 3.45 The free-body diagram of the wedge system for Example 3.20.

Step 2. Apply the equilibrium equations

$$\sum F_y = 0 \ (\uparrow +ve)$$

$$R_B \cos 25 - 400 \text{ N} = 0$$

$$R_B \cos 25 = 400 \text{ N}$$

$$R_B = 400 \text{ N}/\cos 25$$

$$R_B = 441.4 \text{ N}$$

$$\sum F_x = 0 \quad (\leftarrow +ve)$$

$$R_B \sin 25 - P = 0$$

$$441 \text{ N} \sin 25 - P = 0$$

$$P = 441 \text{ N} \sin 25$$

$$P = 186.5 \text{ N}$$

Answer:

$$P = 186.5 \text{ N}.$$

By now, you have mastered the concept of equilibrium and its application to different types of structures used in engineering. Now, let us combine different types of loading in a single problem (Example 3.21) to refresh and summarize the learning of this chapter.

Example 3.21:

A beam, *AB*, is supported by a pin support, *A*, and a roller support, *B*, as shown in Figure 3.46. It is supporting a concentrated load, a uniformly varying load and a concentrated moment. Ignore the weight of the beam. Determine the support reactions for these conditions.

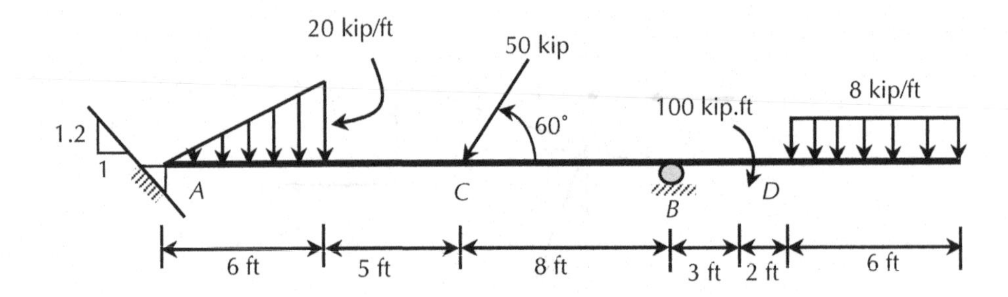

FIGURE 3.46 A beam with an arbitrary loading for Example 3.21.

SOLUTION

Step 1. Draw the free-body diagram as shown in Figure 3.47.

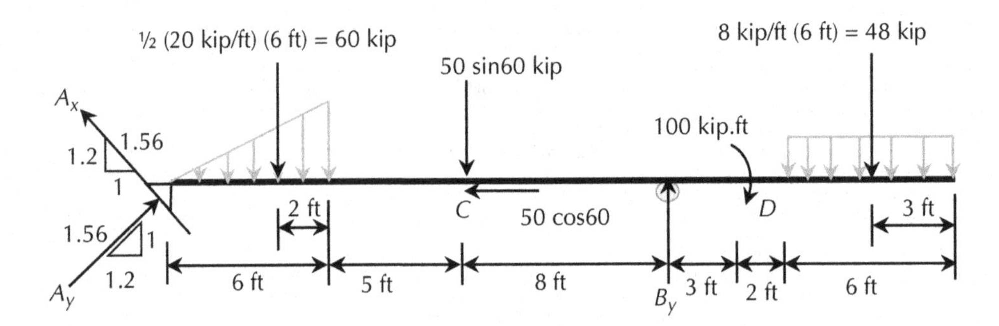

FIGURE 3.47 The free-body diagram of the beam for Example 3.21.

$$\sqrt{(1)^2 + (1.2)^2} = 1.56$$

Note: it is also acceptable to assume A_x and A_y along the horizontal and vertical directions, respectively.

Step 2. Apply the equilibrium equations.

In the beginning $\sum F_x = 0$ and $\sum F_y = 0$ will not help as we will have two unknowns in each direction.

$$\sum M \text{ at } A = 0 \ (\lrcorner + ve)$$

$$60 \text{ kip } (4 \text{ ft}) + 50 \text{ kip } \sin 60 \ (11 \text{ ft}) - B_y \ (19 \text{ ft}) + 100 \text{ kip.ft} + 48 \text{ kip } (27 \text{ ft}) = 0$$

$$240 \text{ kip.ft} + 476.3 \text{ kip.ft} - 19 \ B_y \text{ ft} + 100 \text{ kip.ft} + 1{,}296 \text{ kip.ft} = 0$$

$$-19 \ B_y \text{ ft} + 2{,}109.3 \text{ kip.ft} = 0$$

$$19 \ B_y \text{ ft} = 2{,}109.3 \text{ kip.ft}$$

$$B_y = 111 \text{ kip } (\uparrow)$$

$$\sum F_x = 0(\rightarrow +\text{ve})$$

$$-A_x\left(\frac{1.0}{1.56}\right)+A_y\left(\frac{1.2}{1.56}\right)-50\text{ kip}\cos 60=0$$

$$-0.64A_x+0.77A_y-25=0 \quad (1)$$

$$\sum F_y = 0\ (\uparrow +ve)$$

$$A_x\left(\frac{1.2}{1.56}\right)+A_y\left(\frac{1.0}{1.56}\right)-60\text{ kip}-50\text{ kip}\sin 60+111\text{ kip}-48\text{ kip}=0$$

$$0.77A_x+0.64A_y-40=0 \quad (2)$$

Solving Eq. (1) and (2):

$$A_x = 14.8\text{ kip}$$

$$A_y = 44.7\text{ kip}$$

Answers:

$$A_x = 14.8\ kip$$

$$A_y = 44.7\ kip$$

$$B_y = 111.2\ kip$$

PRACTICE PROBLEMS

PROBLEM 3.1

A 90-kg man is hanging using a cable system shown in Figure 3.48. Determine the tensions at the cables AC and AB?

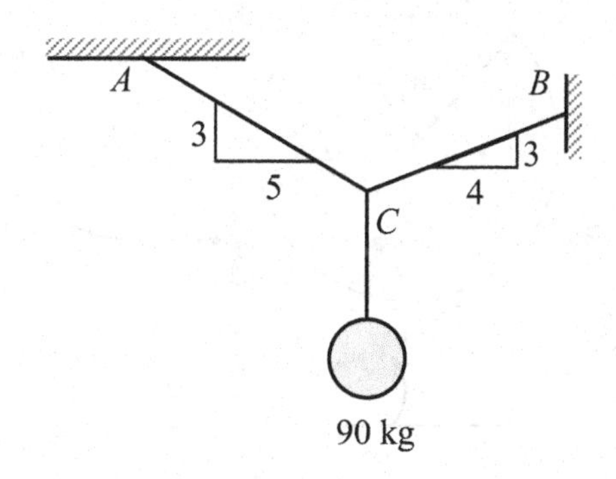

FIGURE 3.48 An idealized structure for Problem 3.1.

PROBLEM 3.2

A 60-ft uniform cylindrical pipe that has the mass of 900 kg is to be lifted by a crane as shown in Figure 3.49. While lifting, the pipe makes an angle of 60° with the ground surface. Assume the mass of the body is uniformly distributed and concentrated along the centerline. Determine the forces at the hooks (*A* and *B*) located at the ends of the pipe.

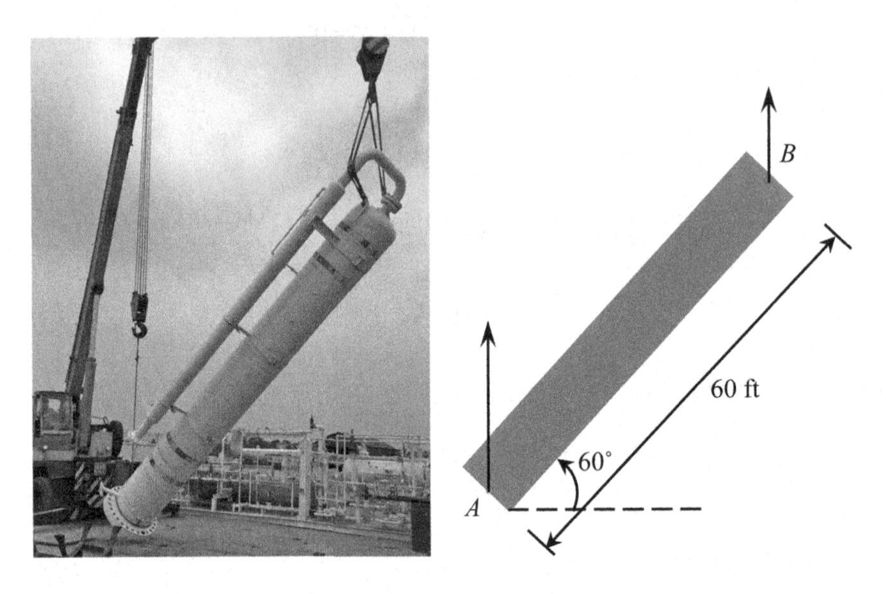

FIGURE 3.49 Lifting of an equipment for Problem 3.2. *(Photo by Bayezid taken in Bangladesh)*

PROBLEM 3.3

A 30-kip sphere is supported by two rigid planes as shown in Figure 3.50. The plane, *AB* is inclined with the horizon at an angle of θ-degrees and the other plane *AC* is vertical. The joint of *AB* and *AC* are pin connected. If the contact surfaces are frictionless, what is the maximum value of θ so that the exerted force on the *AB* does not exceed 30 kip?

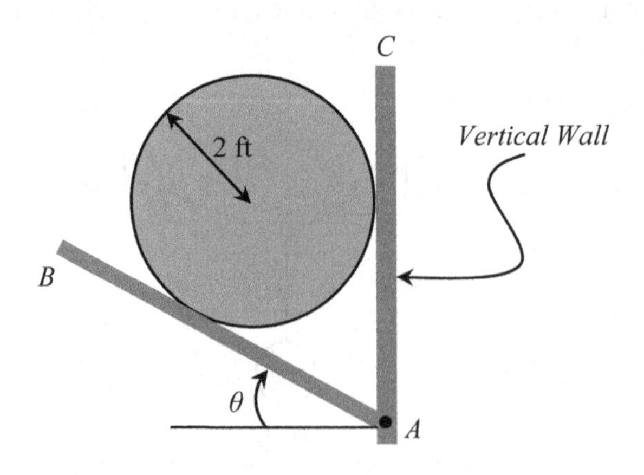

FIGURE 3.50 A sphere for Problem 3.3.

PROBLEM 3.4

A 4-ft diameter cylindrical solid block has the weight of 10 kip per ft length and is to be raised along a vertical wall with the help of a wedge as shown in Figure 3.51. Assume all the contacts are frictionless. If the wedge angle, $\theta = 15°$, what is the minimum amount of force, *F*, required to lift the body?

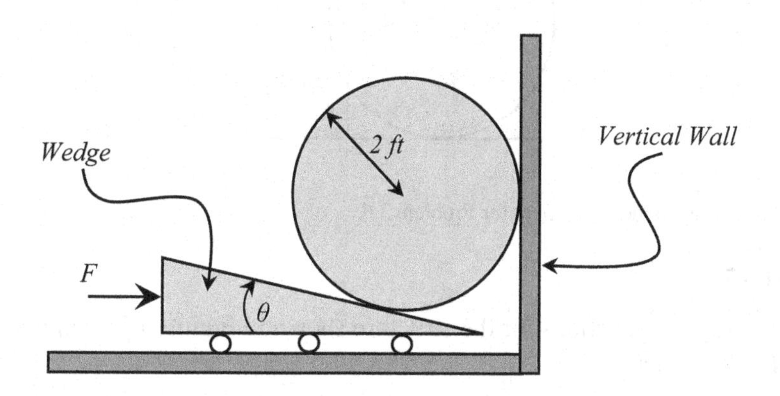

FIGURE 3.51 A wedge system for Problem 3.4.

PROBLEM 3.5

A 2-ft diameter wheel is to be rolled over an obstacle of 8-in. in height as shown in Figure 3.52. The wheel weighs 30 kg. If a push force is applied at an angle of 20° with the ground, calculate the minimum amount of force, *F*, required to roll over the obstacle?

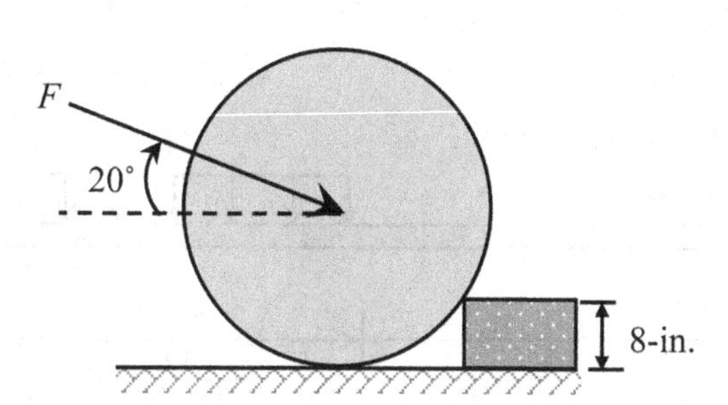

FIGURE 3.52 A wheel to be rolled over for Problem 3.5.

PROBLEM 3.6

A 4-ft diameter wheel is to be rolled over an obstacle of 8-in. in height as shown in Figure 3.53. The wheel weighs 30 kg. If a pull force is applied at an angle of 20° with the ground, determine the minimum amount of force, *F*, required to roll over the obstacle.

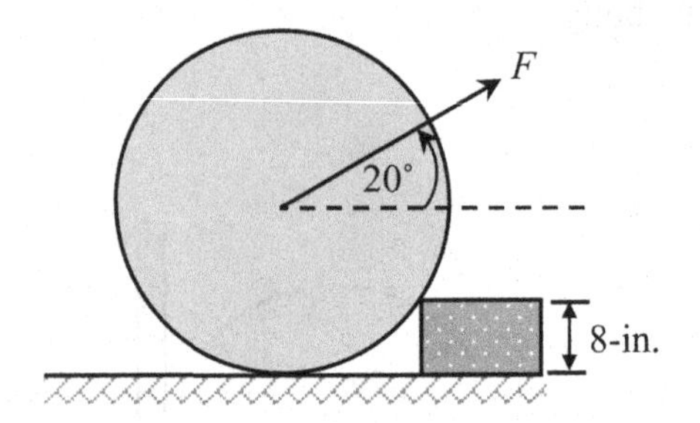

FIGURE 3.53 A wheel to be rolled over for Problem 3.6.

PROBLEM 3.7

Determine the support reactions for the structure shown in Figure 3.54. Support *A* is a pin, and support *B* is a roller.

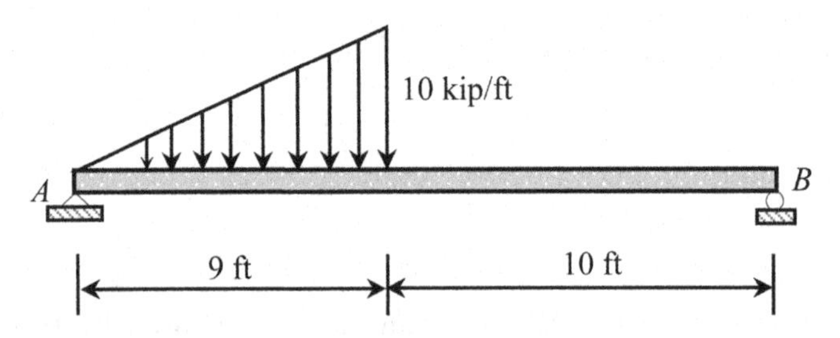

FIGURE 3.54 A beam with an arbitrary loading for Problem 3.7.

PROBLEM 3.8

Determine the support reactions for the structure shown in Figure 3.55. Support *A* is a pin, and support *B* is a roller.

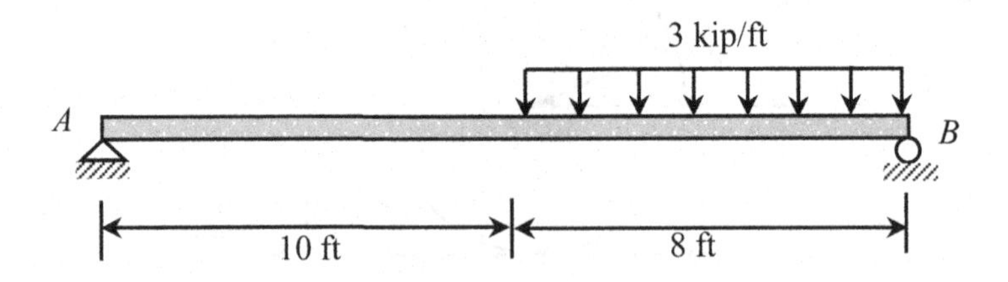

FIGURE 3.55 A beam with an arbitrary loading for Problem 3.8.

PROBLEM 3.9

Determine the support reactions for the structure shown in Figure 3.56. Support *A* is a pin, and support *B* is a roller.

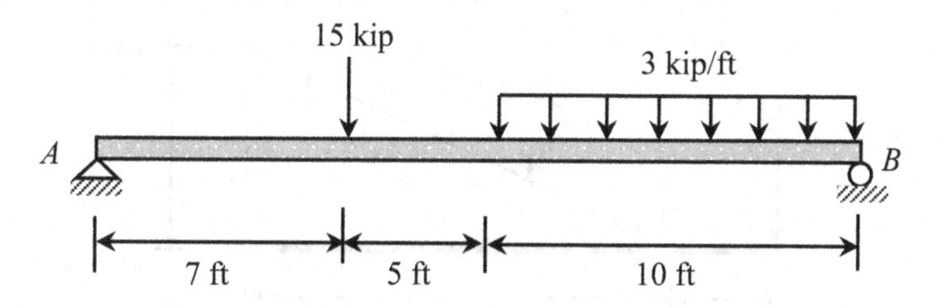

FIGURE 3.56 A beam with arbitrary loading for Problem 3.9

PROBLEM 3.10

Determine the support reactions for the structure shown in Figure 3.57. Support A is a pin, and support B is a roller.

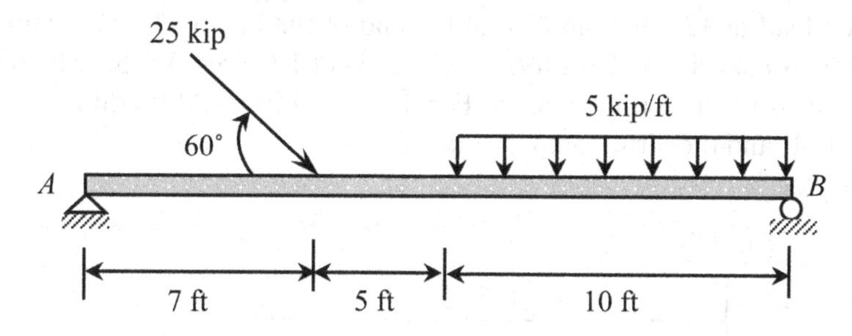

FIGURE 3.57 A beam with arbitrary loading for Problem 3.10.

PROBLEM 3.11

Determine the support reactions for the structure shown in Figure 3.58. Support A is a pin, and support B is a roller.

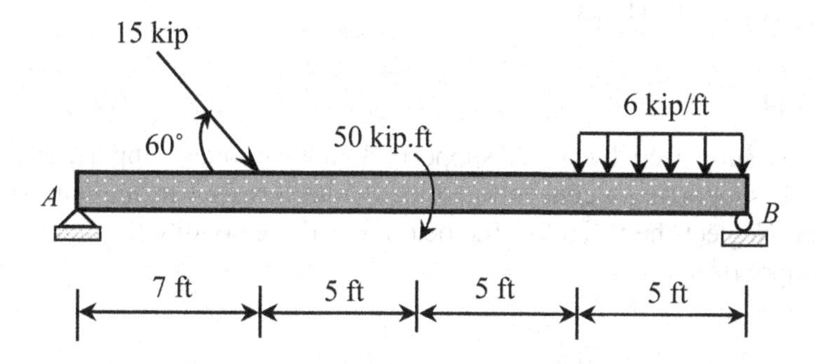

FIGURE 3.58 A beam with arbitrary loading for Problem 3.11.

PROBLEM 3.12

A beam, AB, is supported by a cable, CD, and a pin support, A, as shown in Figure 3.59. A concentrated load of 600 lb is applied at the end of the beam, B. The beam is made up of homogeneous material with a uniform self-weight of 20 lb per ft. The elongation and the self-weight of cable are negligible. For these conditions, determine the reactions at the pin support, A, and the cable tension. The cable can be considered a link support.

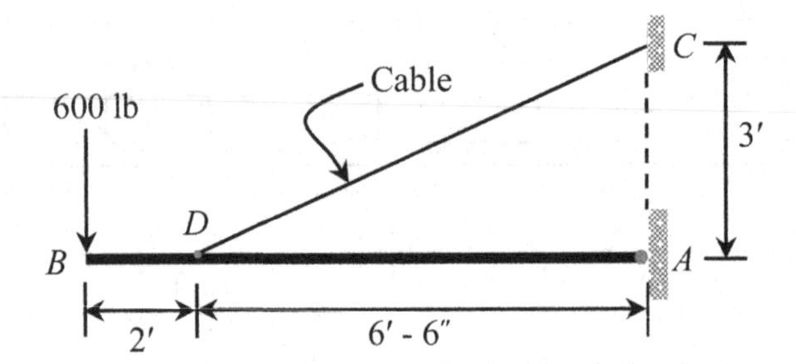

FIGURE 3.59 A beam for Problem 3.12.

PROBLEM 3.13

A beam, *AB*, is supported by a 7-ft strut, *CD,* and a pin support, *A,* as shown in Figure 3.60. A concentrated load of 120 lb is applied at the end of the beam, *B*. The beam is made up with homogeneous material with uniform self-weight of 10 lb per ft. The self-weight of strut is negligible and works as a link support. For these conditions, determine the reactions at the pin support, *A,* and the strut compression.

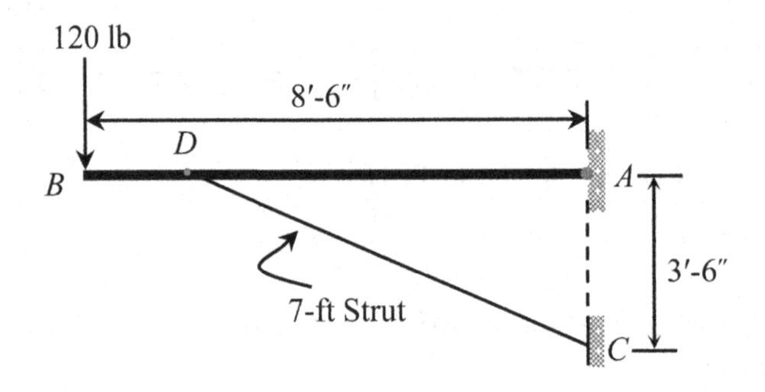

FIGURE 3.60 A beam for Problem 3.13.

PROBLEM 3.14

A beam, *AB,* is supported by a pin support, *A,* and a roller support, *B,* as shown in Figure 3.61. It is supporting a concentrated load, a uniformly varying load and a concentrated moment. Neglect the weight of the beam. For these conditions, determine the reactions at both supports.

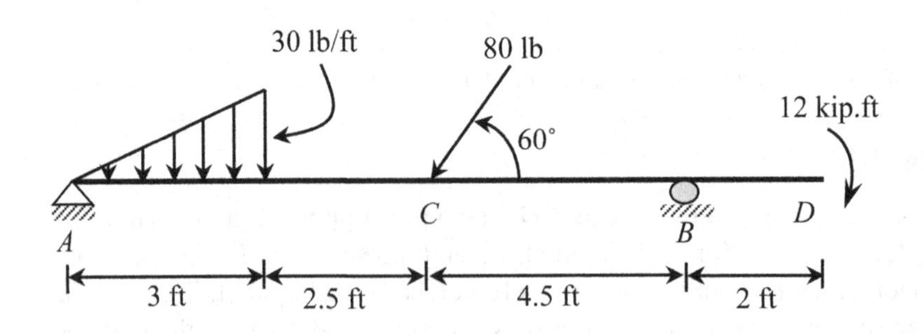

FIGURE 3.61 A beam with arbitrary loading for Problem 3.14.

PROBLEM 3.15

A beam, *AB*, is supported by a pin support, *A*, and a roller support, *B*, as shown in Figure 3.62. It is supporting a concentrated load, a uniformly varying load and a concentrated moment. Neglect the weight of the beam. For these conditions, determine the reactions at both supports.

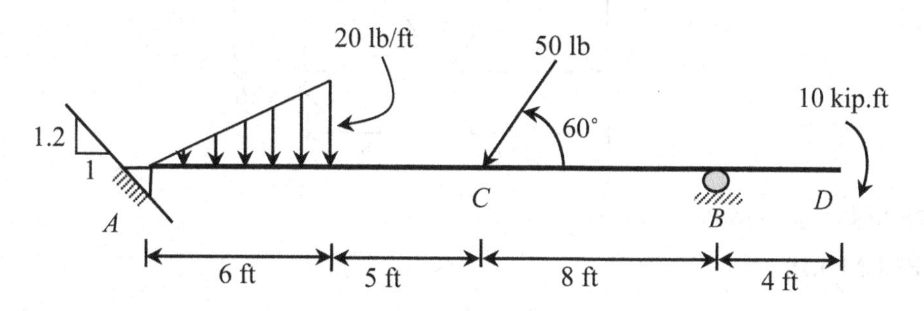

FIGURE 3.62 A beam with arbitrary loading for Problem 3.15.

PROBLEM 3.16

A member, *AB*, is supported by a pin support, *B*, and a roller support, *A*, as shown in Figure 3.63. It is supporting a concentrated load and a uniformly varying load. Neglect the weight of the beam. For these conditions, determine the reactions at both supports.

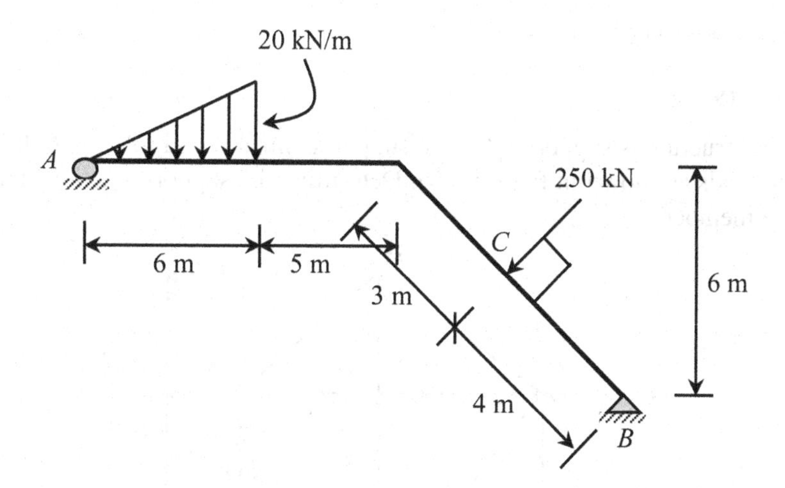

FIGURE 3.63 A complex with arbitrary loading for Problem 3.16.

PROBLEM 3.17

A truss is supported by a pin support, *A*, and a roller support, *B*. It is supporting two concentrated loads as shown in Figure 3.64. Neglect the weight of the members. For these conditions, determine the reactions at the supports.

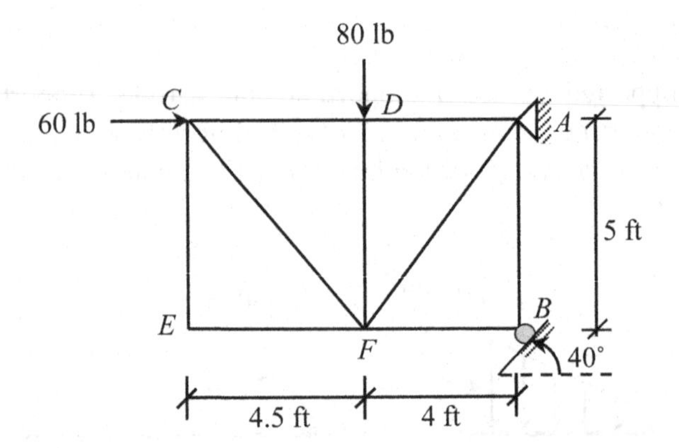

FIGURE 3.64 A truss for Problem 3.17.

PROBLEM 3.18

A truss is supported by a pin support, *A*, and a roller support, *F*. It is supporting two concentrated loads as shown in Figure 3.65. Neglect the weight of the members. For these conditions, determine the reactions at both supports.

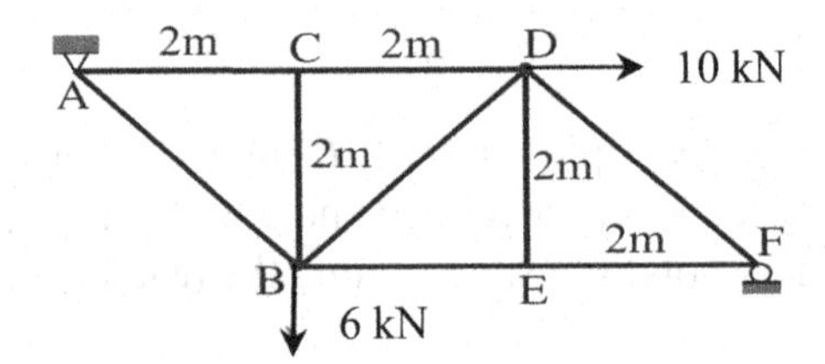

FIGURE 3.65 A truss for Problem 3.18.

PROBLEM 3.19

The following structure is supported by two supports: pin at *A* and roller at *B*. It is supporting two joint loads as shown in Figure 3.66. Determine the support reactions ignoring the weight of the members.

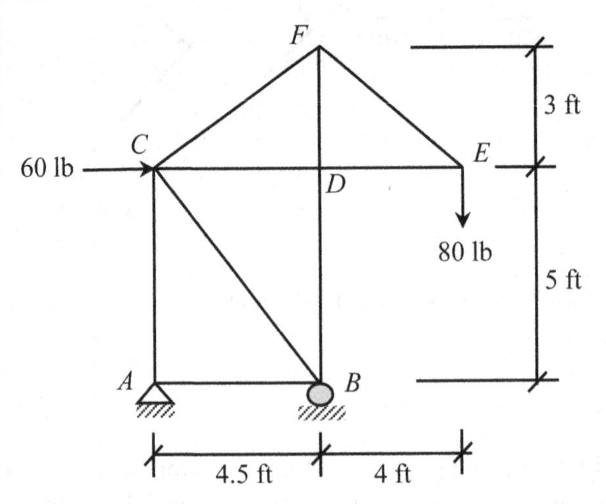

FIGURE 3.66 A truss for Problem 3.19.

PROBLEM 3.20

In Figure 3.67, a 10-lb horse-shaped sign is supported by a beam *AC* and a link member *BD*. Support *A* can be considered a hinge and *BD* is a link member that applies reaction along the member axis. Consider the beam and the link member to be weightless. Calculate the support reactions at *A* and the reaction force at the link member *BD*.

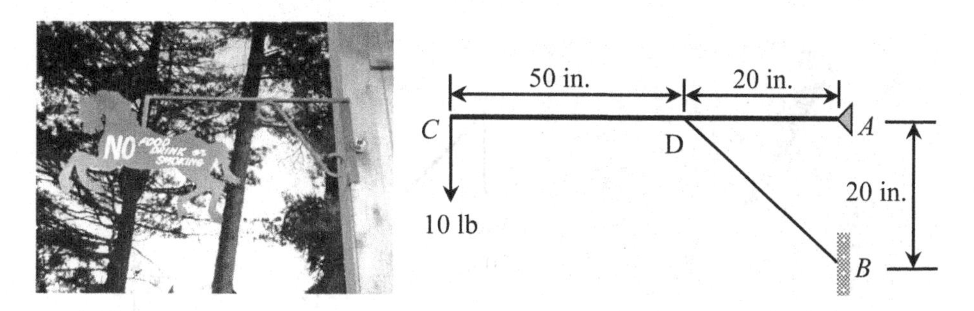

FIGURE 3.67 A beam member for Problem 3.20.

PROBLEM 3.21

A park equipment can be idealized as shown in Figure 3.68. The maximum load in the case including the equipment's self-weight is 600 lb. If the cable and beam are weightless, determine the pin reactions at *A* and the cable tension. The cable can be considered a link and the support A can be considered a hinge.

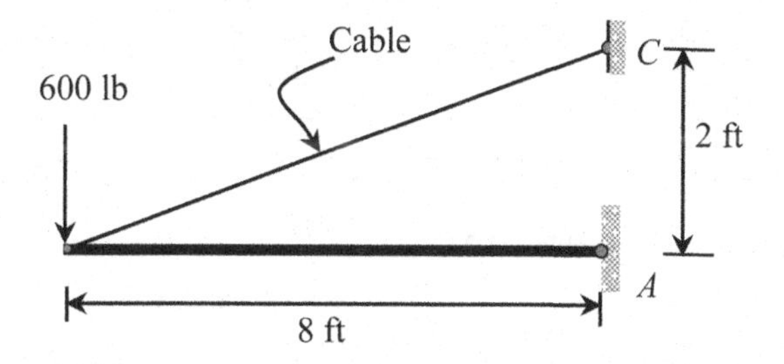

FIGURE 3.68 A beam member for Problem 3.21. *(Photo taken in Pueblo, Colorado)*

PROBLEM 3.22

The crane shown in Figure 3.69 is lifting a concrete vault of 20-kip weight and it can be idealized as shown. Support, *A*, can be considered a pin and *BC* is a link member. Neglect the weight of the crane boom, *ABD*, and the link support, *BC*. Determine the support reactions at *A* and then find out the reaction force at the link *BC*.

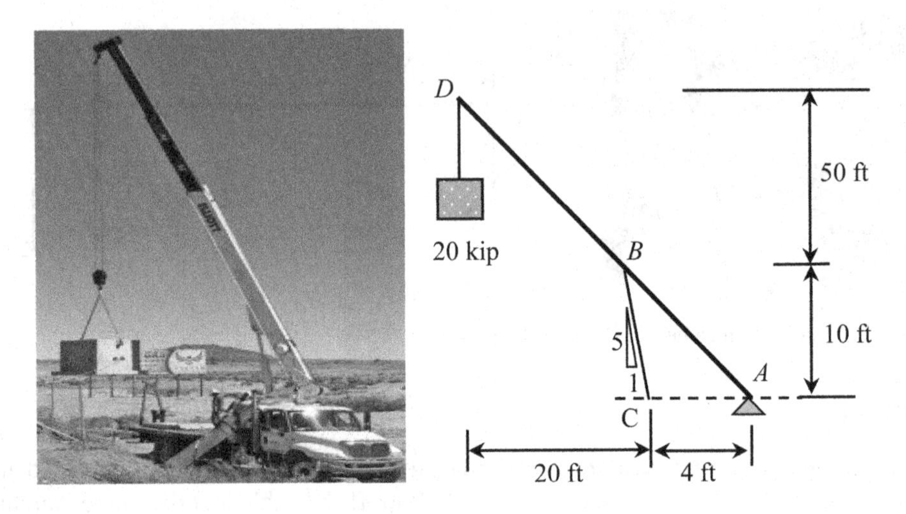

FIGURE 3.69 A crane lifting a concrete vault for Problem 3.22. *(Photo taken near Albuquerque, New Mexico)*

PROBLEM 3.23

A beam, *ABC*, is carrying a load of 5 kip at its center and is supported by a pin support at *A* and cable at *B* and *C* as shown in Figure 3.70. The cable rolls over a frictionless pulley. Determine the tension in the cable and the reactions at the pin support, *A*.

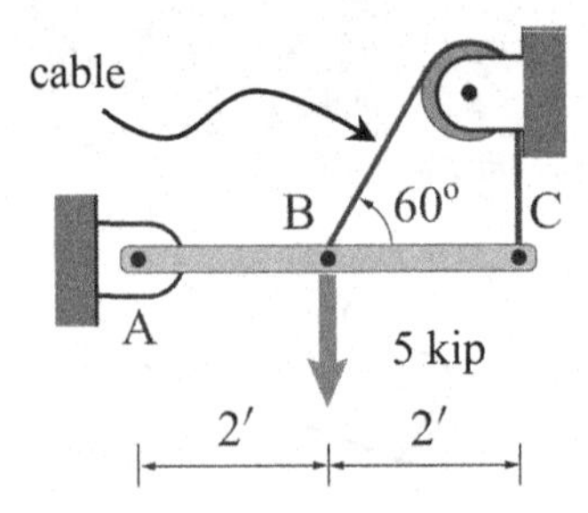

FIGURE 3.70 A beam member for Problem 3.23.

PROBLEM 3.24

A structure is supported by a pin support, A, and a roller support, B. It is supporting two concentrated loads as shown in Figure 3.71. Neglect the weight of the members. For these conditions, determine the reactions at both supports.

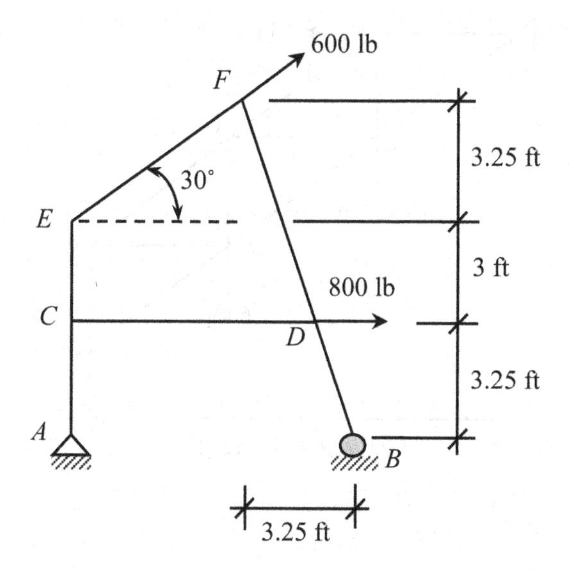

FIGURE 3.71 A structure for Problem 3.24.

PROBLEM 3.25

A truss structure is supporting four joint loads as shown in Figure 3.72. It is supported by a pin support, A, and a roller support, B. Determine the support reactions.

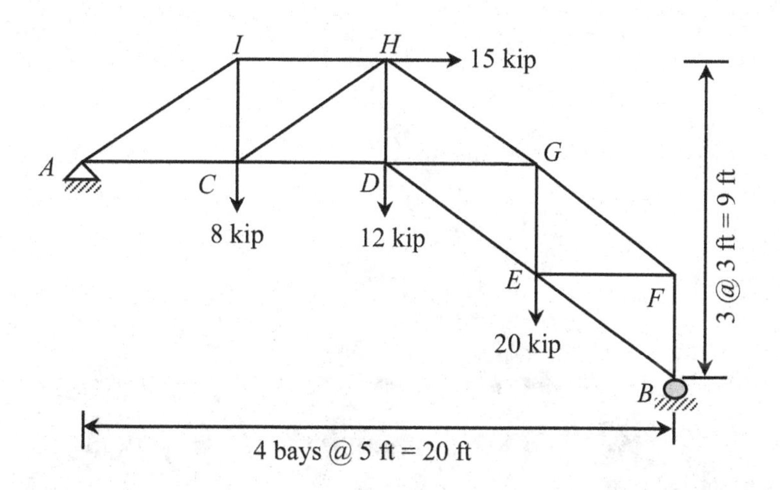

FIGURE 3.72 A truss for Problem 3.25.

PROBLEM 3.26

A truss structure commonly used in stairs, is supporting four joint loads as shown in Figure 3.73. The truss is supported by a pin support, *A*, and a roller support, *B*. Determine the support reactions at *A* and *B*.

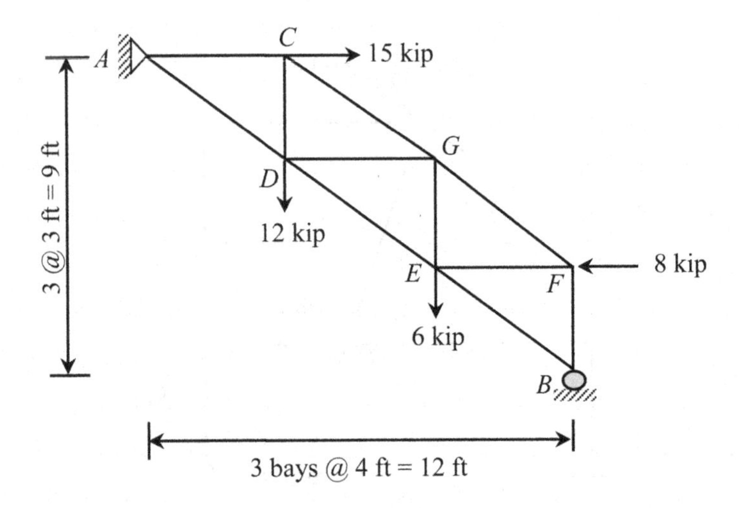

FIGURE 3.73 A truss for Problem 3.26.

PROBLEM 3.27

A construction equipment is being supported by four cables as shown in Figure 3.74. The planview of the cabling system is rectangular, and the total load that is to be supported by the four-cable system is 30 kip. The slant length of each cable is 32 ft. Determine the force in each of the four cables.

(a) Actual structure

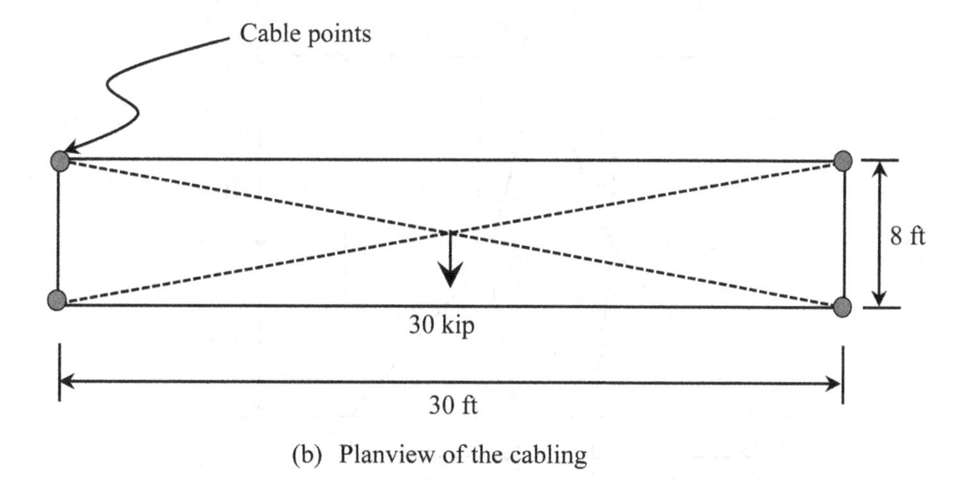

(b) Planview of the cabling

FIGURE 3.74 A construction equipment for Problem 3.27. *(Photo by Armando Perez taken in Pueblo, Colorado)*

PROBLEM 3.28

A concrete wall is being lifted and placed by six cables as shown in Figure 3.75. The profile of the cabling system is also shown. The total load that is to be supported by the cable system is 12 kip and the cable system is symmetrical about the vertical centerline. Determine the force in each of the six cables.

(a) Actual structure

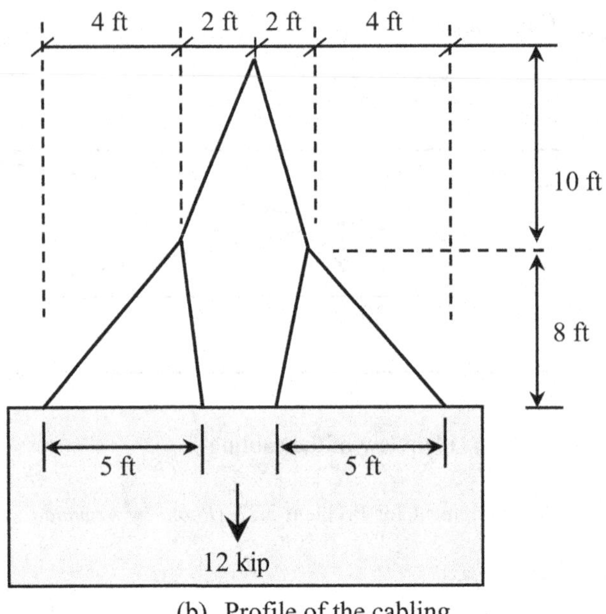

(b) Profile of the cabling

FIGURE 3.75 A crane lifting a concrete wall for Problem 3.28. *(Photo by Armando Perez taken in Pueblo, Colorado)*

PROBLEM 3.29

Wood strut is commonly used to support a roadside signboard as shown in Figure 3.76. At this stage, the wind exerts a force of 280 lb/ft at the height of 14 ft from the ground. The self-weight of the signboard is 400 lb/ft. The foundation of the signboard can be considered a hinge support.

Calculate the developed force in the wood strut if the system is in equilibrium.

Calculate the support reactions.

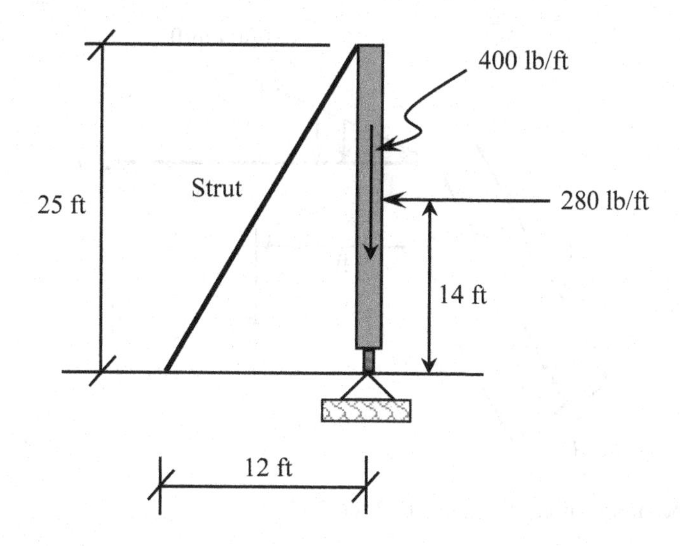

FIGURE 3.76 A wooden signage for Problem 3.29. *(Photo taken in Pueblo, Colorado)*

PROBLEM 3.30

A 1 m × 1 m × 1 m block is resting on an inclined surface as shown in Figure 3.77. Calculate the factor of safety against the tipping of the block.

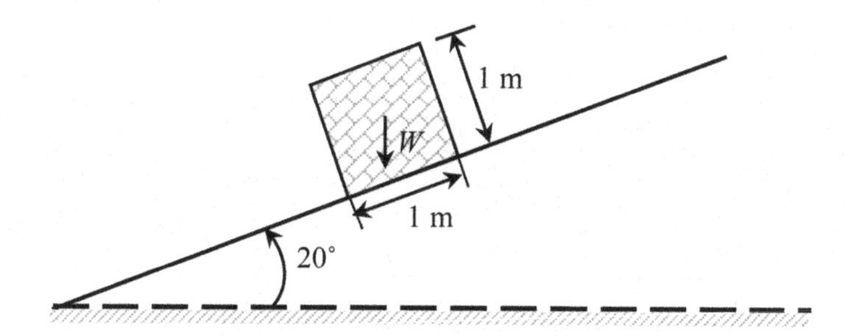

FIGURE 3.77 A block in an inclined plane for Problem 3.30.

PROBLEM 3.31

A member, *AB*, is supported by a pin support, *A*, and a roller support, *B*, as shown in Figure 3.78. It is supporting a uniformly varying load as shown below. Neglect the weight of the beam. The member, *AC*, is inclined at an angle of 70° with the surface. Determine the support reactions for these conditions.

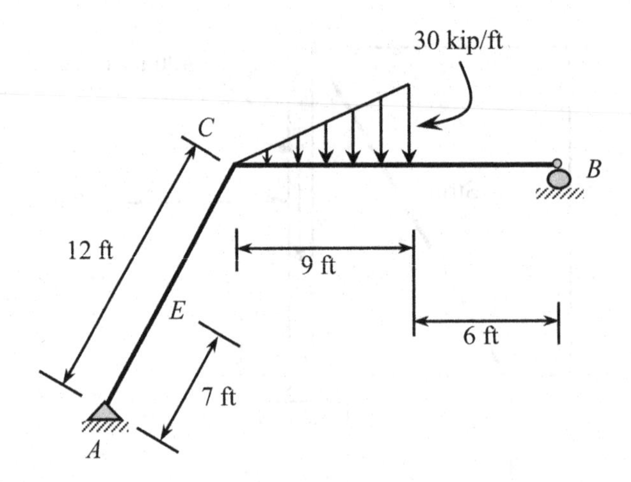

FIGURE 3.78 A beam with arbitrary loading for Problem 3.31.

4 Trusses

4.1 TRUSS STRUCTURE

A truss is a structure (Figure 4.1) where members are pin-connected at their ends, and loads act only at the joints. Due to this configuration, all members in a truss are two-force members meaning that they carry only tension or compression. This is the theory of truss. In fact, these theoretical assumptions are used during analysis and design. The members are organized in a way in which the external loads only act at the joints only. In the real world, we do not see pure pin connections rather some rigidity in the joints as shown in the Figure 4.1. This partially rigid connection makes trusses stronger and safer compared to the calculations.

There is no universal way of classifying trusses. Broadly, they can be divided into two ways:

a. Roof truss
b. Bridge truss

Roof Truss. A roof truss is designed to bridge the space above a room and to provide support for a roof. Trusses usually occur at regular intervals, linked by longitudinal timbers such as purlins. The space between each truss is known as a bay. Several roof trusses are shown in Figure 4.2.

Bridge Truss. The truss used in bridges to support the load from traffic is called a bridge truss. The connected elements (typically straight) may be stressed from tension, compression, or sometimes both in response to dynamic loads. Several bridge trusses are shown in Figure 4.3.

Depending on the dimensions, a truss may be further classified as planar or space truss as shown in Figure 4.4. A planar truss is one where all members and nodes lie within a two-dimensional plane, while a space truss has members and nodes that extend into three dimensions. The top beams in a truss are called top chords and are typically in compression, the bottom beams are called bottom chords and are typically in tension. The interior beams are called webs, and the areas inside the webs are called panels. A space truss is three-dimensional (x, y, and z) and constructed from interlocking struts in a geometric pattern. Space frames can be used to span large areas with few interior supports. Figure 4.4 shows a planar truss and a space truss.

A truss can be classified based on the materials used to construct it. For example, in Figure 4.4, the planar truss is a wood truss, whereas the space truss is a steel truss. In engineering statics, the types of trusses are not the major concern. What matters is the analysis of a truss to determine the member forces (tension or compression). Now the analysis procedure will be discussed.

FIGURE 4.1 A truss in Pueblo Riverwalk in Colorado.

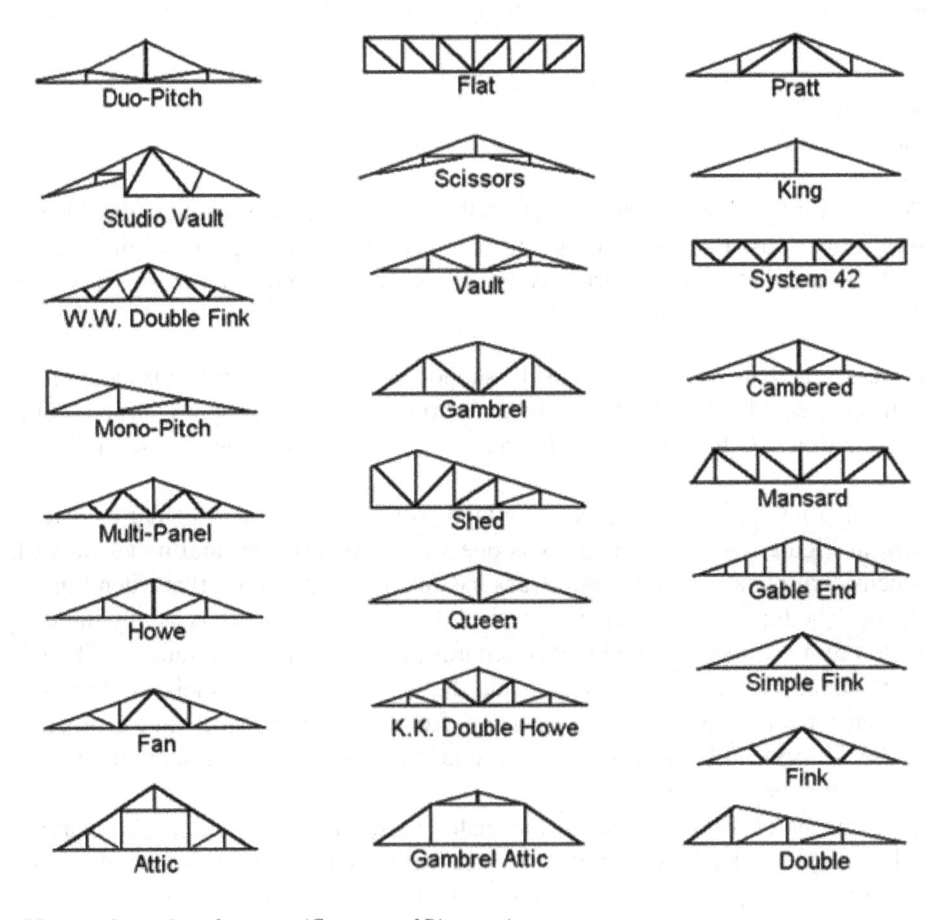

FIGURE 4.2 Several roof trusses. *(Courtesy of Pinterest)*

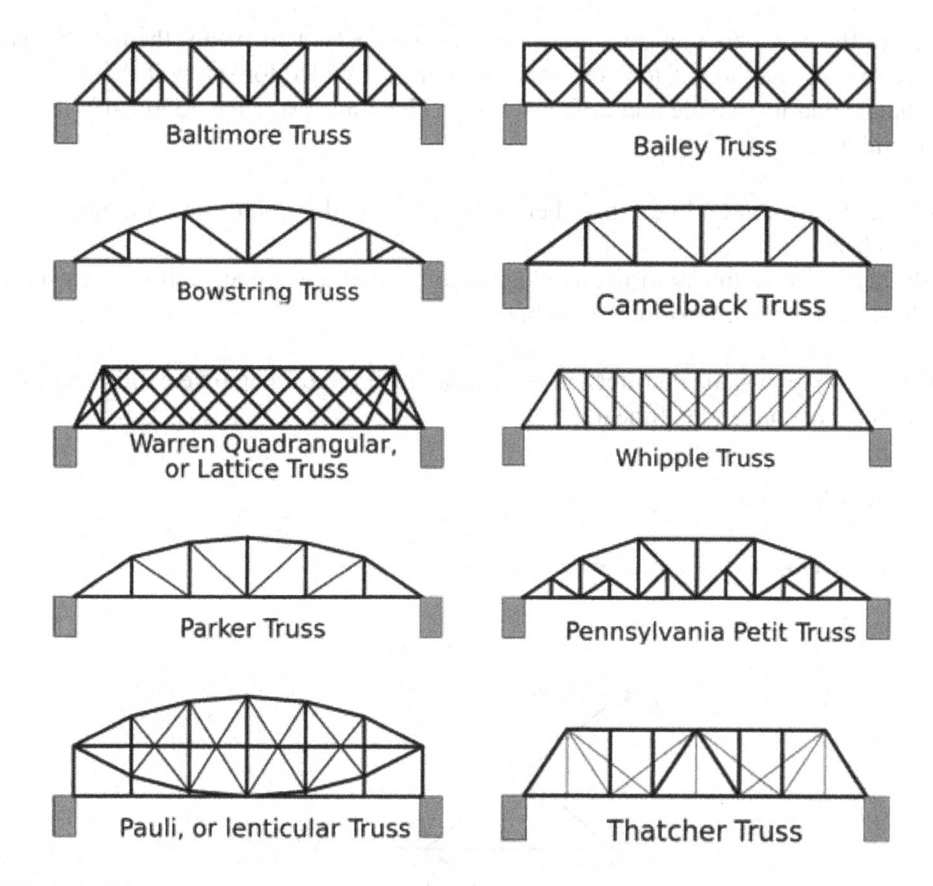

FIGURE 4.3 Several bridge trusses. *(Courtesy of Pinterest)*

Planar truss
(Photo taken in Pueblo)

Space truss
(Photo taken in Niagara Falls)

FIGURE 4.4 Planar and space trusses.

4.2 ZERO-FORCE MEMBERS

A truss is a structure consisting of pin-connected two-force members. A two-force member sustains either tension or compression, not a moment. The self-weight of the member is either neglected or assumed, acting at the ends only. As pin-connected members, the members cannot sustain the moment.

Some of the members in a truss are zero-force members. This means the member is not designed to carry any load. However, this member is provided for safety if the load conditions change and to provide additional stability. There are two rules to identify the zero-force members:

Rule 1. If two non-collinear members form an unloaded joint, both are zero-force members.
Rule 2. If three members form an unloaded joint of which two are collinear, then the third member is a zero-force member.

A two-force body in static equilibrium has two applied forces that are equal in magnitude, opposite in direction, and collinear.

Example 4.1:

Find out the zero-force member(s) of the truss shown in Figure 4.5.

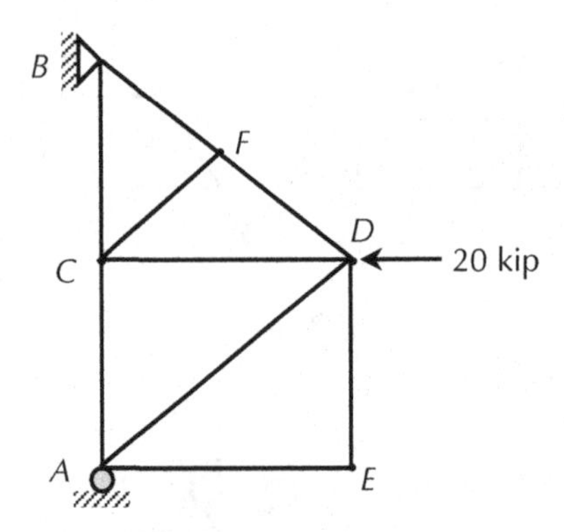

FIGURE 4.5 A truss for Example 4.1.

SOLUTION

AE and *ED* are zero-force members, according to Rule 1.
CF is zero-force member, according to Rule 2. Once *CF* is a zero-force member, *CD* is also a zero-force member too, according to Rule 2.

Example 4.2:

Find out the zero-force member(s) of the truss shown in Figure 4.6.

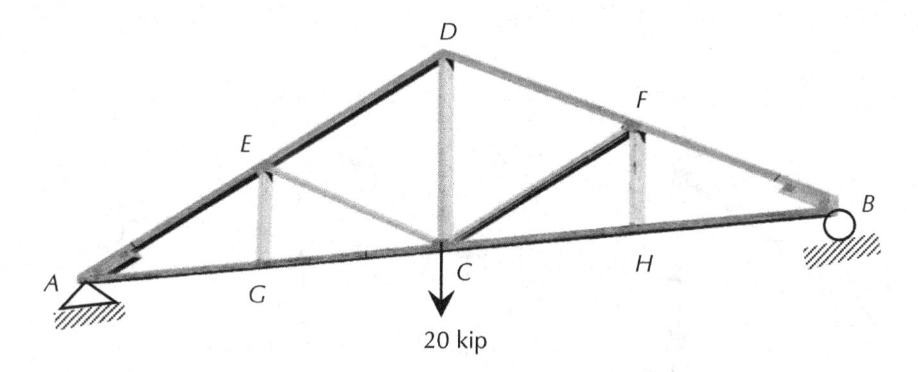

FIGURE 4.6 A truss for Example 4.2.

SOLUTION

GE and *HF* are zero-force members according to Rule 2. Once they are zero-force members, *CE* and *CF* are also zero-force members according to Rule 2.

Example 4.3:

Find out the zero force member(s) of the truss shown in Figure 4.7.

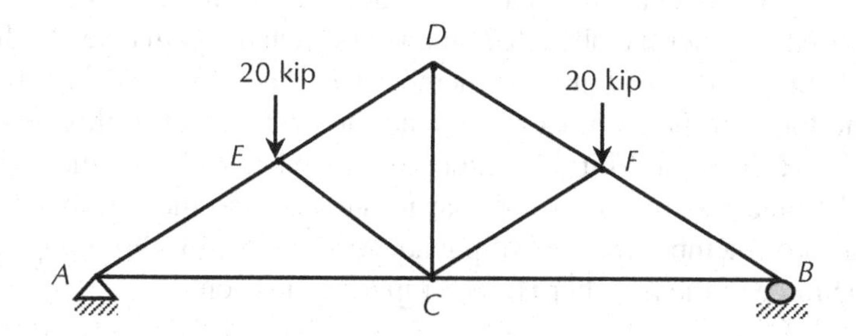

FIGURE 4.7 A truss for Example 4.3.

SOLUTION

None of the members is a zero-force member.

Joints *A*, *B*, *E*, and *F* have a load. Therefore, members associated with these joints might have forces. Joint *D* has three members with no load; however, no two members are linear. So, the associated three members may not be zero-force members. Joint *C* has five members; nothing can be decided from this joint.

4.3 DIRECT-FORCE MEMBERS

The concept of the direct-force member is not well known; it is merely proposed by the authors. There may be some members that carry the applied load directly without distributing it to other members. This happens when a force is applied at a joint along the member

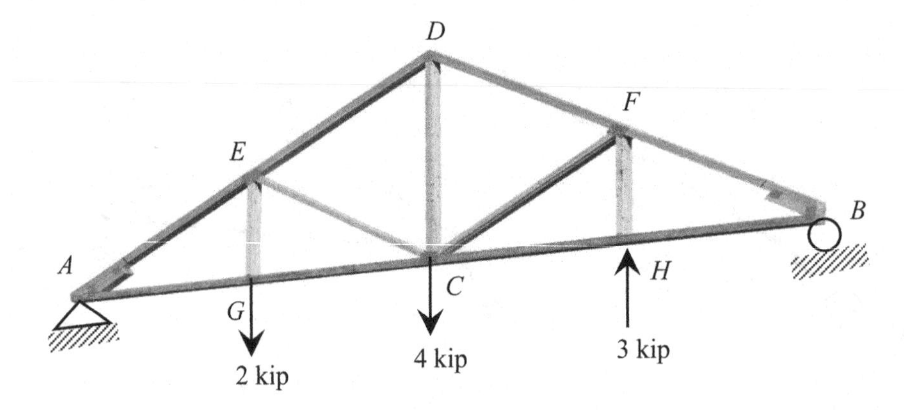

FIGURE 4.8 A truss to clarify the direct-force members.

and other member(s) connected to the joint are perpendicular to the load and the original member. For example, see Figure 4.8. The truss has three loads at three joints, *G*, *C* and *H*. Let us analyze them one by one. The load in the joint *G* is along the member *GE*, and the other two members (*GA* and *GC*) connected to the joint *G* are perpendicular to the load and the member GE. Therefore, *GE* is the direct-force member for the 2-kip load. As the force (2-kip) is applied away from the member *GE*, the force in the member *GE* is 2 kip (tension).

Now, let us consider the joint *C*. It has a 4-kip load along the member *CD*. Two members connected to the joint *C* (*CG* and *CH*) are perpendicular to the load and the member *CD*. However, there are two other members (*CE* and *CF*) which are connected to the joint *C* and are not perpendicular to the load and the member *CD*. Therefore, *CD* is not a direct-force member and the force in the member *CD* cannot be determined in this way. Now, let us consider the joint H. It has a 3-kip load along the member *HF*. Two members connected to the joint *C* (*HC* and *HB*) are perpendicular to the load and the member *HF*. Therefore, HF is the direct-force member for the 3-kip load. As the force (3-kip) is applied toward the member HF, the force in the member HF is 3 kip (compression).

4.4 METHOD OF JOINTS

In the method of joints, each joint is considered as an equilibrium problem, which involves a concurrent force system. A free-body diagram of each joint is constructed and two force equations of equilibrium are written to solve for the two unknown member forces. More clearly, a free-body diagram of the joint is drawn to determine the forces in members connected to a joint. While drawing the free-body diagram, the force in each member is commonly assumed tension. An arrow away from member represents the tension force. After the computation, if the force yields a negative sign, it means a compression force. The equilibrium equations $\left(\sum F_x = 0 \text{ and } \sum F_y = 0\right)$ are applied. Note that if the joint has more than two members, then the forces cannot be determined as only two equilibrium equations $\left(\sum F_x = 0 \text{ and } \sum F_y = 0\right)$ can be applied. In this case, a neighbor joint is considered to determine the connected member's force. Note that the reactions of the support may not be required, but determining such may help for certain cases. Let us review the steps for the joint method:

Step 1. Isolate a joint from the truss that contains no more than two unknowns.
Step 2. Draw a free-body diagram of the joint and solve for the two unknown member forces by satisfying the horizontal and vertical conditions of equilibrium.
Step 3. Select another joint (usually adjacent to the joint just completed) that contains no more than two unknowns and solve the two conditions of equilibrium.
Step 4. Continue this process until all joints have been solved.

Example 4.4:

Determine the member forces of the plane truss shown in Figure 4.9.

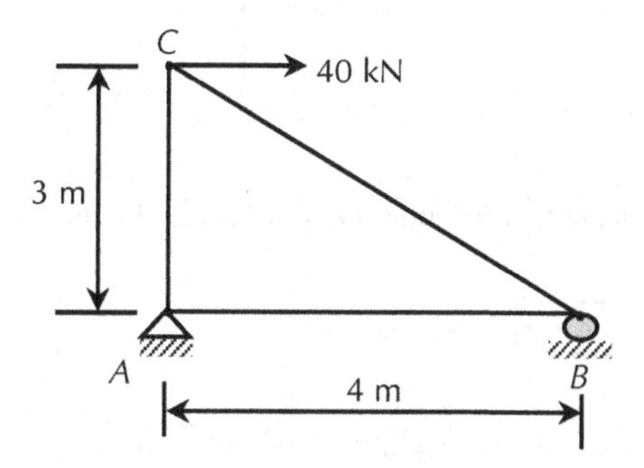

FIGURE 4.9 A truss for Example 4.4.

SOLUTION

The problem did not specify which method to use. Commonly, no specific method is recommended. We will use the joint method here to explain this method. Once you learn the section method, you can use any or mix of both methods. Whatever method you use, the results must be the same.

Consider the free-body diagram of the joint C as shown in Figure 4.10. The reason for choosing this joint is that it has only two unknown forces (two members). If we select the other joints, they have two unknown forces each and the support reactions. It is assumed in the free-body diagram that all the members are in tension. An arrow away from member represents the tension force.

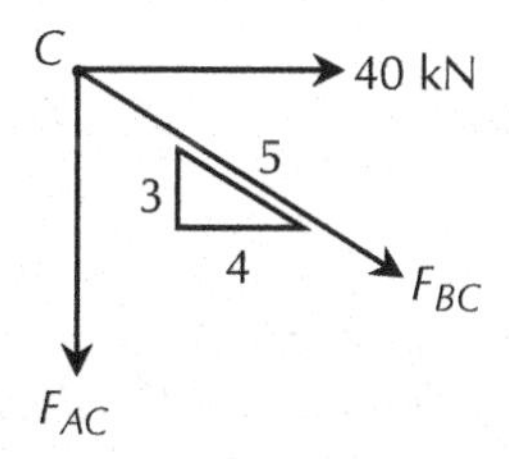

FIGURE 4.10 Free-body diagram of the joint C of the truss for Example 4.4.

$$\sum F_x = 0 \quad (\rightarrow +ve)$$

$$F_{BC}\left(\frac{4}{5}\right) + 40\text{ kN} = 0$$

$$F_{BC} = -50\text{ kN}$$

$$\sum F_y = 0 \quad (\downarrow +ve)$$

$$F_{BC}\left(\frac{3}{5}\right) + F_{AC} = 0$$

$$F_{AC} = -(-50)\left(\frac{3}{5}\right)\text{kN} = 30\text{ kN}$$

The force F_{BC} is negative, meaning it is in compression.

Consider the joint B as shown in Figure 4.11. We already obtained the force F_{BC}. There are only two unknowns left at joint B, i.e., F_{AB} and B_y. We do not need to find B_y.

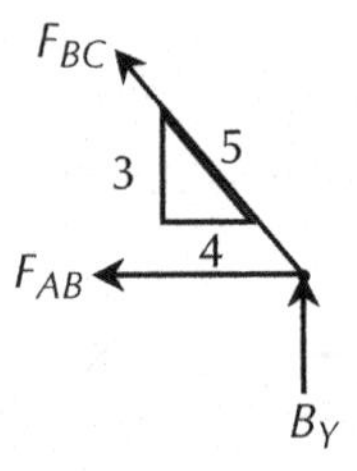

FIGURE 4.11 Free-body diagram of the joint B of the truss for Example 4.4.

$$\sum F_x = 0 \quad (\leftarrow +ve)$$

$$F_{BC}\left(\frac{4}{5}\right) + F_{AB} = 0$$

$$F_{AB} = -F_{BC}\left(\frac{4}{5}\right)$$

$$F_{AB} = -(-50)\left(\frac{4}{5}\right) = 40\text{ kN}$$

Answers:

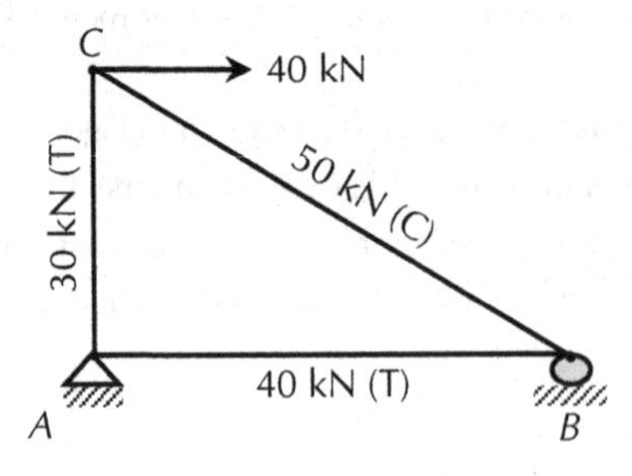

FIGURE 4.12 Answer for Example 4.4.

Remember: the symbols F_{AB} and F_{BA} carry the same meaning. It is the force at AB member, tension or compression. The symbol F_{AB} does not mean the force from A to B or vice versa.

Example 4.5:

Determine the member forces of the plane truss shown in Figure 4.13.

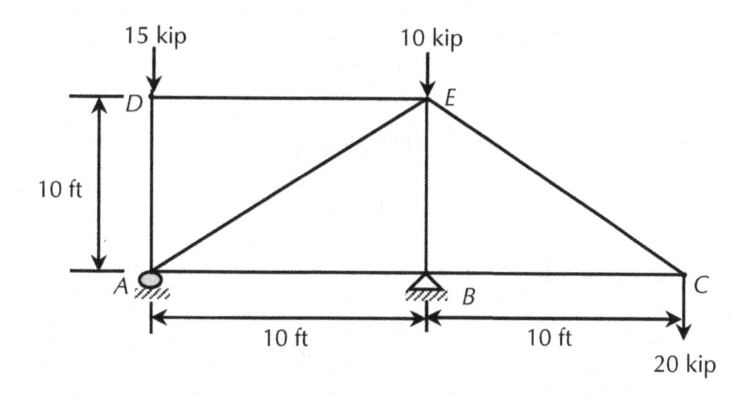

FIGURE 4.13 A truss for Example 4.5.

SOLUTION

We will use the joint method.

Considering the joint C as shown in Figure 4.14. The reason for choosing this joint is that it has only two unknown forces (two members). If we select the other joints, they have more than two unknown forces each except the joint D. We could also consider joint D at the beginning, as it has only two unknown forces and then we proceed from that joint.

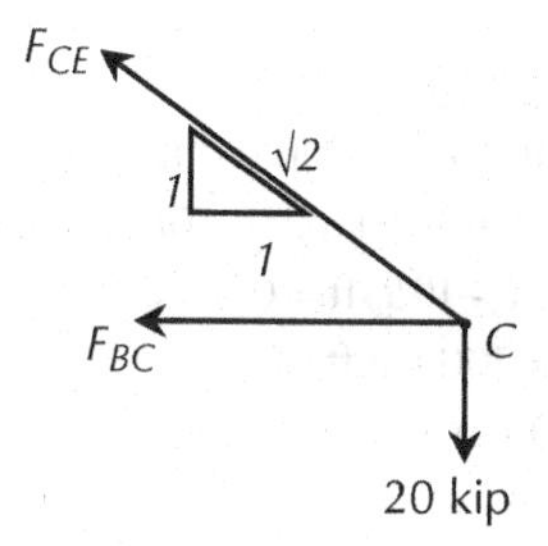

FIGURE 4.14 Free-body diagram of the joint C of the truss for Example 4.5.

$$\sum F_y = 0 \ (\uparrow +ve)$$

$$F_{CE}\left(\frac{1}{\sqrt{2}}\right) - 20 \text{ kip} = 0$$

$$F_{CE} = 20\sqrt{2} \text{ kip}$$

$$F_{CE} = 28.3 \text{ kip (Tension)}$$

$$\sum F_x = 0\ (\rightarrow +ve)$$

$$-F_{CB} - F_{CE}\left(\frac{1}{\sqrt{2}}\right) = 0$$

$$F_{CB} = -F_{CE}\left(\frac{1}{\sqrt{2}}\right)$$

$$F_{CB} = (-28.3\text{ kip})\left(\frac{1}{\sqrt{2}}\right)$$

$$F_{CB} = -20\text{ kip}$$

Now, consider the whole truss (Figure 4.15).

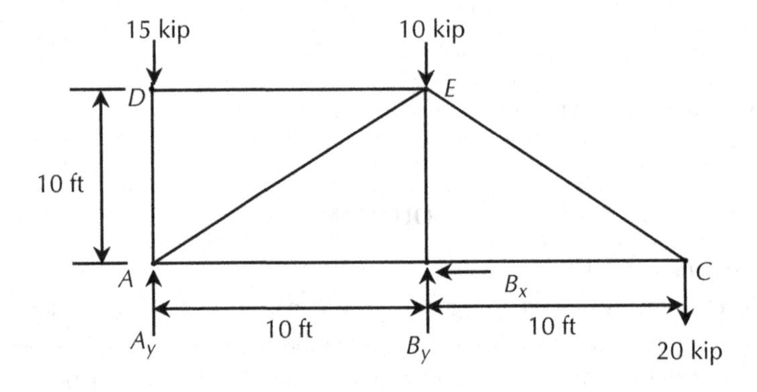

FIGURE 4.15 Free-body diagram of the whole truss for Example 4.5.

$$\sum M \text{ at } A = 0\ (\circlearrowright + ve)$$

$$10\text{ kip }(10\text{ ft}) + 20\text{ kip }(20\text{ ft}) - B_y\,(10\text{ ft}) = 0$$

$$100\text{ kip.ft} + 400\text{ kip.ft} - 10B_y\text{ ft} = 0$$

$$500\text{ kip.ft} - 10B_y\text{ ft} = 0$$

$$10B_y\text{ ft} = 500\text{ kip.ft}$$

$$B_y = 50\text{ kip}$$

$$\sum F_y = 0\ (\uparrow +ve)$$

$$A_y + B_y - 15\text{ kip} - 10\text{ kip} - 20\text{ kip} = 0$$

$$A_y = -B_y + 15\text{ kip} + 10\text{ kip} + 20\text{ kip}$$

$$A_y = -50\text{ kip} + 15\text{ kip} + 10\text{ kip} + 20\text{ kip}$$

$$A_y = -5\text{ kip}$$

Consider the joint B as shown in Figure 4.16:

$$\sum F_x = 0\ (\rightarrow +ve)$$

$$-F_{AB} + F_{CB} = 0$$

$$F_{AB} = F_{CB}$$

$$F_{AB} = -20\text{ kip}$$

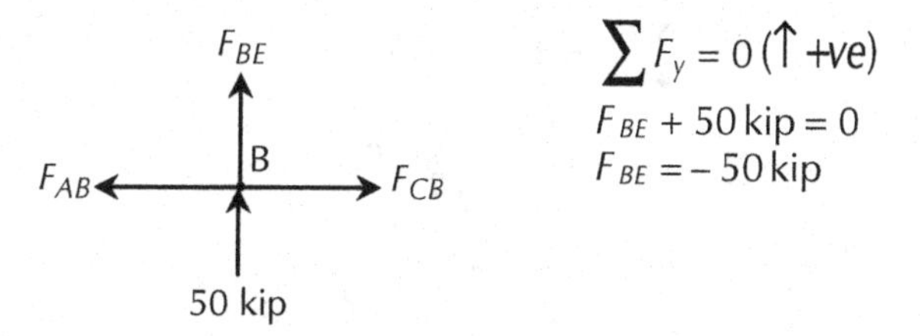

FIGURE 4.16 Free-body diagram of the joint *B* of the truss for Example 4.5.

Consider the joint *D* as shown in Figure 4.17:

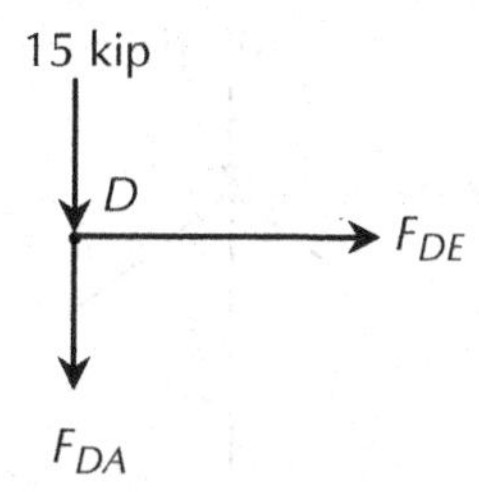

FIGURE 4.17 Free-body diagram of the joint *D* of the truss for Example 4.5.

$$\sum F_y = 0 \ (\uparrow +ve)$$

$$F_{DA} + 15 \text{ kip} = 0$$

$$F_{DA} = -15 \text{ kip}$$

We can also find F_{DA} using visual inspection, $F_{DA} = -15$ kip (direct force)

$$\sum F_x = 0 \ (\rightarrow +ve)$$

$$F_{DE} = 0$$

Consider the joint *A* as shown in Figure 4.18:

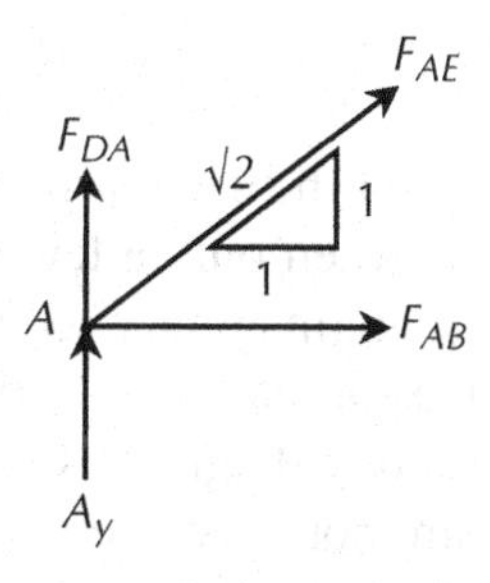

FIGURE 4.18 Free-body diagram of the joint *D* of the truss for Example 4.5.

$$\sum F_x = 0\ (\rightarrow +ve)$$

$$F_{AB} + F_{AE}\frac{1}{\sqrt{2}} = 0$$

$$F_{AE} = -F_{AB}\sqrt{2}$$

$$F_{AE} = -(-20\text{ kip})\sqrt{2}$$

$$F_{AE} = 28.3\text{ kip}$$

Answers:

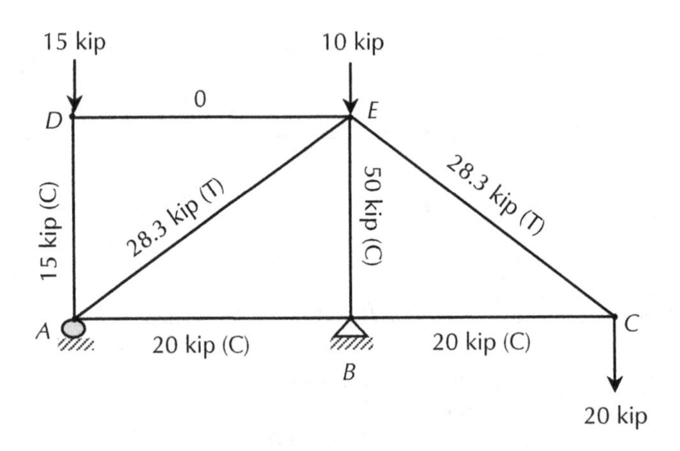

FIGURE 4.19 Answer for Example 4.5.

Remember: Finding out the support reactions is not always required. For many cases, member forces can be determined without finding the support reactions. For example, in Example 4.5, two member forces are already found out without solving for the support reactions.

4.5 METHOD OF SECTIONS

The second method of analysis is called the section method. The method of sections is a particularly useful analysis technique when only a few selected member forces in a truss are desired. In the method of sections, an imaginary section cut is passed through the truss, cutting three members, which includes at least one of the members of interest. The truss is now cut into two parts and a free-body diagram is constructed of either section. Since all forces in the free-body diagram are not concurrent at a common point, this constitutes a rigid body type problem and three equations of equilibrium may be written. These three equations of equilibrium can solve for only three unknown member forces. Note: no more than three unknown members are permitted to be cut for any section passing through the truss.

Example 4.6:

Determine the member forces for the truss shown in Figure 4.20,.

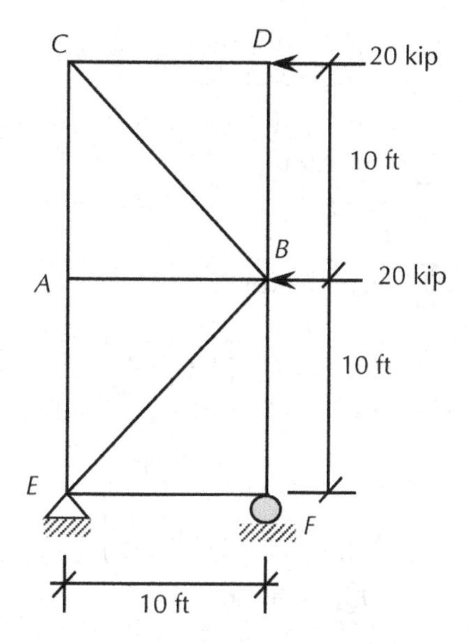

FIGURE 4.20 A truss for Example 4.6.

SOLUTION

By visual inspection:

$F_{CD} = -20$ kip (direct force)
$F_{AB} = 0$ (zero-force member)

It is always a good idea to find out the zero-force members or direct-force members at the beginning of solving the truss. This will reduce the number of unknowns.

Consider the free-body diagram of a section of the truss as shown in Figure 4.21. Note that you can cut any section of the truss. However, the cut should be made in a way that is helpful.

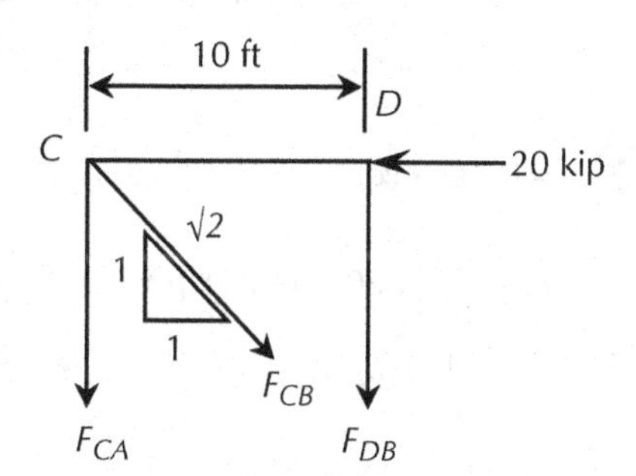

FIGURE 4.21 Free-body diagram of a section of the truss for Example 4.6.

$$\sum F_x = 0 \quad (\rightarrow +ve)$$

$$F_{CB}\left(\frac{1}{\sqrt{2}}\right) - 20\text{ kip} = 0$$

$$F_{CB} = 20\sqrt{2}\text{ kip}$$
$$F_{CB} = 28.28\text{ kip}$$

$$\sum M \text{ at } C = 0 \quad (\lrcorner + ve)$$
$$F_{DB} = 0$$

$$\sum F_y = 0 \quad (\downarrow +ve)$$

$$F_{CA} + F_{DB} + F_{CB}\left(\frac{1}{\sqrt{2}}\right) = 0$$

$$F_{CA} = -F_{DB} - F_{CB}\left(\frac{1}{\sqrt{2}}\right)$$

$$F_{CA} = -0 - 28.28\text{ kip}\left(\frac{1}{\sqrt{2}}\right)$$

$$F_{CA} = -20\text{ kip}$$

Consider the free-body diagram of a section of the truss as shown in Figure 4.22.

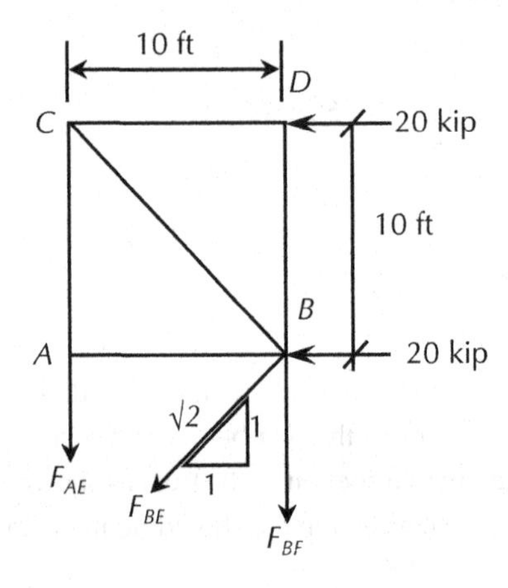

FIGURE 4.22 Free-body diagram of a section of the truss for Example 4.6.

$$\sum F_x = 0 \quad (\leftarrow +ve)$$

$$F_{BE}\left(\frac{1}{\sqrt{2}}\right) + 20\text{ kip} + 20\text{ kip} = 0$$

$$F_{BE}\left(\frac{1}{\sqrt{2}}\right) = -40\text{ kip}$$

$$F_{BE} = -40\sqrt{2}\text{ kip}$$
$$F_{BE} = -56.6\text{ kip}$$

ΣM at $B = 0$ (Assume a counter-clockwise moment is positive)

F_{AE} (10 ft) + 20 kip (10 ft) = 0

F_{AE} (10 ft) + 200 kip.ft = 0

$F_{AE} = -20$ kip

$$\sum F_x = 0 \quad (\leftarrow +ve)$$

$$-F_{AE} - F_{BF} - F_{BE}\left(\frac{1}{\sqrt{2}}\right) = 0$$

$$F_{BF} = -F_{AE} - F_{BE}\left(\frac{1}{\sqrt{2}}\right)$$

$$F_{BF} = -(-20\text{ kip}) - (-56.6\text{ kip})\left(\frac{1}{\sqrt{2}}\right)$$

$$F_{BF} = 60\text{ kip}$$

Consider the joint F as shown in Figure 4.23:

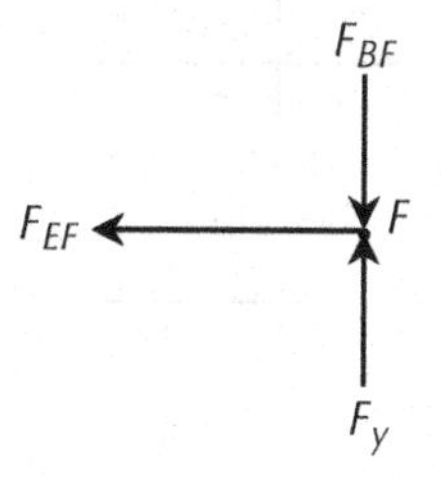

FIGURE 4.23 Free-body diagram of the joint F of the truss for Example 4.6.

$$\sum F_x = 0 \ (\rightarrow +ve)$$

$$F_{EF} = 0$$

Answer:

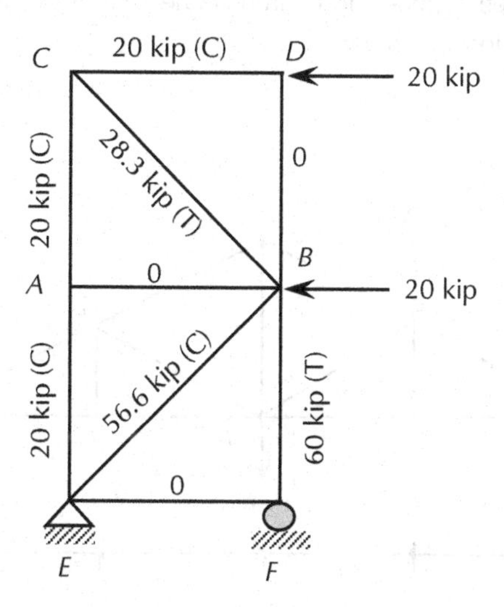

FIGURE 4.24 Answer for Example 4.6.

Remember: there is no restriction on which method (joint or section) to be used to solve a truss unless it is specified in the exam. Whatever method is used, the answer (solution) must be the same.

Example 4.7:

The truss shown in Figure 4.25 is supporting two loads. Determine the member forces at *CD, CF,* and *DF*.

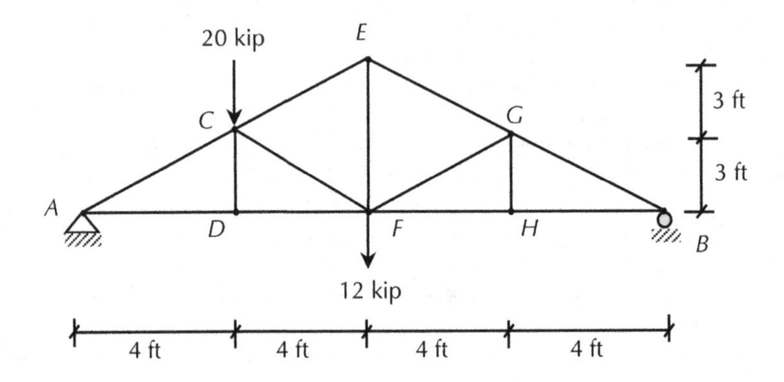

FIGURE 4.25 A truss for Example 4.7.

SOLUTION

First, considering the joint – D as shown in Figure 4.26, and applying $\Sigma F_y = 0$, it is found that CD is a zero-force member. The second rule of the zero-force members also supports that CD is a zero-force member – if three members form an unloaded joint of which two are collinear, then the third member is a zero-force member.

Next, consider the left of the a–a cut section:

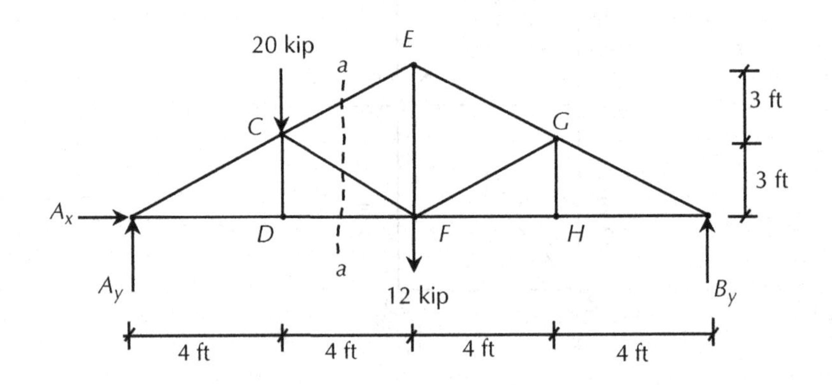

FIGURE 4.26 Free-body diagram of the whole section of the truss for Example 4.7.

Consider the left side of the section *a-a*, as shown in Figure 4.27.

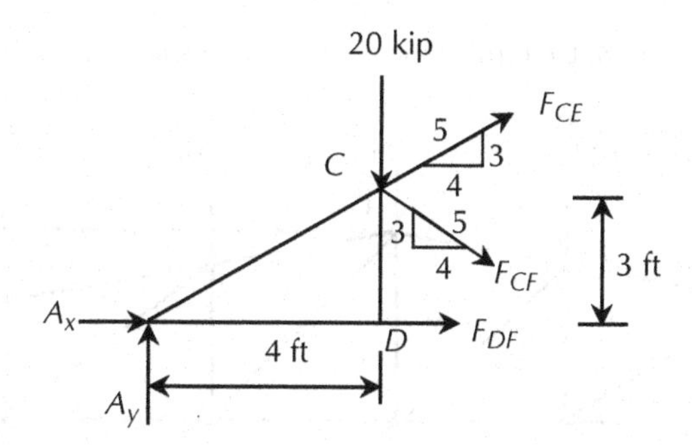

FIGURE 4.27 Free-body diagram of the left of *a-a* section of the truss for Example 4.7.

ΣM at $A = 0$ (assuming clockwise moment is positive)

$$20 \text{ kip } (4 \text{ ft}) + F_{CF}\,(3/5)\,(4 \text{ ft}) + F_{CF}\,(4/5)\,(3 \text{ ft}) = 0$$
$$80 \text{ kip.ft} + 2.4\,F_{CF} + 2.4\,F_{CF} \text{ ft} = 0$$
$$4.8\,F_{CF} \text{ ft} = -\,80 \text{ kip.ft}$$
$$F_{CF} = -16.67 \text{ kip}$$

To find F_{DF}, the support reaction at A must be determined. In the free-body diagram of the whole body, as shown in Figure 4.26, apply $\sum M \text{ at B} = 0$ (assuming counter-clockwise moment is positive) as follows:

$$20 \text{ kip } (12 \text{ ft}) + 12 \text{ kip } (8 \text{ ft}) - A_y\,(16 \text{ ft}) = 0$$
$$240 \text{ kip.ft} + 96 \text{ kip.ft} - 16A_y \text{ ft} = 0$$
$$336 \text{ kip.ft} - 16A_y \text{ ft} = 0$$
$$16A_y \text{ ft} = 336 \text{ kip.ft}$$
$$A_y = 21 \text{ kip}$$

$$\sum F_x = 0 \ (\rightarrow +ve); A_x = 0$$

Now, in the left part of the cut section *a-a* apply $\sum M \text{ at } C = 0$ (assuming clockwise moment is positive) as follows:

$$A_y\,(4 \text{ ft}) - F_{DF}\,(3 \text{ ft}) = 0$$
$$21 \text{ kip } (4 \text{ ft}) - F_{DF}\,(3 \text{ ft}) = 0$$
$$84 \text{ kip.ft} - F_{DF}\,(3 \text{ ft}) = 0$$
$$3F_{DF} = 84 \text{ kip.ft}$$
$$F_{DF} = 28 \text{ kip}$$

Answers:

$F_{CD} = 0$

$F_{CF} = 16.7$ kip (Compression)

$F_{DF} = 28$ kip (Tension)

***Remember:** you can cut anywhere in any shape while using the section method. A wise cut can reduce the solution time. It is always advised to think a lot before cutting a section.*

Example 4.8:

A truss is shown in Figure 4.28. Determine the member forces at the members: *FG*, *CE*, and *EF*.

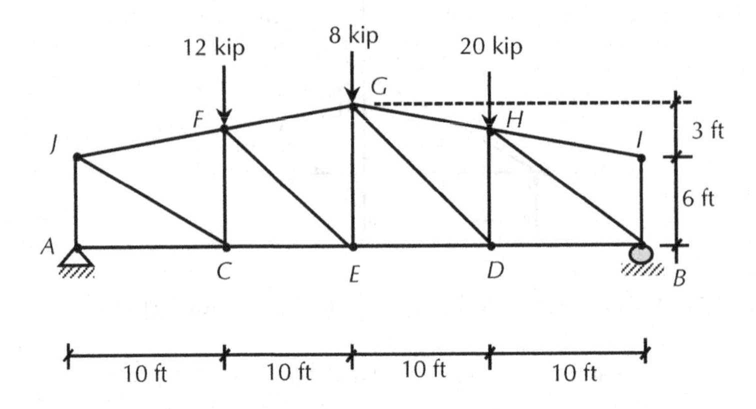

FIGURE 4.28 A truss for Example 4.8.

SOLUTION

We have discussed that the joint method is good for a very simple truss. To determine the forces in *CE* and *EF*, the joint method may be tedious. Let us try solving this problem by the section method.

Let us cut as shown by the dashed line *a–a*; as shown in Figure 4.29.

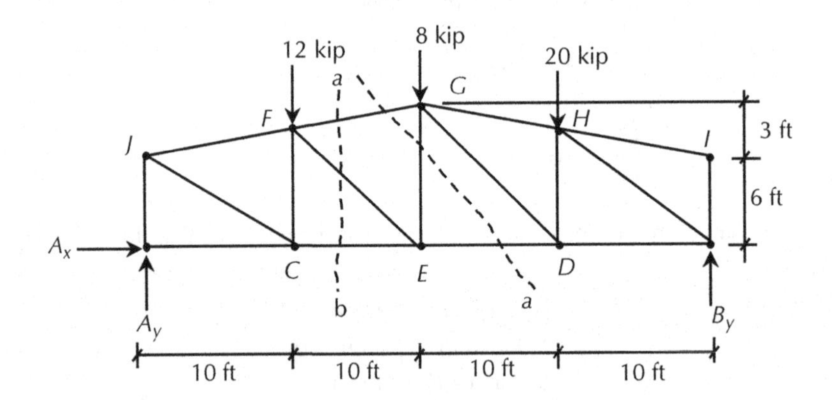

FIGURE 4.29 Free-body diagram of the whole section of the truss for Example 4.8.

The free-body diagram of the left part of the *a–a* cut can be drawn as shown in Figure 4.30.

$\sum M$ at $E = 0$ will determine F_{FG} if A_y is known. Let us find A_y first. In the free-body diagram of the whole body, as shown in Figure 4.29, apply $\sum M$ at $B = 0$ (assuming the counter-clockwise moment is positive) as follows:

$$20 \text{ kip } (10 \text{ ft}) + 8 \text{ kip } (20 \text{ ft}) + 12 \text{ kip } (30 \text{ ft}) - A_y \,(40 \text{ ft}) = 0$$

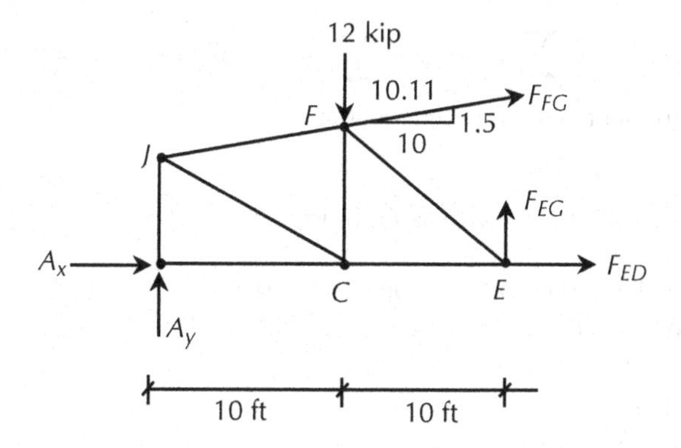

FIGURE 4.30 Free-body diagram of the left part of the cut section *a-a* for Example 4.8.

$$200 \text{ kip.ft} + 160 \text{ kip.ft} + 360 \text{ kip.ft} - 40A_y \text{ ft} = 0$$
$$720 \text{ kip.ft} - 40A_y \text{ ft} = 0$$
$$40A_y \text{ ft} = 720 \text{ kip.ft}$$
$$A_y = 18 \text{ kip}$$

$$\sum F_x = 0 \quad (\rightarrow +ve);\ A_x = 0$$

Now, in the left part of the cut section *a-a* apply $\sum M \text{ at } E = 0$ (assuming the clockwise moment is positive) as follows:

$$18\,\text{kip}(20\,\text{ft}) - 12(10\,\text{ft}) + F_{FC}\left(\frac{10}{10.11}\right)(7.5\,\text{ft}) + F_{FC}\left(\frac{1.5}{10.11}\right)(10\,\text{ft}) = 0$$

$$360 \text{ kip.ft} - 120 \text{ kip.ft} + 7.418\ F_{FC} \text{ ft} + 1.48\ F_{FC} \text{ ft} = 0$$
$$240 \text{ kip.ft} + 8.9\ F_{FC} \text{ ft} = 0$$
$$8.9\ F_{FC} \text{ ft} = -240 \text{ kip.ft}$$
$$F_{FC} = -26.96 \text{ kip}$$

Then, let us cut the section again at *a-b* lines to bring the desired F_{CE} and F_{EF} out, as shown in Figure 4.31.

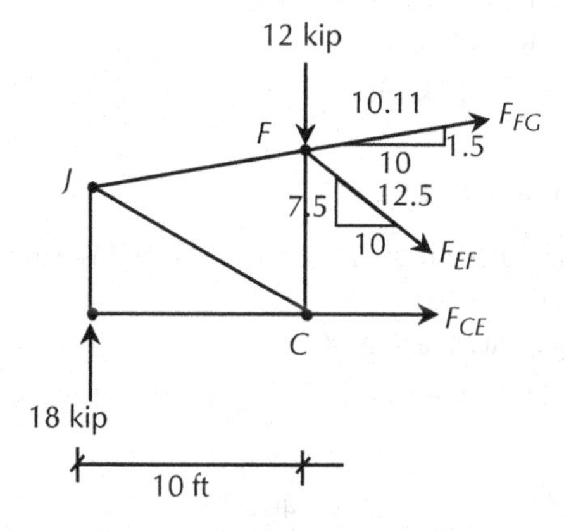

FIGURE 4.31 Free-body diagram of the left part of the cut section a-b of the truss for Example 4.8.

It can be again seen that $\sum F_x = 0$ and $\sum F_y = 0$ will not work here as well as multiple unknown forces are present here in each direction.

$\sum M$ at $F = 0$ will determine F_{CE} as follows:

18 kip (10 ft) – F_{CE} (7.5 ft) = 0 (assuming clockwise moment is positive)
180 kip.ft – 7.5F_{CE} ft = 0
7.5F_{CE} ft = 180 kip.ft
F_{CE} = 24 kip

In section *a-b* now only one force is unknown, F_{EF}. It can be found out applying $\sum F_x = 0$ or $\sum F_y = 0$. Let us apply $\sum F_x = 0$ $(\rightarrow +ve)$ as follows:

$$F_{CE} + F_{FG}\left(\frac{10}{10.11}\right) + F_{EF}\left(\frac{10}{12.5}\right) = 0$$

$$24\text{ kip} + (-26.96\text{ kip})\left(\frac{10}{10.11}\right) + F_{EF}\left(\frac{10}{12.5}\right) = 0$$

– 2.67 kip + 0.8 F_{EF} = 0
0.8F_{EF} = 2.67 kip
F_{EF} = 3.33 kip

Now, see that to find out F_{EF} we need to determine F_{CE} and F_{FG} first. If the problem asks you to determine F_{EF} only, there is a direct way to do it which is discussed later.
Check your answer by applying $\sum F_x = 0$ $(\uparrow +ve)$ in the *a-b* section:

$$18\text{ kip} - 12\text{ kip} + F_{FG}\left(\frac{1.5}{10.11}\right) - F_{EF}\left(\frac{7.5}{12.5}\right) = 0$$

6 kip + 0.1484 F_{FG} – 0.6 F_{EF} = 0
6 kip + 0.148 (–26.96 kip) – 0.6 F_{EF} = 0
2 kip – 0.6 F_{EF} = 0
0.6 F_{EF} = 2 kip
F_{EF} = 3.33 kip (matches)

Alternative way and direct way to find F_{EF}:

A point is to be identified through which F_{FG} and F_{CE} pass. Let us extend *JF* and *AC* backward to intersect at *O* as shown in Figure 4.32.

From similar angle triangles Δ*OAJ* and Δ*OCF*

$$\frac{x}{6} = \frac{x+10}{7.5}$$

$$x = 40\text{ ft}$$

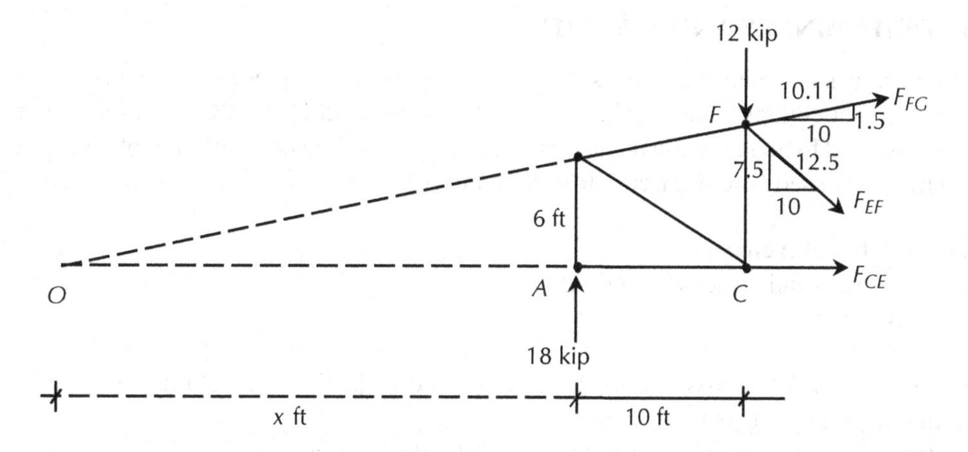

FIGURE 4.32 Free-body diagram of the left part of the cut section *a-b* of the truss for Example 4.8.

Now in the section *a-b*, apply ΣM at $O = 0$

$$18\,\text{kip}(40\,\text{ft}) - 12\,\text{kip}(50\,\text{ft}) - F_{EF}\left(\frac{10}{12.5}\right)(7.5\,\text{ft}) - F_{EF}\left(\frac{7.5}{12.5}\right)(50\,\text{ft}) = 0$$

$$720\ \text{kip.ft} - 600\ \text{kip.ft} - 6\,F_{EF} - 30\,F_{EF} = 0$$

$$120\ \text{kip.ft} - 36\,F_{EF} = 0$$

$$36\,F_{EF} = 120\ \text{kip.ft}$$

$$F_{EF} = 3.33\ \text{kip}$$

Answers:

F_{FG} = 27 kip (Compression)
F_{CE} = 24 kip (Tension)
F_{EF} = 3.3 kip (Tension)

Remember:

- *The first step in solving a truss is to find out the zero-force or direct-force members. If you could identify the zero-force or direct-force members, you can delete these members to make the solution easier.*
- *The method to be used depends on the type of problem, member organizations and complexity. A general rule of thumb is that the joint method is good for a simple and easy truss. The section method is more appropriate for a complex truss.*
- *You can use any single method or a combination unless it is specified in the examination.*
- *For vertical downward loading, it is common to have tensile forces at the bottom chords, and compressive forces at the top chords.*

4.6 DETERMINACY AND STABILITY

A structure is said to be stable if its support restraints and member organization are such that it does not collapse due to external loading of any kind from any direction . Stability can be assessed by mathematical calculation, and by visual inspection. The following symbols have been used here to discuss this phenomenon.

m = number of members
r = number of independent reactions
j = number of joints

For a plane truss (2D truss), Table 4.1 can be used to justify the stability (external) and determinacy of a plane truss.

A truss must be assessed visually to justify its internal stability. Both the support restraints and member arrangement must be able to resist both horizontal and vertical loads either anticipated or accidental. The bar arrangement must be such that it must have at least one diagonal (i.e., sum of triangles) so that the truss does not squeeze. Stability is commonly studied in a higher course such as structural analysis. The truss shown in Figure 4.33 is internally unstable, as one bay of the truss has no diagonal.

TABLE 4.1
Stability and Determinacy Analysis of a Plane Truss

Analysis	Classification
$m + r < 2j$	Unstable
$m + r = 2j$	Stable and statically determinate
$m + r > 2j$	Stable and statically indeterminate

FIGURE 4.33 Trusses in a residential building. *(Photo taken in Long Island, New York)*

Example 4.9:

Determine the determinacy and the stability of the truss shown in Figure 4.34.

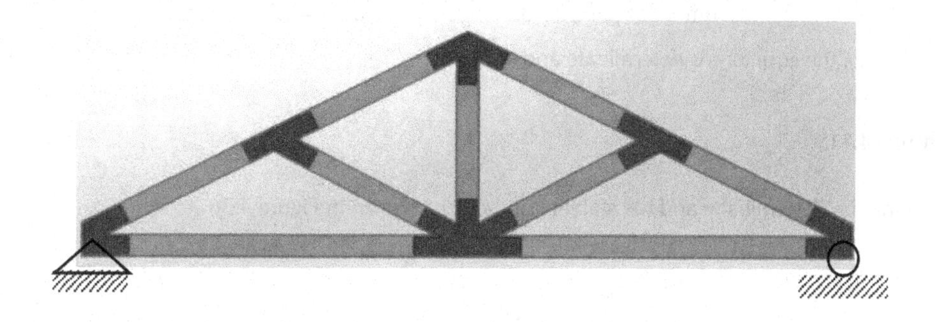

FIGURE 4.34 A truss for Example 4.9.

SOLUTION

Number of member, $m = 9$	$m + r = 12$
Number of reaction, $r = 3$	$2j = 12$
Number of joint, $j = 6$	$m + r = 2j$

As $m + r = 2j$, the structure is determinate and stable.

Example 4.10:

Determine the determinacy and the stability of the truss shown in Figure 4.35.

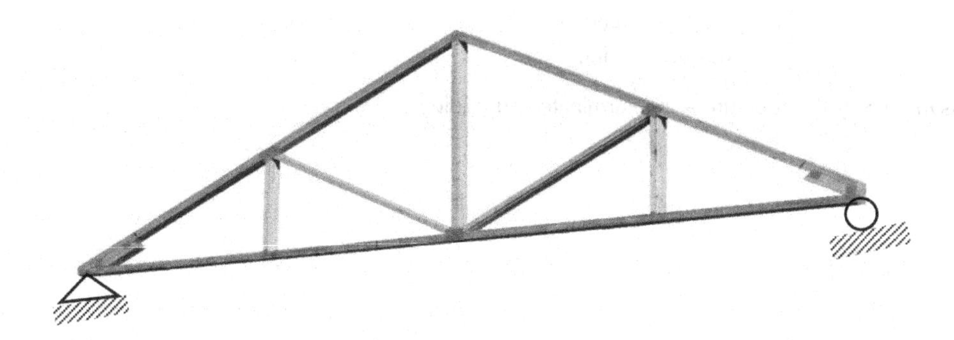

FIGURE 4.35 A truss for Example 4.10.

SOLUTION

Number of member, $m = 13$	$m + r = 16$
Number of reaction, $r = 3$	$2j = 16$
Number of joint, $j = 8$	$m + r = 2j$

As $m + r = 2j$, the structure is determinate and stable.

Example 4.11:

Determine the determinacy and the stability of the truss shown in Figure 4.36.

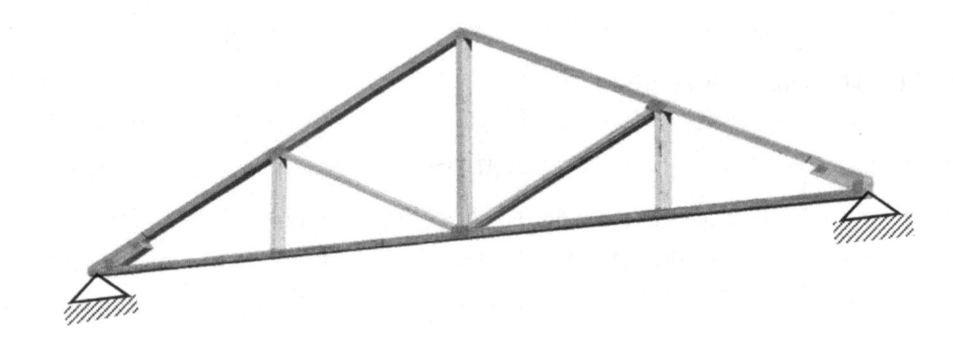

FIGURE 4.36 A truss for Example 4.11.

SOLUTION

Number of member, $m = 13$	$m + r = 17$
Number of reaction, $r = 4$	$2j = 16$
Number of joint, $j = 8$	$m + r > 2j$

As $m + r > 2j$, the structure is indeterminate and stable.

PRACTICE PROBLEMS

PROBLEM 4.1

Find out the zero-force member(s) for the trusses shown in Figure 4.37.

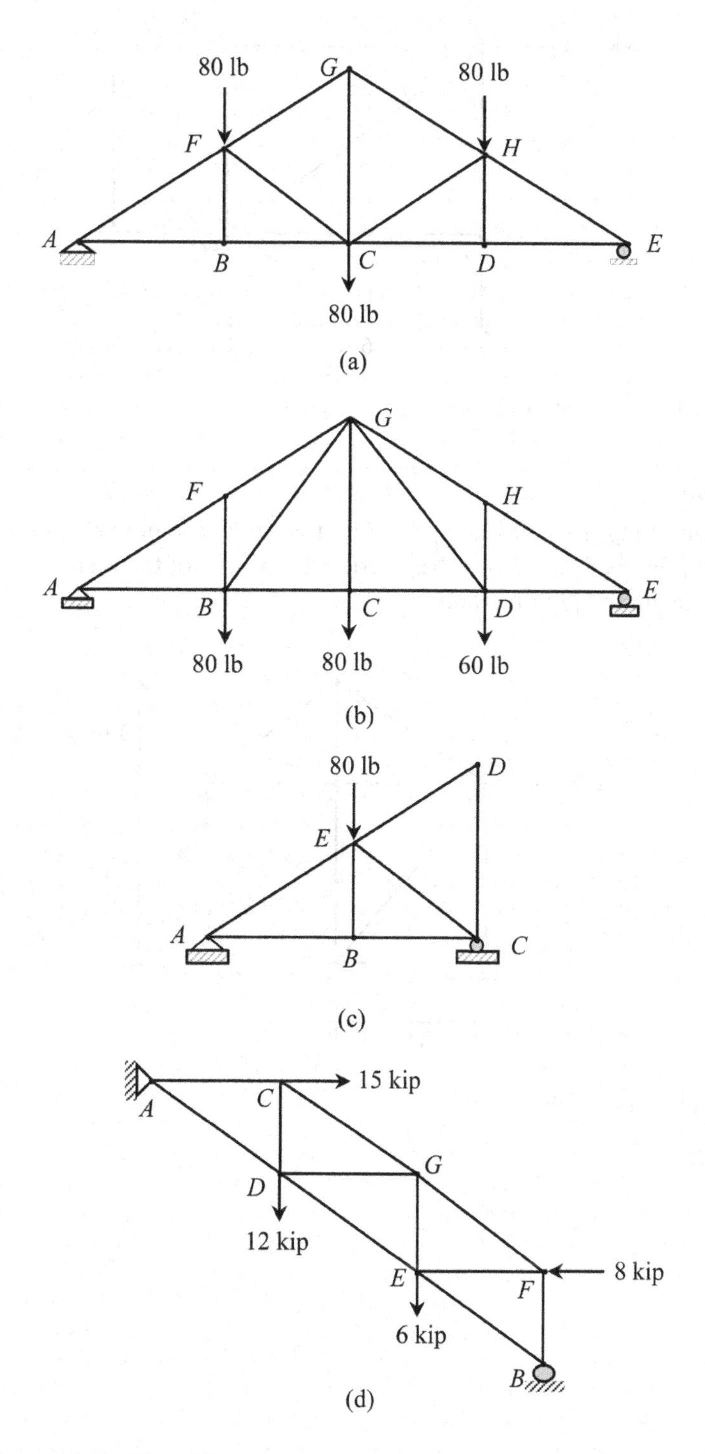

FIGURE 4.37 Trusses for Problem 4.1.

PROBLEM 4.2

Find out all the member forces using the joint method for the truss shown in Figure 4.38.

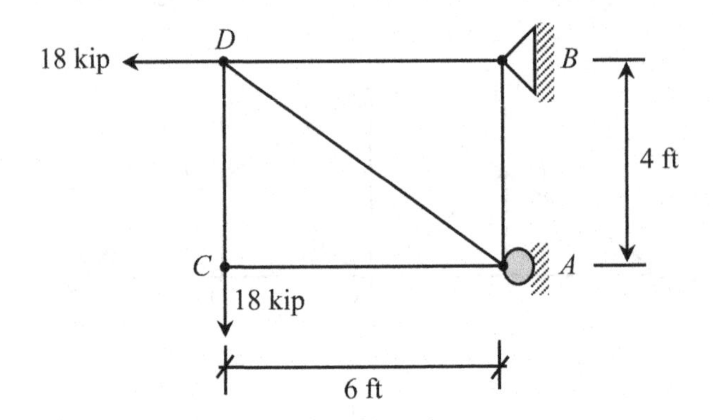

FIGURE 4.38 A truss for Problem 4.2.

PROBLEM 4.3

The truss shown in Figure 4.39 is supported by two supports: pin at A and roller at B. It is supporting two joint loads as shown. Neglecting the weight of the members, determine the member forces using the joint method.

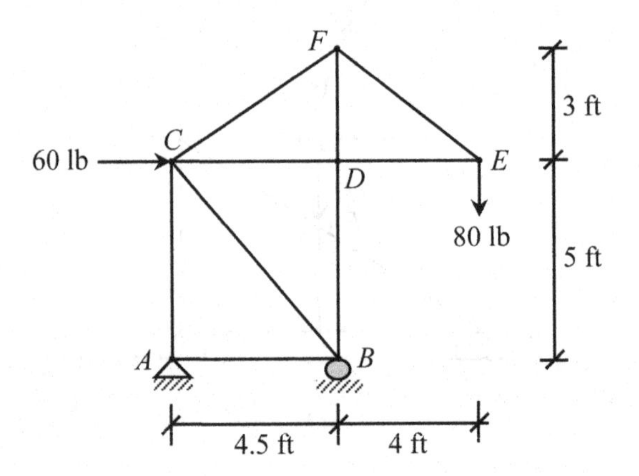

FIGURE 4.39 A truss for Problem 4.3.

PROBLEM 4.4

Find out all the member forces for the truss shown in Figure 4.40.

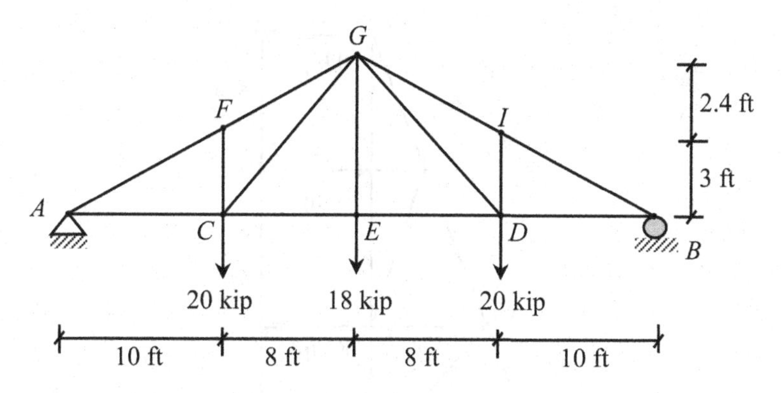

FIGURE 4.40 A truss for Problem 4.4.

PROBLEM 4.5

Find out all the member forces for the following truss as shown in Figure 4.41.

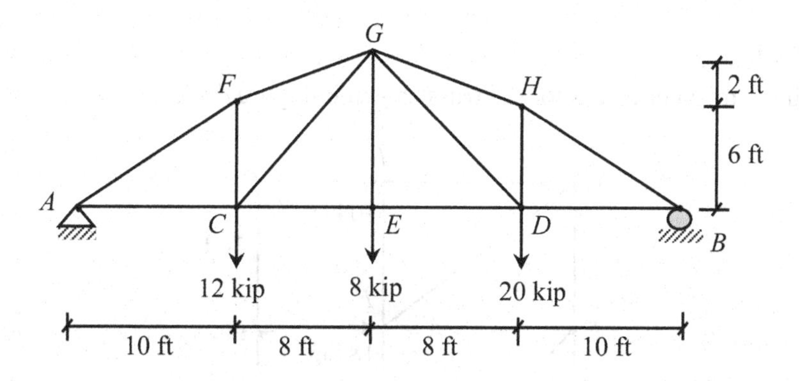

FIGURE 4.41 A truss for Problem 4.5.

PROBLEM 4.6

Find out all the member forces for the truss shown in Figure 4.42.

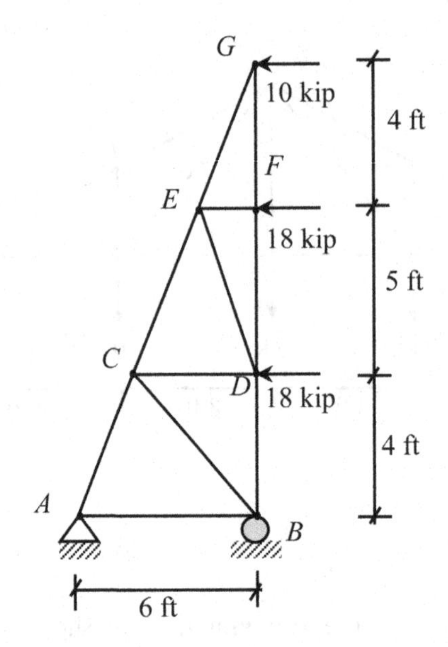

FIGURE 4.42 A truss for Problem 4.6.

PROBLEM 4.7

Find out all the member forces for the truss shown in Figure 4.43.

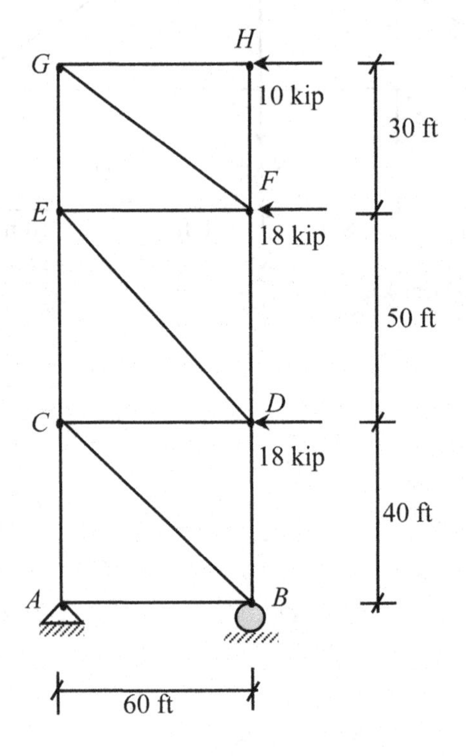

FIGURE 4.43 A truss for Problem 4.7.

PROBLEM 4.8

Find out all the member forces for the truss shown in Figure 4.44. Use $CG = 8$ ft.

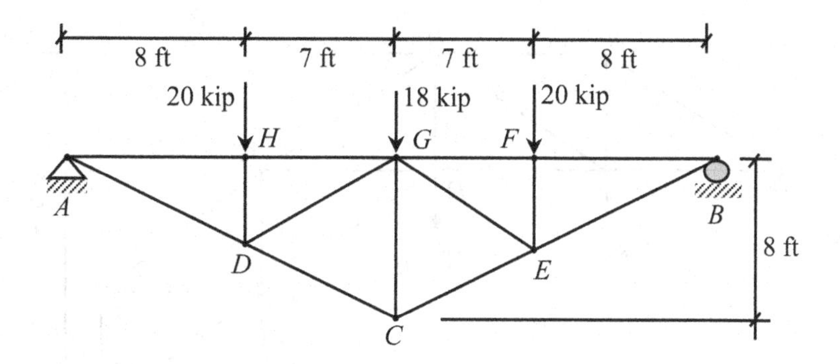

FIGURE 4.44 A truss for Problem 4.8.

PROBLEM 4.9

Find out all the member forces for the truss shown in Figure 4.45. The applied loads are perpendicular to the roof surface.

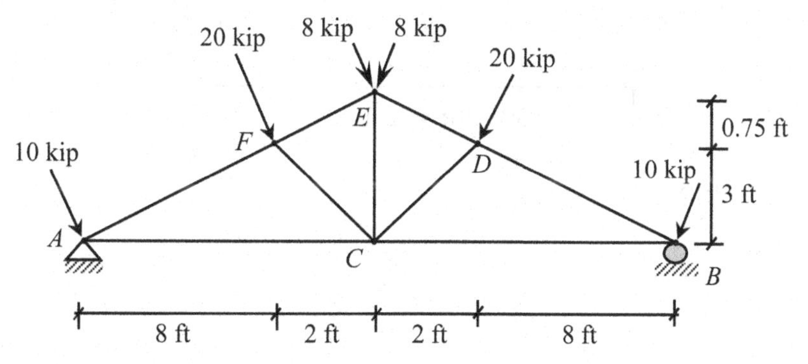

FIGURE 4.45 A Truss for Problem 4.9. *(Photo taken in Denver, Colorado)*

PROBLEM 4.10

A truss structure is supporting four joint loads as shown in Figure 4.46. It is supported by a pin support, *A* and a roller support, *B*. Determine the member forces.

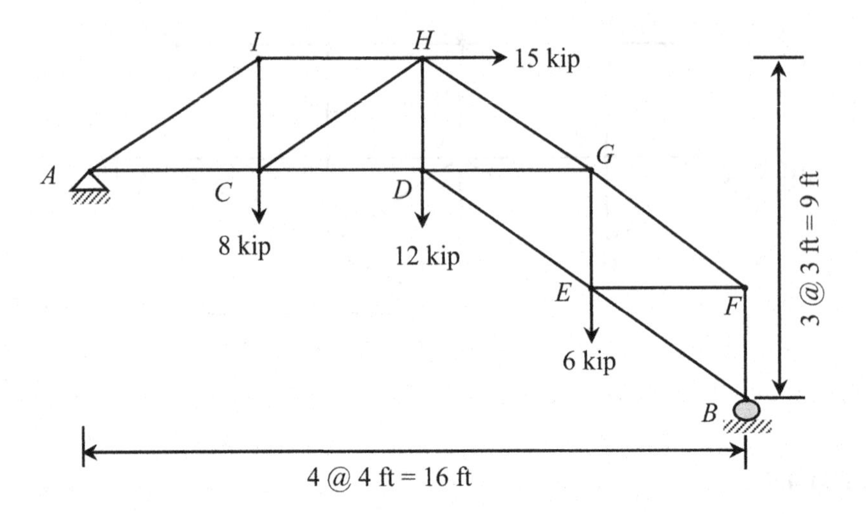

FIGURE 4.46 A truss for Problem 4.10.

PROBLEM 4.11

A truss structure is supporting four joint loads as shown in Figure 4.47. It is supported by a pin support, *A*, and a roller support, *B*. Determine the member forces.

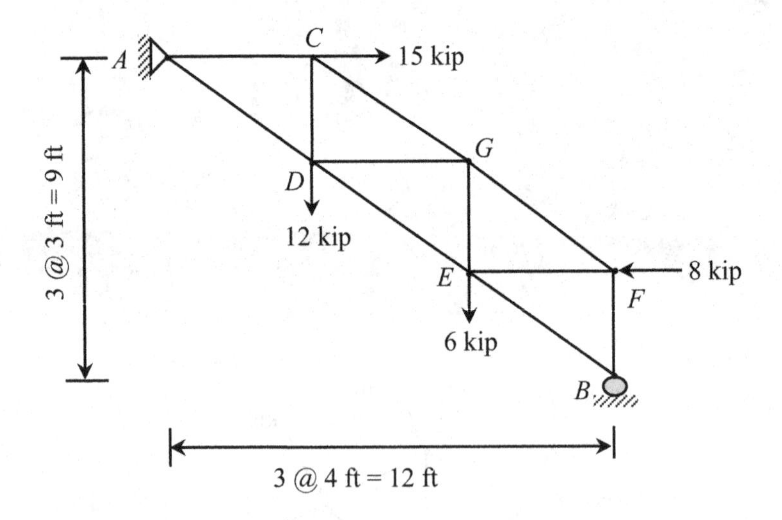

FIGURE 4.47 A truss for Problem 4.11.

PROBLEM 4.12

Idealize the truss shown in Figure 4.48. The truss is located at the Ellis Island National Museum of Immigration (New Jersey, USA). Assume reasonable loads at the truss joints and determine the member forces.

FIGURE 4.48 A truss for Problem 4.12. *(Photo taken in Ellis Island, New York)*

5 Arches, Cables and Pulleys

5.1 GENERAL

Chapters 3 and 4 discussed how equilibrium equations can be used to find out different unknown reactions in particles, beams and trusses. This chapter will discuss how the equilibrium equations can be used for several other types of structures such as arches, cables and frictionless pulleys. To remind you, determination of forces/reactions is the first step of designing a structural member. Once forces/reactions are calculated, the possible deformations, twist, bending, distortion, etc. are then determined to examine whether the target member will be structurally serviceable in the real world.

5.2 ARCHES

An arch is a curved structure spanning a space and may or may not support the weight above it, as shown in Figure 5.1. An arch may be synonymous with a vault, but a vault may be distinguished as a continuous arch forming a roof. A dome can be considered a three-dimensional arch.

An arch is in a pure compression form. It can span a long distance by distributing forces into compressive stresses and, in turn, eliminating tensile stresses. This is sometimes referred to as arch action. As the forces in the arch are carried to the ground, the arch will push outward at the base, called thrust. As the rise, or height, of the arch decreases, the outward thrust increases. The thrust needs to be restrained, either with internal ties or external bracing, such as abutments, to maintain arch action and prevent the arch from collapsing.

The most common true arch configurations are the fixed arch, two-hinged arch, and three-hinged arch. The fixed arch is most often used in reinforced concrete bridge and tunnel construction, where the spans are short. Because it is subject to additional internal stress caused by thermal expansion and contraction, this type of arch is considered to be statically indeterminate.

The two-hinged arch is most often used to bridge long spans. This type of arch has pinned connections at the bases only and no interior hinge as shown in Figure 5.2. Unlike the fixed arch, the pinned base can rotate, allowing the structure to move freely and compensate for the thermal expansion and contraction caused by changes in outdoor temperature. However, this can result in additional stresses, so the two-hinged arch is also statically indeterminate, although not to the degree of the fixed arch.

The three-hinged arch is not only hinged at its base, like the two-hinged arch, but at the mid-span as well, as shown in Figure 5.3. The additional connection at the mid-span allows the three-hinged arch to move in two opposite directions and compensate for any expansion and contraction. This type of arch is thus not subject to additional stress caused by thermal change. The three-hinged arch is therefore said to be statically determinate. It is most often used for medium-span structures, such as roofs of large buildings.

FIGURE 5.1 An arch structure. *(Courtesy of Wikipedia)*

FIGURE 5.2 Two-hinged arches. *(Photo taken in the Southern State Parkway, New York)*

Another advantage of the three-hinged arch is that the pinned bases are more easily developed than fixed ones, allowing for shallow, bearing-type foundations in medium-span structures. In the three-hinged arch, thermal expansion and contraction of the arch will cause vertical movements at the peak pin joint but will have no appreciable effect on the bases, further simplifying the foundation design.

FIGURE 5.3 A three-hinged arch on Interstate 25 (I-25). *(Photo taken in Denver, Colorado)*

The analysis procedure of three-hinged arches can be summarized as:

Step 1. Draw the free-body diagram of the whole arch.

Step 2. Determine the support reactions in the vertical direction using the equilibrium equations $\left(\sum M = 0 \text{ and } \sum F_y = 0\right)$.

Step 3. Detach at the interior hinge; draw the free-body diagram of either section and determine the unknown reactions using the three equilibrium equations.

Step 4. Consider the free-body diagram of the whole arch and determine the remaining unknown reaction using the equilibrium equation, $\sum F_x = 0$.

In this text, arches with both supports located at the same elevation are discussed. Arches with supports located at different elevations are discussed in higher-level courses such as structural analysis.

Example 5.1:

A three-hinged arch is supporting two loads of 20 kip and 30 kip as shown in Figure 5.4. The supports *A* and *B* are pinned supports, and *C* is a pin joint. Determine the reactions (B_x, and B_y) at the support *B*, the reactions (C_x, and C_y) at the interior hinge, *C*, and the reactions (A_x, and A_y) at the support *A*.

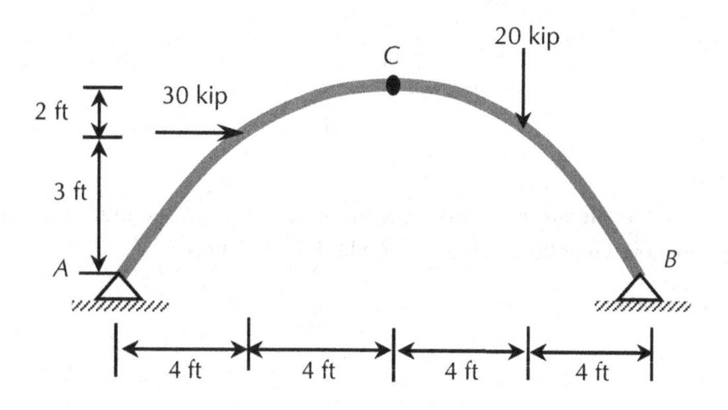

FIGURE 5.4 A three-hinged arch for Example 5.1.

SOLUTION

Step 1. Draw the free-body diagram of the whole arch. Consider the whole structure as shown in Figure 5.5. Let the support reactions at the support *A* be A_x and A_y along the horizontal and the vertical directions respectively. The horizontal and the vertical reactions at *B* be B_x and B_y, respectively. The directions of A_x, A_y, B_x and B_y are assumed with the best guess.

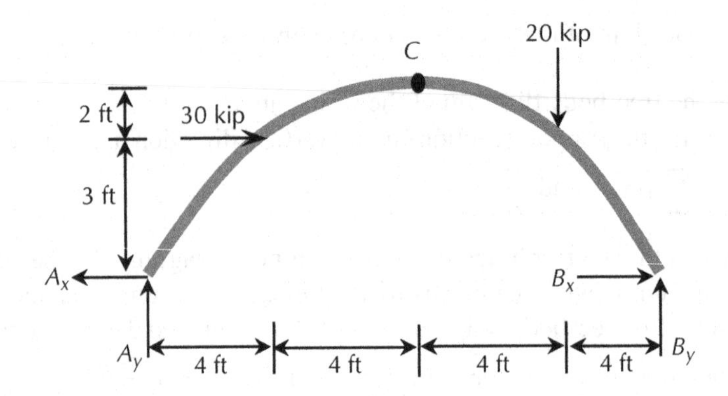

FIGURE 5.5 Free-body diagram of the arch for Example 5.1.

Step 2. Determine the support reactions in the vertical direction using the equilibrium equations.

$$\sum M \text{ at A} = 0 \ (\circlearrowright + ve)$$

$$30 \text{ kip } (3 \text{ ft}) + 20 \text{ kip } (12 \text{ ft}) - B_y \, (16 \text{ ft}) = 0$$
$$90 \text{ kip.ft} + 240 \text{ kip.ft} - 16 \, B_y \text{ ft} = 0$$
$$330 \text{ kip.ft} - 16 \, B_y \text{ ft} = 0$$
$$16 \, B_y \text{ ft} = 330 \text{ kip.ft}$$
$$B_y = 20.63 \text{ kip}$$

Now, $\sum F_y = 0 \ (\uparrow +ve)$

$$A_y + B_y - 20 \text{ kip} = 0$$
$$A_y = 20 \text{ kip} - B_y$$
$$A_y = 20 \text{ kip} - 20.63 \text{ kip}$$
$$A_y = -\,0.63 \text{ kip}$$

Step 3. Detach at the interior hinge; draw the free-body diagram of either section and determine the unknown reactions using the equilibrium equations.

Cut at C and consider the right part as shown in Figure 5.6

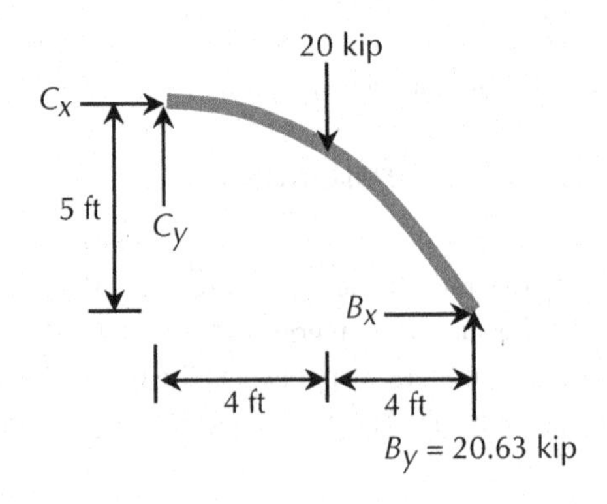

FIGURE 5.6 Free-body diagram of the right side of the arch for Example 5.1.

$\sum M$ at $C = 0$ (↲+ve)

$$20 \text{ kip } (4 \text{ ft}) - B_x (5 \text{ ft}) - 20.63 \text{ kip } (8 \text{ ft}) = 0$$
$$80 \text{ kip.ft} - 5B_x \text{ ft} - 165 \text{ kip.ft} = 0$$
$$-5B_x \text{ ft} - 85 \text{ kip.ft} = 0$$
$$5B_x \text{ ft} + 85 \text{ kip.ft} = 0$$
$$5B_x \text{ ft} = -85 \text{ kip.ft}$$
$$B_x = -17 \text{ kip}$$

$\sum F_x = 0$ (→ +ve)

$$C_x + B_x = 0$$
$$C_x = -B_x$$
$$C_x = -(-17 \text{ kip})$$
$$C_x = 17 \text{ kip}$$

$\sum F_y = 0$ (↑ +ve)

$$C_y + B_y - 20 \text{ kip} = 0$$
$$C_y = 20 \text{ kip} - B_y$$
$$C_y = 20 \text{ kip} - 20.63 \text{ kip}$$
$$C_y = -0.63 \text{ kip}$$

Step 4. Consider the free-body diagram of the whole arch and determine the remaining unknown reaction.

Consider the whole structure shown in Step 1 and apply $\Sigma F_x = 0$.

$\sum F_x = 0$ (→ +ve)

$$-A_x + B_x + 30 \text{ kip} = 0$$
$$A_x = B_x + 30 \text{ kip}$$
$$A_x = -17 \text{ kip} + 30 \text{ kip}$$
$$A_x = 13 \text{ kip}$$

Answers:

C_x = 17 kip (You cannot tell which way; it depends on which side, AC or BC you are considering)
C_y = 0.6 kip (You cannot tell which way; it depends on which side, AC or BC you are considering)
B_x = 17 kip (←)
B_y = 20.6 kip (↑)
A_x = 13 kip (←)
A_y = 0.6 kip (↓)

Example 5.2:

A three-hinged arch is supporting nine concentrated loads of 20 kip each as shown in Figures 5.7 and 5.8. The supports A and B are pinned supports, and C is a pin joint. Determine the reactions (B_x, and B_y) at the support B, the reactions (C_x, and C_y) at the interior hinge C, and the reactions (A_x, and A_y) at the support A.

FIGURE 5.7 A three-hinged arch. *(Courtesy of Wikipedia)*

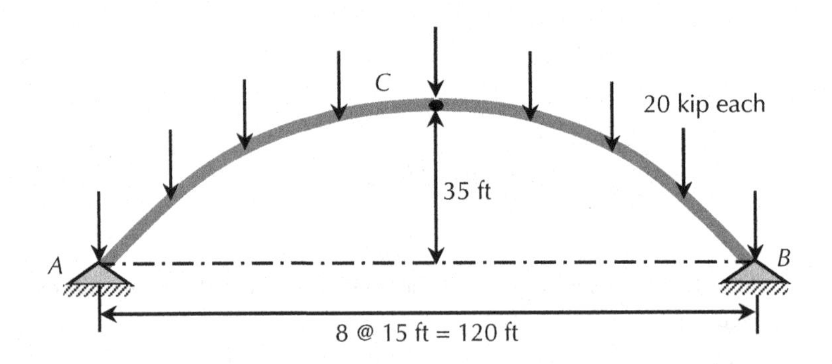

FIGURE 5.8 The idealized structure of the three-hinged arch for Example 5.2.

Step 1. Draw the free-body diagram of the whole arch. Consider the whole structure as shown in Figure 5.9.

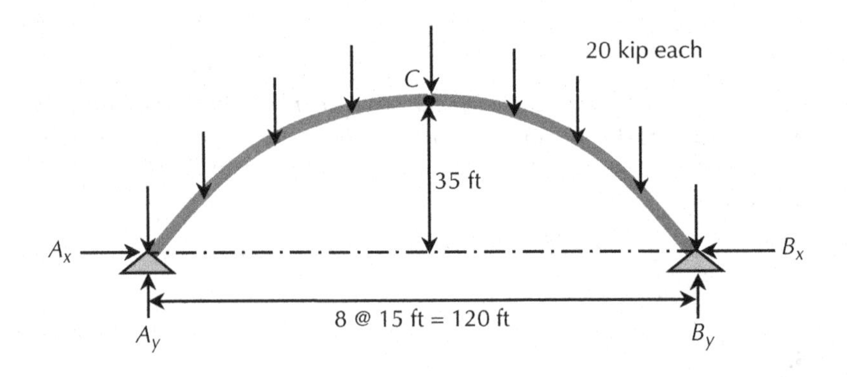

FIGURE 5.9 Free-body diagram of the arch for Example 5.2.

Step 2. Determine the support reactions in the vertical direction using the equilibrium equations.

$$\sum M \text{ at A} = 0\ (\lrcorner + ve)$$

$$20\text{ kip}(0\text{ ft}) + 20\text{ kip}(15\text{ ft}) + 20\text{ kip}(30\text{ ft}) + 20\text{ kip}(45\text{ ft}) + 20\text{ kip}(60\text{ ft}) + 20\text{ kip}(75\text{ ft}) + 20\text{ kip}(90\text{ ft}) + 20\text{ kip}(105\text{ ft}) + 20\text{ kip}(120\text{ ft}) - B_y\,(120\text{ ft}) = 0$$

$$20\text{ kip }(0 + 15 + 30 + 45 + 60 + 75 + 90 + 105 + 120)\text{ ft} - B_y\,(120\text{ ft}) = 0$$
$$20\text{ kip }(540\text{ ft}) - B_y\,(120\text{ ft}) = 0$$
$$10{,}800\text{ kip.ft} - B_y\,(120\text{ ft}) = 0$$
$$B_y = 90\text{ kip}$$

(Apply common sense to check the answer; symmetric loading.)

$$\sum F_y = 0\ (\uparrow +ve)$$

$$A_y + B_y - 20\text{ kip} - 20\text{ kip} - 20\text{ kip} - 20\text{ kip} - 20\text{ kip} - 20\text{ kip} - 20\text{ kip} - 20\text{ kip} - 20\text{ kip} = 0$$
$$A_y = 180\text{ kip} - B_y$$
$$A_y = 180\text{ kip} - 90\text{ kip}$$
$$A_y = 90\text{ kip}$$

Step 3. Detach at the interior hinge; draw the free-body diagram of the either section and determine the unknown reactions using the equilibrium equations.

Cut at C and consider the right part as shown in Figure 5.10.

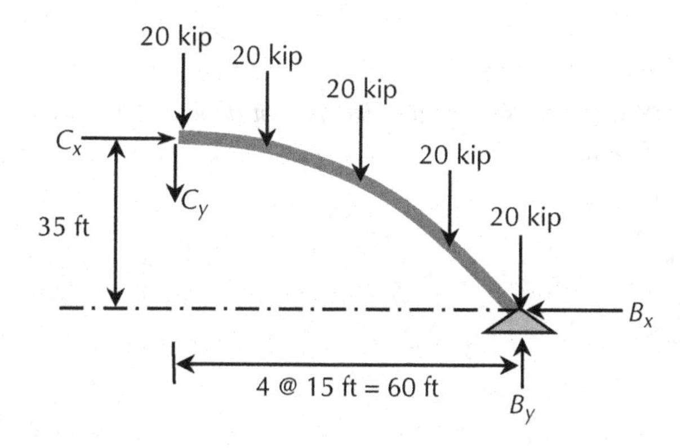

FIGURE 5.10 Free-body diagram of the right side of the arch for Example 5.2.

$$\sum M \text{ at } C = 0\ (\lrcorner + ve)$$

$$20\text{ kip }(0\text{ ft}) + 20\text{ kip }(15\text{ ft}) + 20\text{ kip }(30\text{ ft}) + 20\text{ kip }(45\text{ ft}) + 20\text{ kip }(60\text{ ft}) + B_x\,(35\text{ ft}) - B_y\,(60\text{ ft}) = 0$$
$$3{,}000\ 0\text{ kip.ft} + B_x\,(35\text{ ft}) - 90\text{ kip }(60\text{ ft}) = 0$$
$$B_x\,(35\text{ ft}) - 2{,}400\text{ kip.ft} = 0$$
$$B_x = 68.6\text{ kip}$$

$$\sum F_x = 0\ (\rightarrow +ve)$$
$$C_x - B_x = 0$$

$C_x = B_x$
$C_x = 68.6$ kip

$$\sum F_x = 0\ (\uparrow +ve)$$

$-C_y + B_y - 20$ kip $- 20$ kip $- 20$ kip $- 20$ kip $- 20$ kip $= 0$
$C_y = B_y - 100$ kip
$C_y = 90$ kip $- 100$ kip
$C_y = -10$ kip

Note that the concentrated force applied at C is considered in the free-body diagram of the right side of the arch shown in Figure 5.10. As the problem did not clarify the right or leftside of C, we can take this freedom of choosing any side. If there is a concentrated load at any point of a member, the transverse reaction (shear force) gets an abrupt change at this point. If we do not consider the 20-kip load at C, then C_y will be 10 kip.

Step 4. Consider the whole structure shown in Step 1 and apply $\Sigma F_x = 0$.

$$\sum F_x = 0\ (\rightarrow +ve)$$

$A_x - B_x = 0$
$A_x = B_x$
$A_x = 68.6$ kip

Answers:

$C_x = 68.6$ kip (You cannot specify the direction; it depends on which side, AC or BC, you are considering)
$C_y = 10$ kip
$A_x = 68.6$ kip ($\rightarrow$)
$B_y = 90$ kip ($\uparrow$)
$B_x = 68.6$ kip ($\leftarrow$)
$A_y = 90$ kip ($\uparrow$)

Remember: although there is no horizontal force in arches, the supports experience horizontal reactions to balance the outward thrust.

5.3 CABLES

A cable is a flexible member that can only support tension. Cables can be made of practically any material, such as wool or any other natural or artificial fibers, but in most engineering applications cables are made of steel. Cables have a few properties that are worth noting. First, while it is possible to pull on something with a cable or to have a cable take a tensile load, cables cannot push something or take compressional loads. Another property of cables is a consequence of Newton's third law – every action has an equal and opposite reaction. In the case of cables, this means that if a tensile load is applied at one end of the cable that same force will be experienced throughout the length of the cable all the way to the opposite end. One benefit of cable structures is that the tension forces in the cables can be used as points of references in the analysis of structures.

One of the major benefits of cable structures is that they allow you to displace loads from one part of a structure to another. In contrast, the loads from a structure can be displaced from its extremities to a central pillar. Such is the case for cable-stayed bridges and suspension bridges which will be discussed in greater detail later. It is important to note that mixtures or hybrids of structures can be used as well.

5.3.1 Uniformly Loaded Cable

A uniformly loaded cable means the cable which supports no other load but its self-weight. Consider the following uniformly distributed loaded cable system with both supports at the same elevation as shown in Figures 5.11 and 5.12.

FIGURE 5.11 Electric cable with uniform loading (self-weight) *(Courtesy of Enincon)*

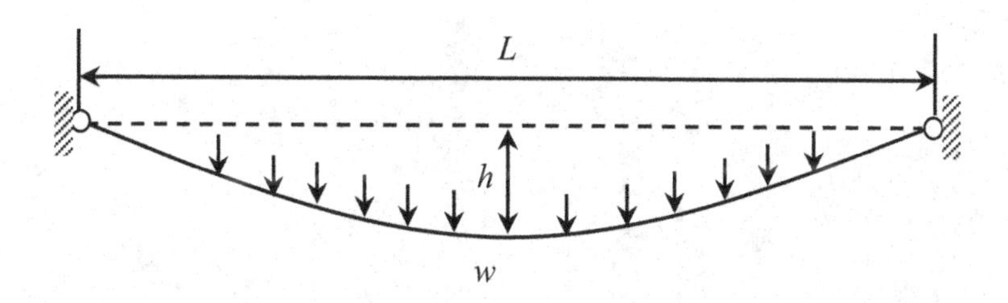

FIGURE 5.12 A cable with uniformly distributed load.

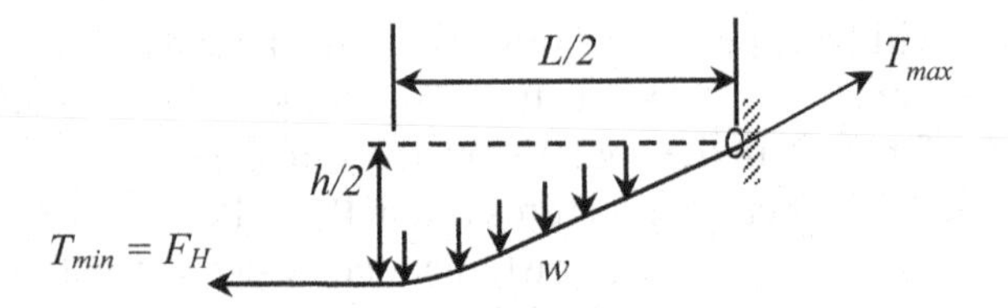

FIGURE 5.13 Free-body diagram of the right side of the cable with uniform load.

where

L = span of the cable system, m or ft
h = maximum sag at the middle of the cable, m or ft
w = uniformly distributed load, commonly the self-weight, lb/ft or N/m

The maximum tension occurs at the supports in the direction of the cable. The minimum tension occurs at the middle of the cable in horizontal direction. The right side of a uniformly loaded cable is shown in Figure 5.13:

The maximum tension at the supports (T_{max}) and the minimum tension at the lowest point (T_{min}) of cable can be determined as follows:

$$T_{\max} = \frac{wL}{2}\sqrt{1+\left(\frac{L}{4h}\right)^2}$$

$$T_{\min} = F_H = \frac{wL^2}{8h}$$

If the elevations of the cable supports are at two different positions, then the above-mentioned equations (T_{max} and T_{min}) will not be appropriate. Such advanced topics are commonly covered in a higher-level course such as structural analysis.

Example 5.3:

The cable system in the Royal Gorge Bridge in Colorado has a span of 880 ft and the maximum sag of 20 ft, as shown in Figure 5.14. If the uniformly distributed load on the cable can be considered as 0.20 kip/ft, determine the maximum and the minimum tensions developed at the cable.

FIGURE 5.14 The cable in the Royal Gorge Bridge in Colorado.

SOLUTION

Given,

Span, $L = 880$ ft

Sag, $h = 20$ ft

Uniformly distributed load, $w = 0.20$ kip/ft

Maximum tension, $T_{\max} = \dfrac{wL}{2}\sqrt{1+\left(\dfrac{L}{4h}\right)^2}$

$$T_{\max} = \frac{\left(0.20\,\dfrac{\text{kip}}{\text{ft}}\right)(880\,\text{ft})}{2}\sqrt{1+\left(\frac{880\text{ ft}}{4(20\text{ ft})}\right)^2}$$

Therefore, $T_{\max} = 972$ kip

Minimum tension, $T_{\min} = F_H = \dfrac{wL^2}{8h}$

$$T_{\min} = \frac{\left(0.20\,\dfrac{\text{kip}}{\text{ft}}\right)(880\text{ ft})^2}{8(20\,\text{ft})}$$

Therefore, $T_{\min} = 968$ kip

Answers:

Maximum tension = 972 kip

Minimum tension = 968 kip

Example 5.4:

A cable system with a uniformly distributed load on the cable is 1.0 kip/ft as shown in Figure 5.15. The maximum allowable sag in this cable is 10 ft, and the maximum tensile capacity of the cable is 75 kip. Calculate the maximum possible span length of the cable.

FIGURE 5.15 A cable with the self-weight as the uniformly distributed load.

SOLUTION

Given,

Span, $L = ?$

Sag, $h = 10$ ft

Uniformly distributed load, $w = 1.0$ kip/ft

Maximum tension, $T_{max} = 75$ kip

$$\text{Known: } T_{\max} = \frac{wL}{2}\sqrt{1+\left(\frac{L}{4h}\right)^2}$$

$$\text{Therefore, } 75 \text{ kip} = \frac{\left(1.0\frac{\text{kip}}{\text{ft}}\right)L}{2}\sqrt{1+\left(\frac{L}{4(10 \text{ ft})}\right)^2}$$

$$150 = L\sqrt{1+\left(\frac{L}{4(10)}\right)^2}$$

$$(150)^2 = (L^2)\left(1+\left(\frac{L}{40}\right)^2\right)$$

$$22{,}500 = L^2 + \frac{L^4}{1600}$$

$$L^4 + 1600L^2 - 360{,}000{,}00 = 0$$

Solving $L = 72.48$ ft

Answer:

Maximum possible span = 72.5 ft.

5.3.2 Non-Uniformly Loaded Cable

Cables may support concentrated loading spaced certain intervals. In that case, each segment of cable becomes straight. This type of cable can be solved using the procedure used in truss analysis (joint method or section method) except that cable can only take tension.

Example 5.5:

A cable system supported by two supports *A* and *B* is in equilibrium as shown in Figure 5.16. The length of the *BD* cable is 80 ft. The difference in elevation between *A* and *B* is 15 ft.

Calculate the following:

- The reaction at both supports *A* and *B*
- The cable tension at each segment
- The total length of the cable

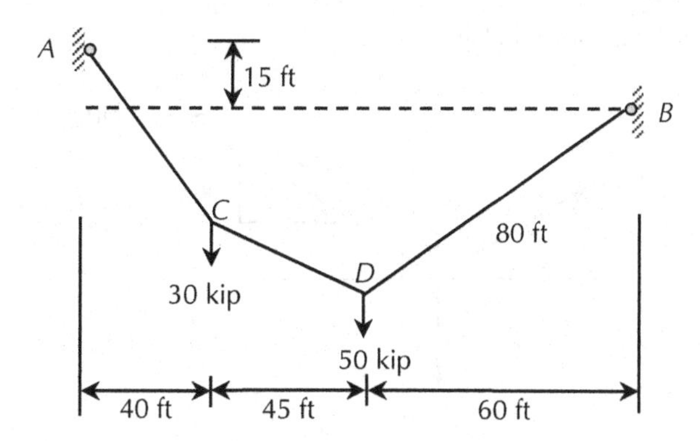

FIGURE 5.16 A cable with non-uniform load for Example 5.5.

SOLUTION

The reaction at B can be determined very easily as the slope of the BD segment is known.

The free-body diagram of the whole cable is shown in Figure 5.17:

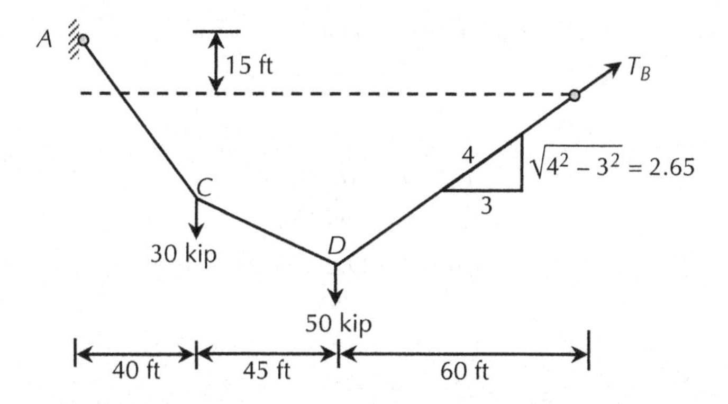

FIGURE 5.17 Free-body diagram of the cable for Example 5.5.

$$\sum M \text{ at A} = 0 \ (\lrcorner + ve)$$

$$30 \text{ kip } (40 \text{ ft}) + 50 \text{ kip } (85 \text{ ft}) - T_B \,(2.65/4)\,(145 \text{ ft}) - T_B \,(3/4)\,(15 \text{ ft}) = 0$$
$$1{,}200 \text{ kip.ft} + 4{,}250 \text{ kip.ft} - 96.06\ T_B \text{ ft} - 11.25\ T_B \text{ ft} = 0$$
$$5{,}450 \text{ kip.ft} - 107.31\ T_B \text{ ft} = 0$$
$$T_B = 50.8 \text{ kip}$$

The cable tension at segment BD is equal to the reaction at B, i.e., $T_{BD} = 50.8$ kip.

Reaction at A cannot be determined very easily as the slope of the segment AC is not known. We will have to proceed from the segment BD.

Consider the joint D, as shown in Figure 5.18.

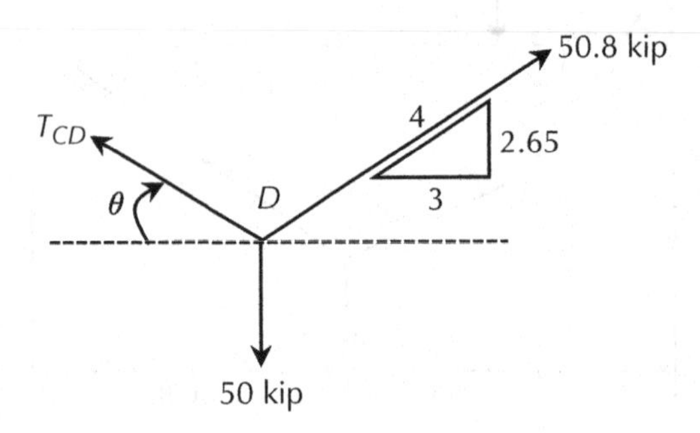

FIGURE 5.18 The joint D of the cable of Example 5.5.

$$\sum F_y = 0\ (\uparrow +ve)$$

$$T_{CD}\sin\theta + 50.8\ (2.65/4)\ \text{kip} - 50\ \text{kip} = 0$$
$$T_{CD}\sin\theta = 16.35\ \text{kip} \ldots\ldots\ldots\ldots\ldots\ldots(1)$$

Now, $\sum F_x = 0\ (\rightarrow +ve)$

$$-T_{CD}\cos\theta + 50.8\ (3/4)\ \text{kip} = 0$$
$$T_{CD}\cos\theta = 38.1\ \text{kip} \ldots\ldots\ldots\ldots\ldots\ldots(2)$$

From (1)/(2):

$$\frac{T_{CD}\sin\theta}{T_{CD}\cos\theta} = \frac{16.35\ \text{kip}}{38.1\ \text{kip}}$$

$$\tan\theta = 0.429$$
$$\theta = \tan^{-1}(0.429)$$
$$\theta = 23.23^\circ$$

From (1):

$$T_{CD} = 16.35\ \text{kip}/\sin 23.23$$
$$T_{CD} = 41.5\ \text{kip}$$

Consider the joint C, as shown in Figure 5.19.

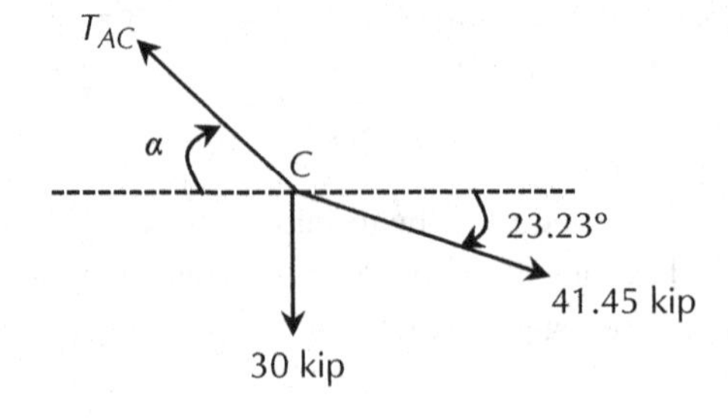

FIGURE 5.19 The joint C of the cable for Example 5.5.

$$\sum F_y = 0 \ (\uparrow +ve)$$

$$T_{AC}\sin\alpha - 41.45 \text{ kip } \sin 23.23 - 30 \text{ kip} = 0$$
$$T_{AC}\sin\alpha = 46.35 \text{ kip} \ \ldots\ldots\ldots\ldots\ldots\ldots\ldots (3)$$

Now, $\sum F_x = 0 \ (\rightarrow +ve)$

$$-T_{CD}\cos\alpha + 41.45 \text{ kip } \cos 23.23 = 0$$
$$T_{CD}\cos\alpha = 38.1 \text{ kip} \ \ldots\ldots\ldots\ldots\ldots\ldots\ldots (4)$$

From (1)/(2):

$$\frac{T_{AC}\sin\alpha}{T_{AC}\cos\alpha} = \frac{46.35 \text{ kip}}{38.1 \text{ kip}}$$

$$\tan\alpha = 0.429$$
$$\alpha = \tan^{-1}(0.429)$$
$$\alpha = 50.6^\circ$$

From (3):

$T_{AC} = 46.35$ kip/sin50.6

$T_{AC} = 60$ kip

The cable tension at segment AC is equal to the reaction at A, i.e., $T_A = 60$ kip.

Now let us find out the cable length.

Length of $BD = 80$ ft.

From the segment AC, as shown in Figure 5.20.

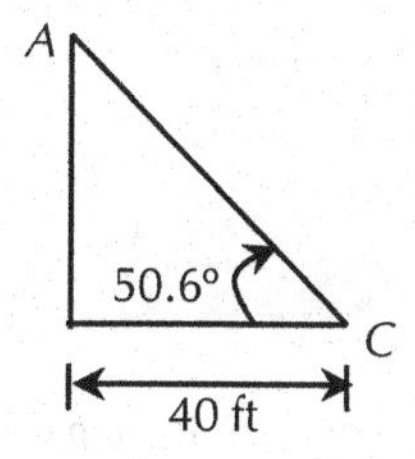

FIGURE 5.20 The segment AC of the cable for Example 5.5.

Length of $AC = 40$ ft/cos50.6 = 63 ft

From the segment CD, as shown in Figure 5.21.

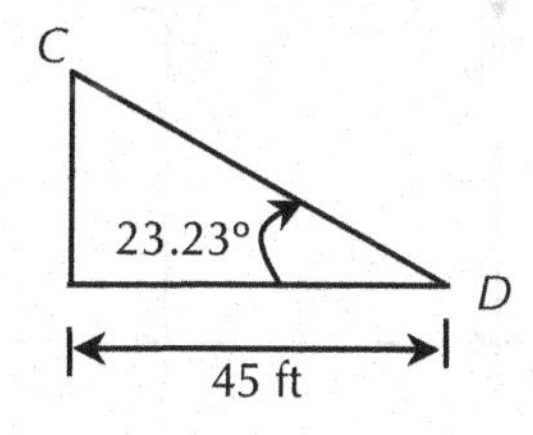

FIGURE 5.21 The segment CD of the cable for Example 5.5.

Length of CD = 45 ft/cos23.23 = 48.97 ft

Total length of cable = 80 ft + 63 ft + 48.97 ft = 191.97 ft ≈ 192 ft

Answers:

The reaction at B = 50.8 kip
The reaction at A = 60 kip
T_{CD} *= 41.5 kip*
T_{BD} *= 50.8 kip*
T_{AC} *= 60 kip*
Total length of cable = 192 ft

Example 5.6:

A highway traffic signal system can be idealized as shown in Figures 5.22 and 5.23. The cable can be assumed weightless and is in equilibrium. Calculate the tensions developed at each segment of the cable and the unknown force.

FIGURE 5.22 A highway traffic signal system for Example 5.6. *(Photo taken in Long Island, New York)*

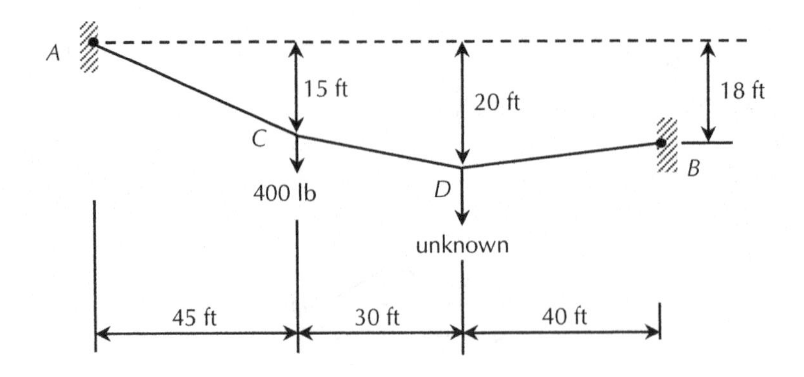

FIGURE 5.23 The idealized traffic signal system for Example 5.6.

SOLUTION

Free-body diagram of a part of the cable system, as shown in Figure 5.24:

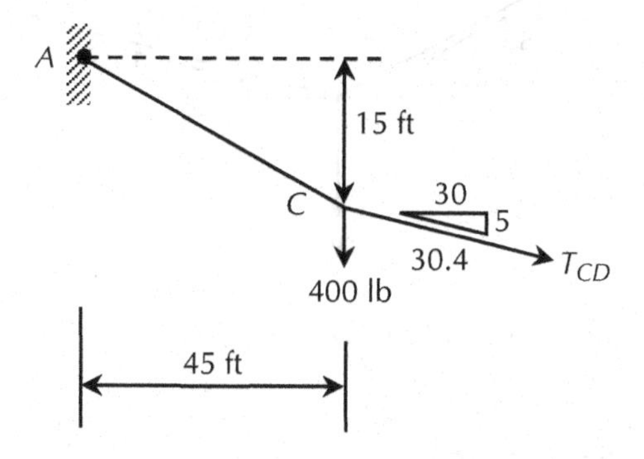

FIGURE 5.24 The free-body diagram of the traffic signal system for Example 5.6.

The vertical difference between C and D = 20 ft – 15 ft = 5 ft
Length of segment $CD = \sqrt{(30\,\text{ft})^2 + (5\,\text{ft})^2} = 30.41\,\text{ft}$

$$\sum M \text{ at A} = 0\ (\circlearrowright + ve)$$

$$(400\text{ lb})(45\text{ ft}) + T_{CD}\left(\frac{5}{30.4}\right)(45\text{ ft}) - T\left(\frac{30}{30.4}\right)(15\text{ ft}) = 0$$

18,000 lb.ft – 7.4 T_{CD} ft = 0
Therefore, T_{CD} = 2,432 lb

Consider the joint C, as shown in Figure 5.25.

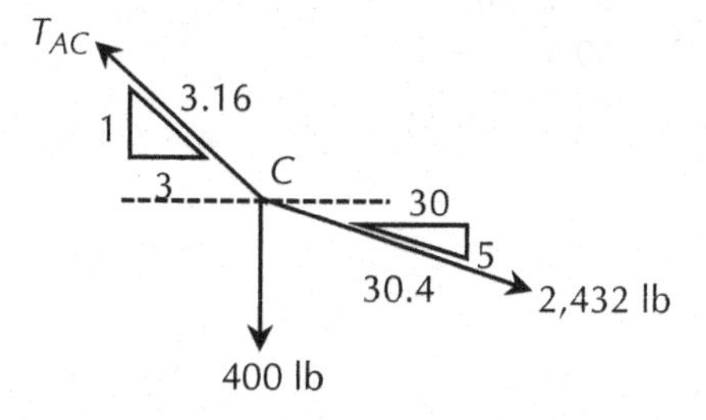

FIGURE 5.25 The free-body diagram of joint C.

$$\sum F_y = 0\ (\uparrow + ve)$$

$$T_{AC}\left(\frac{1}{3.16}\right) - 400\text{ lb} - 2{,}432\text{ lb}\left(\frac{5}{30.4}\right) = 0$$

TAC = 2,528 lb

$\sum F_x = 0\ (\rightarrow + ve)$ will also give you the answer.

Consider the joint D, as shown in Figure 5.26.

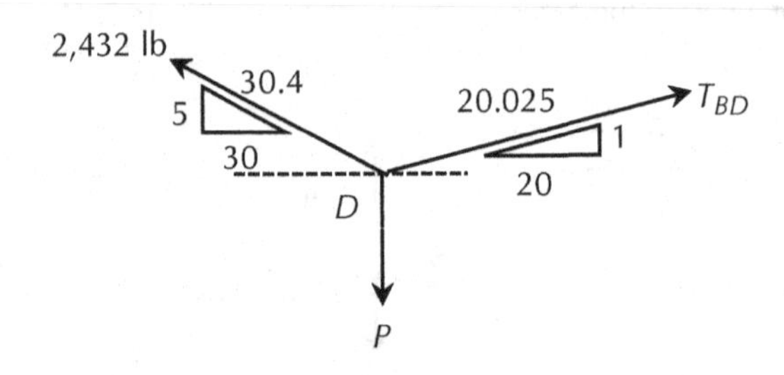

FIGURE 5.26 The free-body diagram of Joint D.

$$\sum F_x = 0\ (\rightarrow +ve)$$

$$2{,}432\,\text{lb}\left(\frac{30}{30.4}\right) - T_{BD}\left(\frac{20}{20.025}\right) = 0$$

$$2{,}400\ \text{lb} - 0.9988\ T_{BD} = 0$$
$$T_{BD} = 2{,}403\ \text{lb}$$

$$\sum F_y = 0\ (\uparrow +ve)$$

$$2{,}432\ \text{lb}\left(\frac{5}{30.4}\right) - P - T_{BD}\left(\frac{1}{20.025}\right) = 0$$

$$2{,}432\ \text{lb}\left(\frac{5}{30.4}\right) - P - (2{,}403\,\text{lb})\left(\frac{1}{20.025}\right) = 0$$

$$400\ \text{lb} - P - 120\ \text{lb} = 0$$
$$280\ \text{lb} - P = 0$$
$$P = 280\ \text{lb}$$

Answers:

$T_{CD} = 2{,}432$ *lb*
$T_{AC} = 2{,}528$ lb
$T_{BD} = 2{,}403$ lb
Unknown force = 280 lb

5.4 PULLEYS

A pulley is a wheel on an axle or shaft that is designed to support movement and change of direction of a cable, rope or belt along its circumference, as shown in Figure 5.27. Pulleys are used in a variety of ways to lift loads, apply forces and to transmit power. Pulley systems may consist of friction between the surfaces or friction may be neglected for well-finished and lubricated ones. In this section, only a frictionless pulley is discussed. Pulleys with friction will be discussed in Chapter 11.

The simplest theory of operation for a frictionless pulley system assumes that the pulleys and lines are weightless, and that there is no energy loss due to friction. It is also assumed

FIGURE 5.27 A pulley.

that the lines do not stretch. In equilibrium, the forces on the moving block must sum to zero. In addition, the tension in the rope must be the same for each of its parts. This means that the two parts of the rope supporting the moving block must each support half the load. The mechanical advantage of a pulley system can be analyzed using free-body diagrams which balance the tension force in the rope with the force of gravity on the load. In an ideal system, the massless and frictionless pulleys do not dissipate energy and allow for a change of direction of a rope that does not stretch or wear; in this case, a force balance on a free body that includes the load and the number of supporting sections of a rope with tension. The following examples clarify the analysis procedure.

Rule: Same cable, same force in a frictionless pulley.

Example 5.7:

Determine the tension at the hangers at *A* and *B* for the frictionless pulley system, as shown in Figure 5.28.

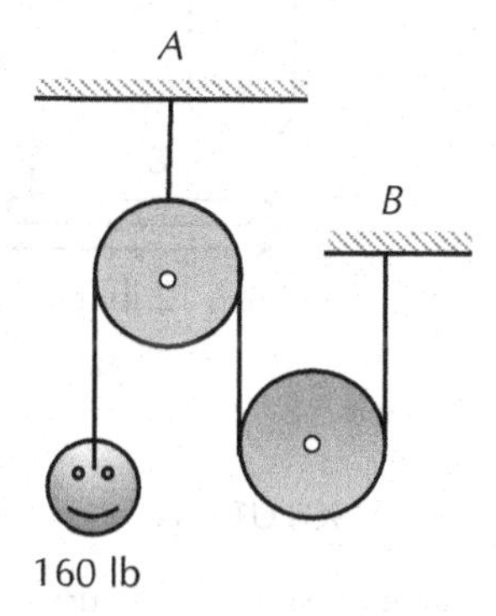

FIGURE 5.28 A pulley system for Example 5.7.

SOLUTION

From the free-body diagram, as shown in Figure 5.29:

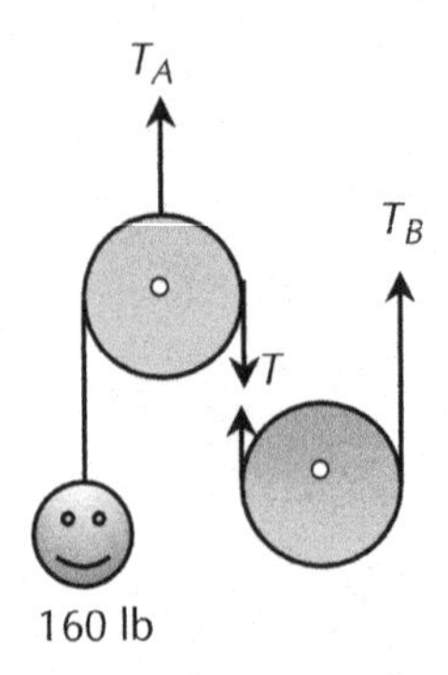

Using the principle, 'Same Cable Same Force'
From the left pulley: $T = 160$ lb
From the right pulley: $T_B = T = 160$ lb

Applying $\Sigma F_y = 0$ in the left pulley
$T_A = T + 160 \text{ lb} = 320 \text{ lb}$

FIGURE 5.29 Free-body diagram of the pulley system for Example 5.7.

Remember: same cable, same force in a frictionless pulley.

Answers:

Tension at A = 320 lb

Tension at B = 160 lb

Example 5.8:

A uniform plank of length 8 ft has mass per unit length of 12 lb/ft and is held up in the air horizontally by light inextensible strings tied to each end of the plank, as shown in Figure 5.30. Each of these strings is attached to a small light pulley which is connected to two large pulleys by a single string. If there is no friction, determine the tensions at supports A and B required to keep the system in equilibrium.

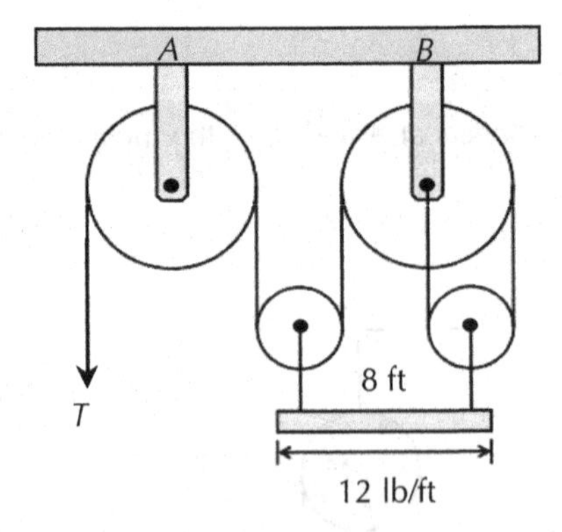

FIGURE 5.30 A pulley system for Example 5.8.

SOLUTION

From the free-body diagram of the plank, as shown in Figure 5.31:

Weight of plank = 12 lb/ft (8 ft) = 96 lb

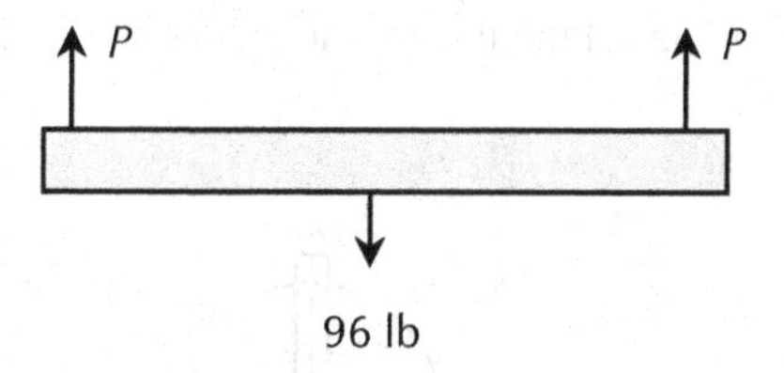

FIGURE 5.31 Free-body diagram of the plank for Example 5.8.

$$\sum F_y = 0\ (\uparrow +ve)$$

$$P + P - 96\text{ lb} = 0$$
$$2P - 96\text{ lb} = 0$$
$$P = 48\text{ lb}$$

Now let us consider the free bodies of the two left pulleys as shown in Figure 5.32.

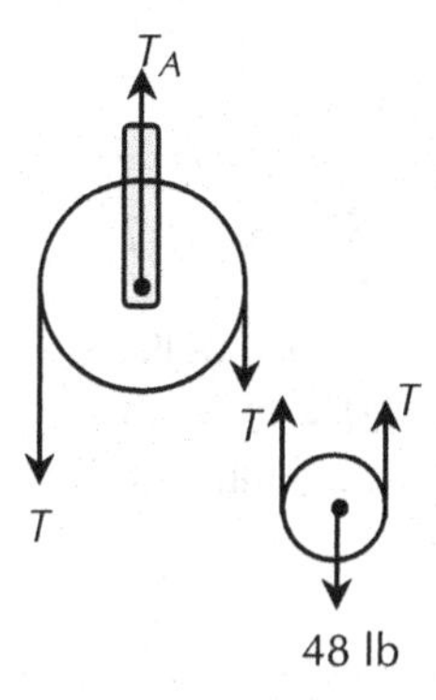

FIGURE 5.32 Free-body diagram of the left two pulleys for Example 5.8.

From the smaller pulley:

$$\sum F_y = 0\ (\uparrow +ve)$$

$$T + T - 48\text{ lb} = 0$$
$$2T - 48\text{ lb} = 0$$
$$T = 24\text{ lb}$$

From the bigger pulley:

$$\sum F_y = 0\ (\uparrow +ve)$$

$$T_A - T - T = 0$$
$$T_A = 2T$$
$$T_A = 2(24\text{ lb})$$
$$T_A = 48\text{ lb}$$

Now let us consider the free bodies of the three right pulleys, as shown in Figure 5.33.

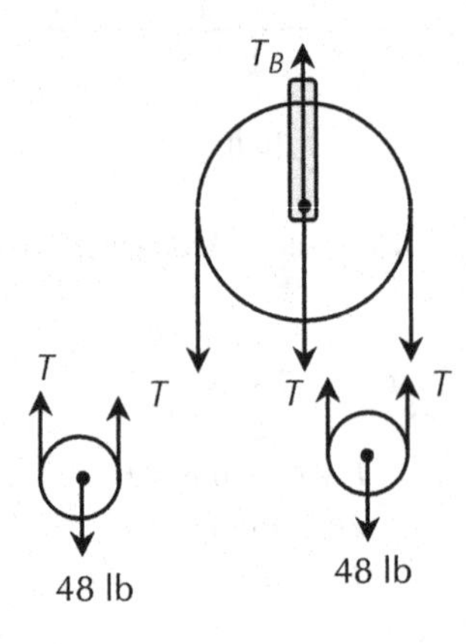

FIGURE 5.33 Free-body diagram of the right-three pulleys for Example 5.8.

From the smaller pulley:

$$\sum F_y = 0\ (\uparrow +ve)$$

$$T + T - 48\ \text{lb} = 0$$
$$2T - 48\ \text{lb} = 0$$
$$T = 24\ \text{lb}$$

From the bigger pulley:

$$\sum F_y = 0\ (\uparrow +ve)$$

$$T_B - T - T - T = 0$$
$$T_B - 3T = 0$$
$$T_B = 3T$$
$$T_B = 3(24\ \text{lb})$$
$$T_B = 72\ \text{lb}$$

Answer:

$T_A = 48\ lb$

$T_B = 72\ lb$

Example 5.9:

A frictionless collar, *C*, is connected to a 200-lb load via a frictionless pulley, *A*, as shown in Figure 5.34. The collar, *C*, can only slide horizontally on the frictionless bar, *DE*, shown below. Determine the amount of force to be applied to the collar to maintain equilibrium. Also, determine the support reaction at *A*.

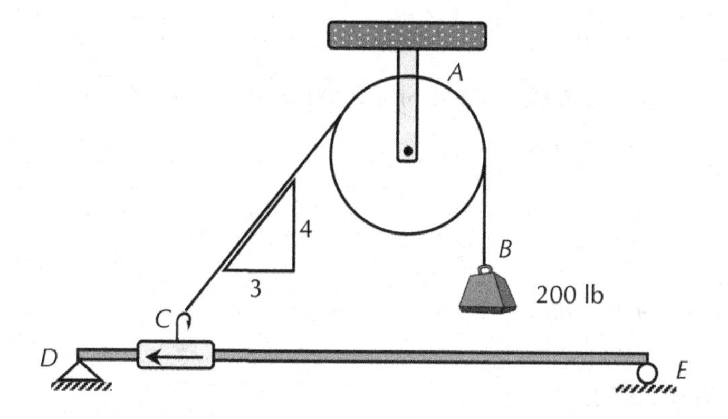

FIGURE 5.34 A pulley system for Example 5.9.

SOLUTION

Remember, same cable, same force in a frictionless pulley. Considering the free-body diagram of the collar as shown in Figure 5.35.

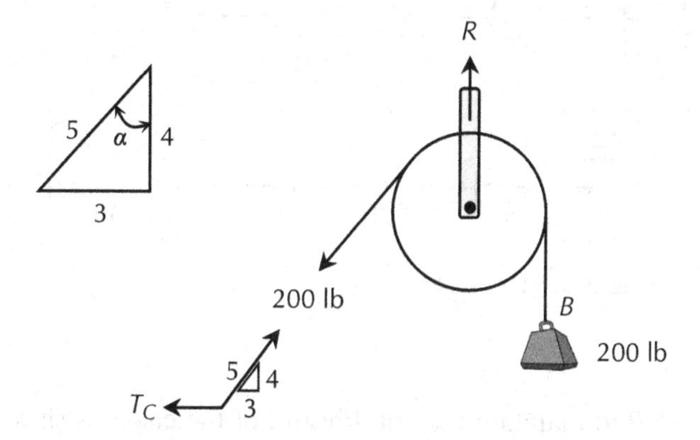

FIGURE 5.35 Free-body diagram of the pulley system for Example 5.9.

$$\sum F_x = 0 \; (\rightarrow +ve)$$

$$200\left(\frac{3}{5}\right) \text{lb} - T_C = 0$$

$$T_C = 200\left(\frac{3}{5}\right) \text{lb}$$

$$T_C = 120 \text{ lb}$$

Considering the free-body diagram of the pulley:

$$\alpha = \tan^{-1}\left(\frac{3}{4}\right) = 36.87^{\circ}$$

$$R = \sqrt{P^2 + Q^2 + 2PQ\cos\alpha}$$

$$R = \sqrt{(200\,\text{lb})^2 + (200\,\text{lb})^2 + 2(200\,\text{lb})(200\,\text{lb})(\cos 36.87)}$$

$$R = 380\,\text{lb}$$

Answer:

TC = 120 lb

Support reaction at A = 380 lb.

PRACTICE PROBLEMS

PROBLEM 5.1

Determine the reactions at *A*, *B*, and *C* for the three-hinged arch as shown in Figure 5.36.

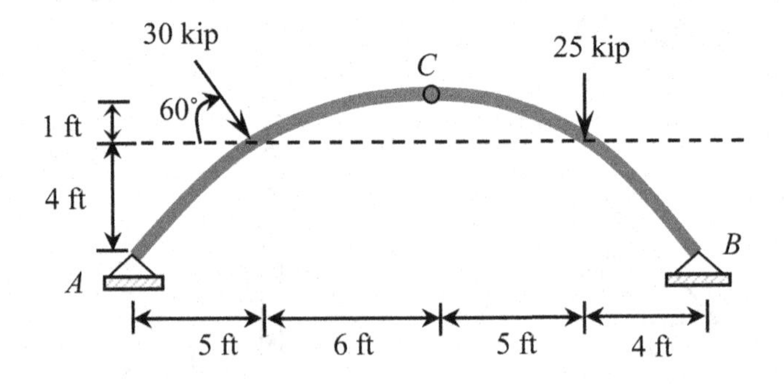

FIGURE 5.36 A three-hinged arch for Problem 5.1.

PROBLEM 5.2

What is the value of *P* to maintain the equilibrium of the cable as shown in Figure 5.37? Determine the cable force at each segment of the cable for the equilibrium condition. Cable segment *DE* is horizontal.

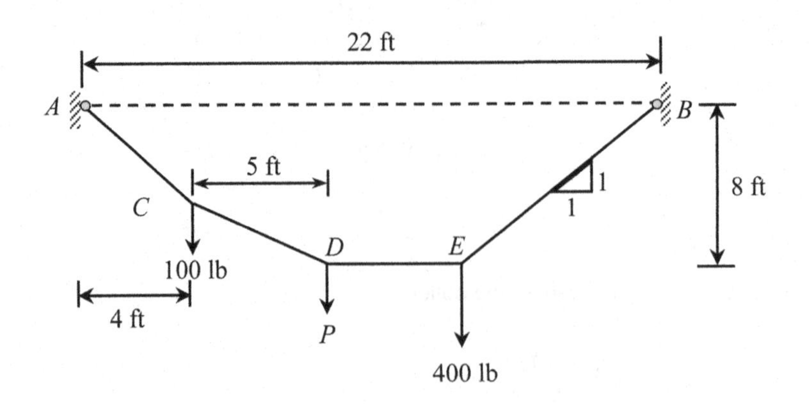

FIGURE 5.37 A cable with non-uniform loading for Problem 5.2.

PROBLEM 5.3

Determine the maximum and the minimum cable forces for the cable with uniformly distributed load as shown in Figure 5.38.

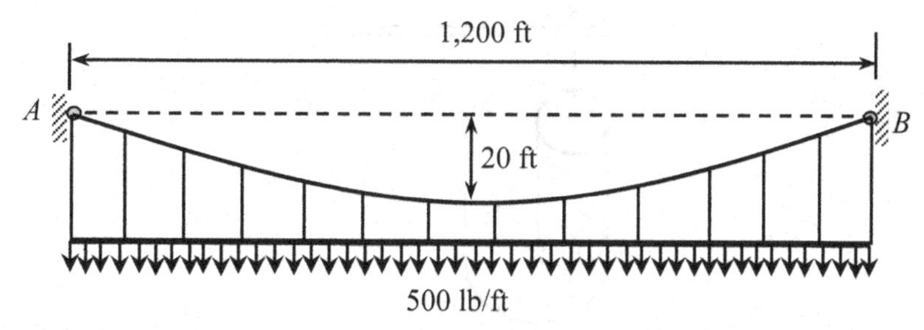

FIGURE 5.38 A cable with uniform loading for Problem 5.3. *(Photo taken in Canyon City, Colorado)*

PROBLEM 5.4

Determine the maximum and the minimum cable forces for the cable with uniformly distributed load, as shown in Figure 5.39.

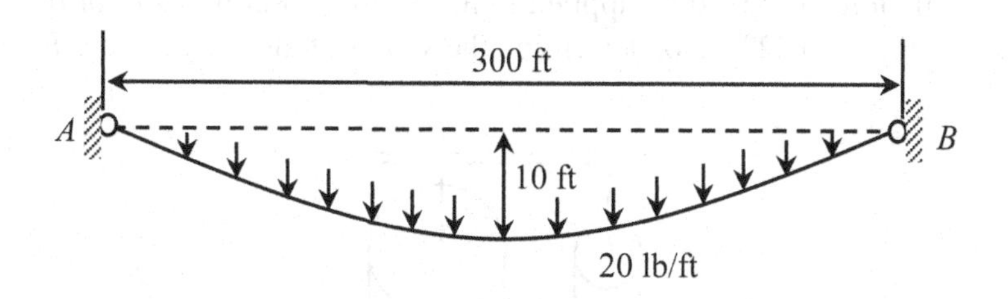

FIGURE 5.39 A cable with uniform loading for Problem 5.4.

PROBLEM 5.5

For the frictionless pulley shown in Figure 5.40 how much load is to be applied at the Pulley D to maintain the equilibrium? Also, determine the support tensions at A and B.

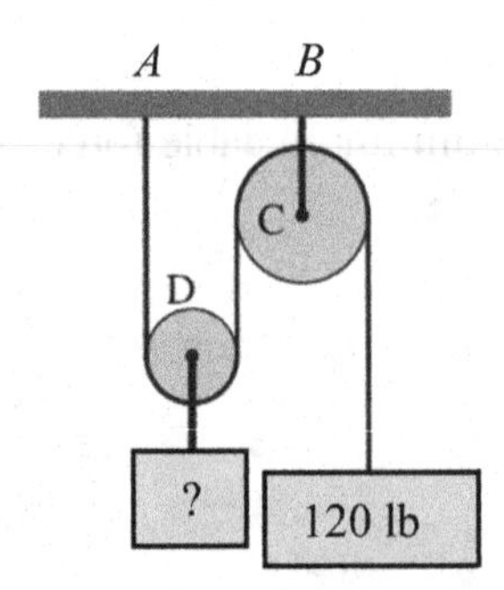

FIGURE 5.40 A pulley system for Problem 5.5.

PROBLEM 5.6

For the frictionless pulley shown in Figure 5.41, how much load is to be applied at the pulley B to maintain the equilibrium? Also, find out the support tension at A.

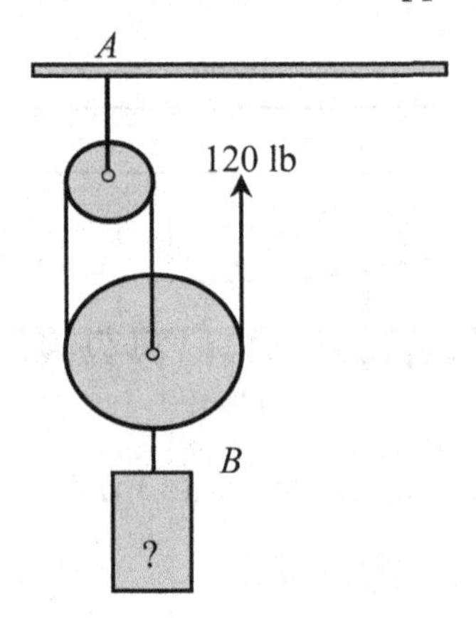

FIGURE 5.41 A pulley system for Problem 5.6.

PROBLEM 5.7

How much tensile force (T) is to be applied to maintain the equilibrium of the frictionless pulley shown in Figure 5.42? Also, determine the support tensions at A and B.

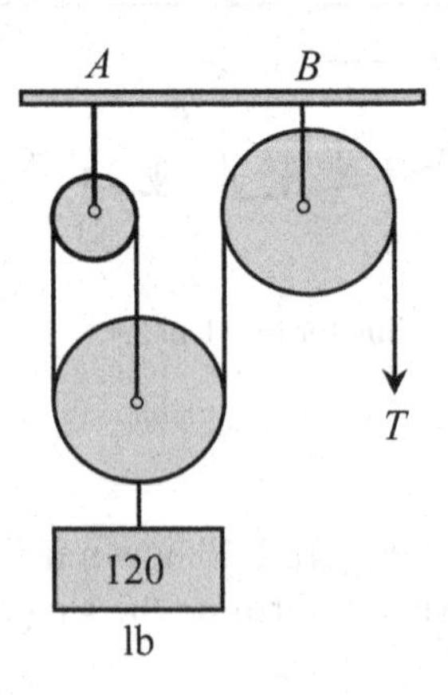

FIGURE 5.42 A pulley system for Problem 5.7.

PROBLEM 5.8

How much tensile force (T) is to be applied to maintain the equilibrium of the frictionless pulley shown in Figure 5.43? Also, determine the support tensions at A, B and C.

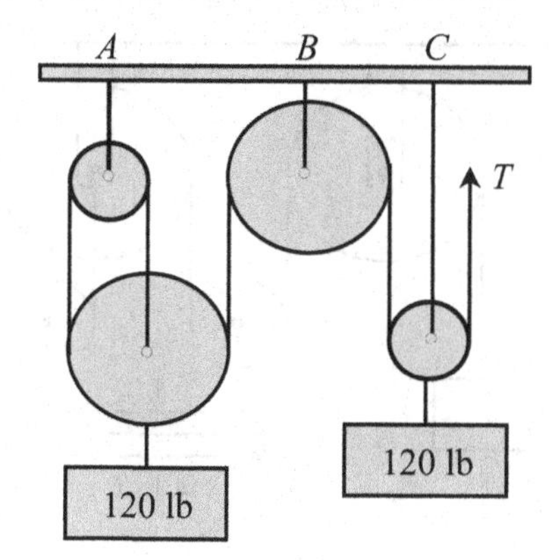

FIGURE 5.43 A pulley system for Problem 5.8.

PROBLEM 5.9

A girl and her baby brother are hanging to a system of two light pulleys with no friction, as shown in Figure 5.44. If the girl weighs 60 lb, determine the weight of the boy to maintain the equilibrium.

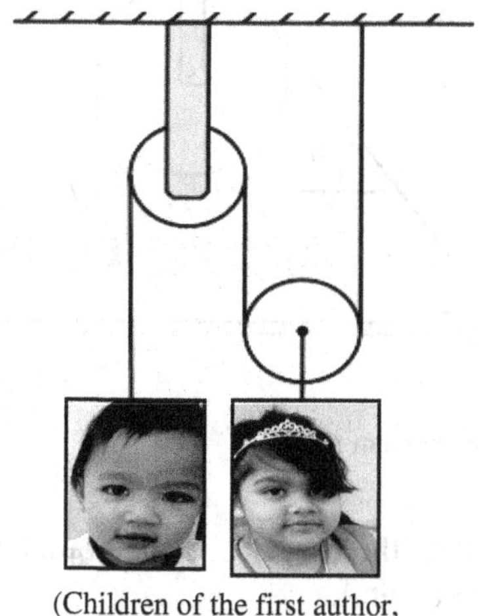

(Children of the first author, M. R. Islam)

FIGURE 5.44 A pulley system for Problem 5.9.

PROBLEM 5.10

A uniform plank of length of 4 ft has mass per unit length 4 lb/ft and is held up in the air horizontally by light inextensible strings tied to each end of the plank, as shown in

Figure 5.45. Each of these strings is attached to a small light pulley which is connected to two large pulleys by a single string. If there is no friction, determine the mass of T required to keep the system in equilibrium.

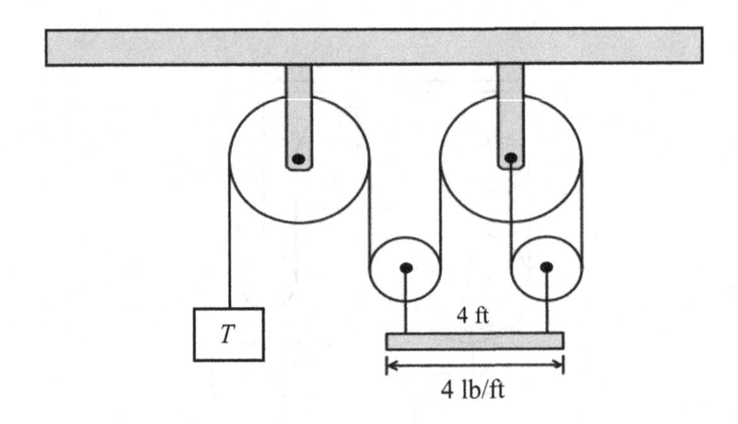

FIGURE 5.45 A pulley system for Problem 5.10.

PROBLEM 5.11

A frictionless collar, C, is connected to a 200-lb load via a frictionless pulley, A. The collar, C, can slide only horizontally on the frictionless bar, DE, as shown in Figure 5.46.

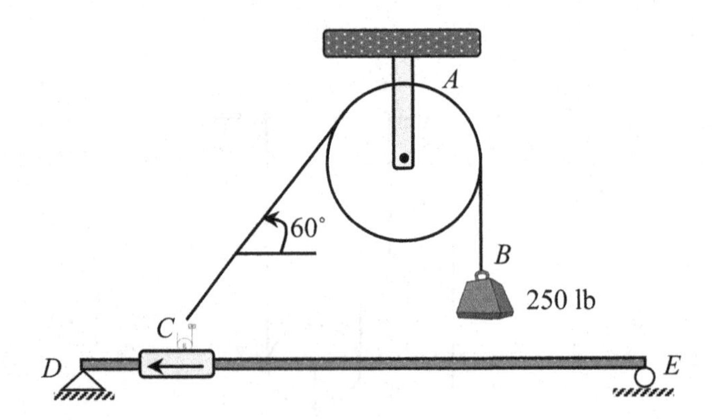

FIGURE 5.46 A Pulley system for Problem 5.11.

Determine the force at C along the bar, DE, to maintain equilibrium.
Determine the support reaction at A.

6 Frame Structures

6.1 TRUSS VERSUS FRAME

A *frame structure is* a structure where at least one of its members is a *multiforce member.* A *multiforce* member is defined as a member with at least three forces acting on it or with at least two forces and at least one couple acting on it. In the analysis of trusses (Figure 6.1), joints are considered pinned and the members are free to rotate about the pin. Therefore, a truss cannot transfer moments. Members are subjected to only axial forces (tensile and compression) only. However, members of frames are joined rigidly at joints by means of welding or monolithic casting. A frame may also be pin connected. However, the members are designed in a way that the member may have axial force, shear force, and moment. Details of axial force, shear force, and moment are discussed in Chapter 7. A load can be applied at the members in a frame structure, whereas a truss can have loads on joints only. In a frame, if a non-loaded member is pin-connected, then that member can be assumed to be a two-force member (similar to a member in a truss). The difference between a truss and frame structure is summarized in Table 6.1:

6.2 FRAME ANALYSIS

The analysis procedure of frame structure is outlined below:

- **Step 1.** Consider the whole structure first to draw the free-body diagram. Then, by applying the equilibrium equations, a few unknown reactions may be determined.
- **Step 2.** Identify the two-force member(s) – the transversely unloaded members that are pin-connected. Remember, in truss, all members are two-force members in a truss. In a frame, there may not be any two-force member. In the free-body diagram of a two-force member, show only the axial force.
- **Step 3.** Isolate each member to draw the free-body diagram. Show the applied forces and the unknown reactions clearly.
- **Step 4.** Use the equilibrium equation(s) to solve for the unknowns.

  Remember:

  - Reactions common to any two contacting members are equal in magnitude but opposite in direction.
  - The joint method used for trusses does not work for frames unless all joint members are two-force members (pin joint with no transverse loading).

Let us see examples for clarification.

Truss Frame

FIGURE 6.1 A truss (left) and a frame structure (right).

TABLE 6.1
The Difference Between Truss and Frame Structure Analysis

Truss	Frame
• All members are pin-connected.	• Joints are mainly rigid; pin connection is also available.
• Load can be applied at joints only.	• Load can be applied at joints and transversely in members.
• Members cannot sustain moment.	• Members can sustain moment.

Case I: Pin joints with joint loads

The structure shown in Figure 6.2a has all joints pin/hinge-connected and there is no transverse load inside the member; there is only one load at the joint. Therefore, all members are two-force members and it is a truss. For truss analysis, both the joint method and the section method are applicable. The free-body diagram of joint C is shown in Figure 6.2b:

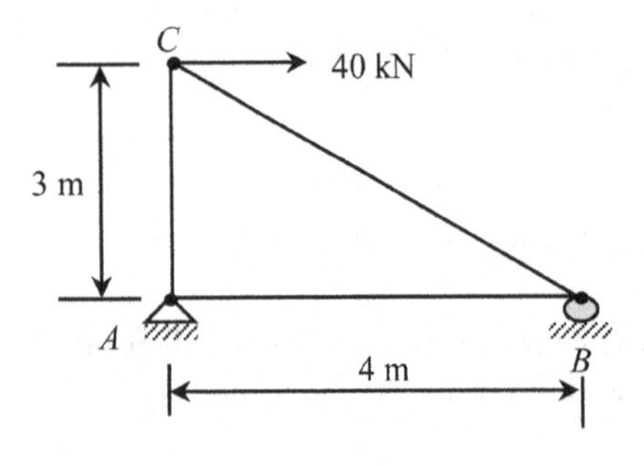

a) The original structure

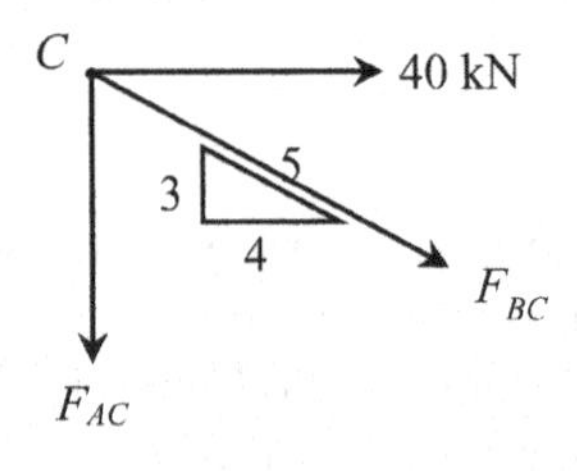

b) The Free-body diagram of joint C

FIGURE 6.2 Drawing of the free-body diagram of a truss (pin joints with joint loads).

Case II: Rigid joints with transverse loads

Assume all joints are rigid and that there are transverse loads on the members as shown in Figure 6.3a. Therefore, we can conclude that not all members are two-force members and thus it is a frame. For a frame, both the joint method and the section methods are applicable. Let us look at the free-body diagram of joint *C*, as shown in Figure 6.3b:

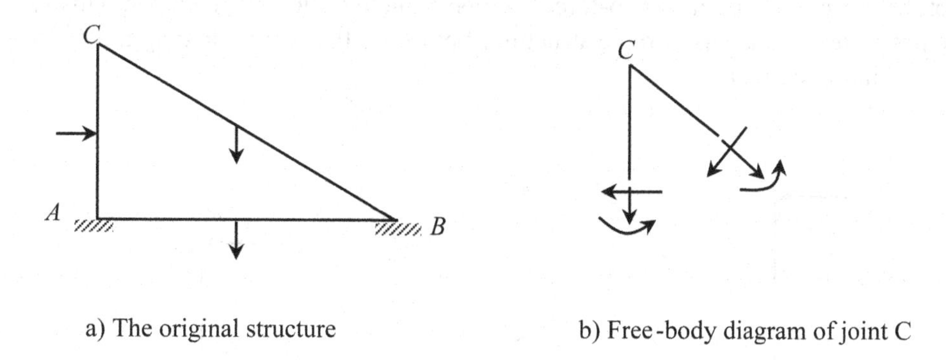

FIGURE 6.3 Free-body diagrams of a rigid joint with member loads.

In the free-body diagram of joint *C*, both members have axial/normal force, shear force and moment. The details of shear force and moment are discussed in Chapter 7. For a truss member (two-force member), there is only axial force (tension or compression). If we take a section of a frame, we would get a similar free-body diagram with shear and moment. In frame structure, analysis is done by considering members. For example, the free-body diagrams of the members of the frame shown in Figure 6.3 can be drawn as shown in Figure 6.4.

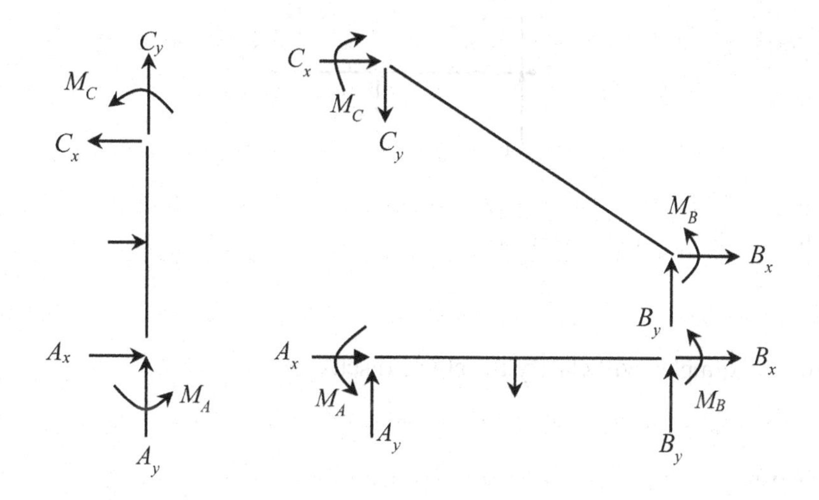

FIGURE 6.4 Free-body diagrams of the members for rigid joints and transversely loaded members.

Case III: Pin joints with joint and member loads

Assume all joints are pin/hinge-connected and that there is a transverse load on member *AB* and a joint load at *C*, as shown in Figure 6.5a. For member AC, the joints are pinned and there is no transverse loading on the member. Therefore, *AC* is a two-force member. Similarly, member BC is a two-force member. Although member AB is pin/hinge-connected, it has a transverse load on. Therefore, member AB is a multiforce member. This structure is a combination of two-force members and a multiforce member. This type of structure is also termed as a frame structure. Let us see the free-body diagram of Joint *C*, as shown in Figure 6.5b.

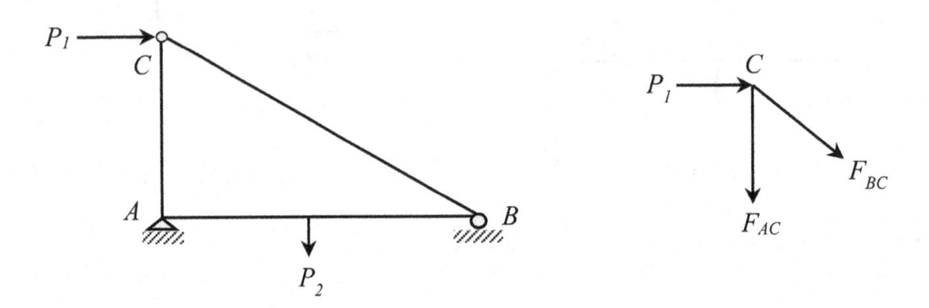

FIGURE 6.5 Drawing of a free-body diagram for frame with member loads.

The free-body diagrams of the *AB* member can be drawn as shown in Figure 6.6.

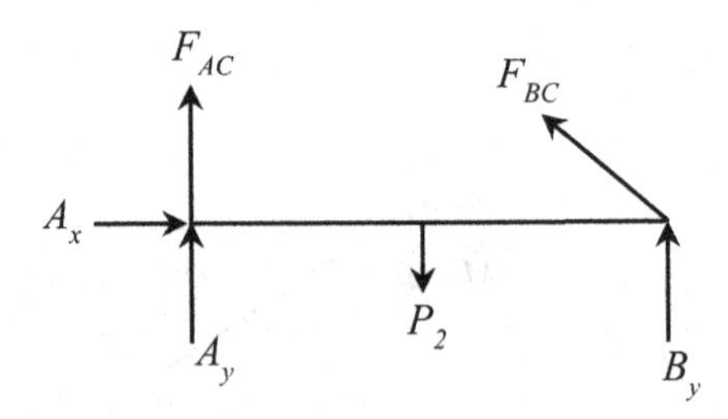

FIGURE 6.6 A Free-body diagram of the member *AB*.

The following examples will clarify the above discussion.

Example 6.1:

A frame structure is supporting three concentrated loads as shown in Figure 6.7. Determine the reactions at the supports *A* and *B*. Assume the support at *A* is a pin, *B* is a roller and *C* is a rigid (fixed-connected) joint.

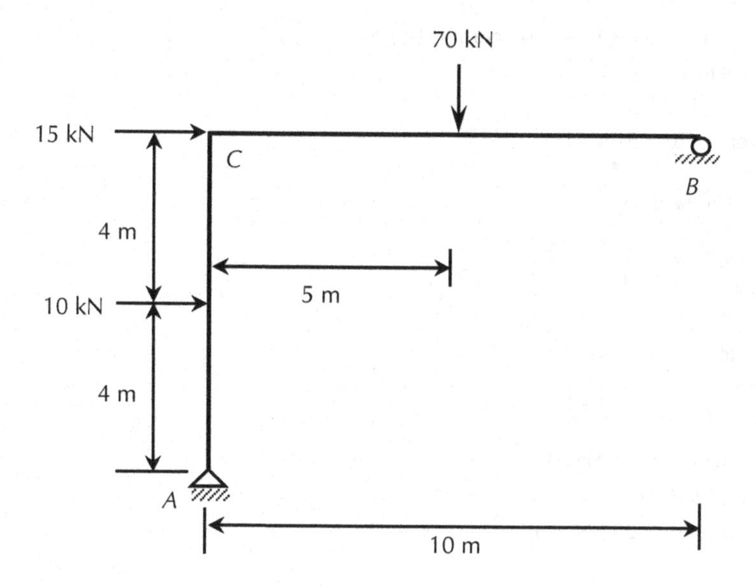

FIGURE 6.7 A frame structure for Example 6.1.

SOLUTION

Let us draw the free-body diagram of the frame as shown in Figure 6.8. Support A has two unknown reactions and B has a single unknown for being pin and roller supports, respectively.

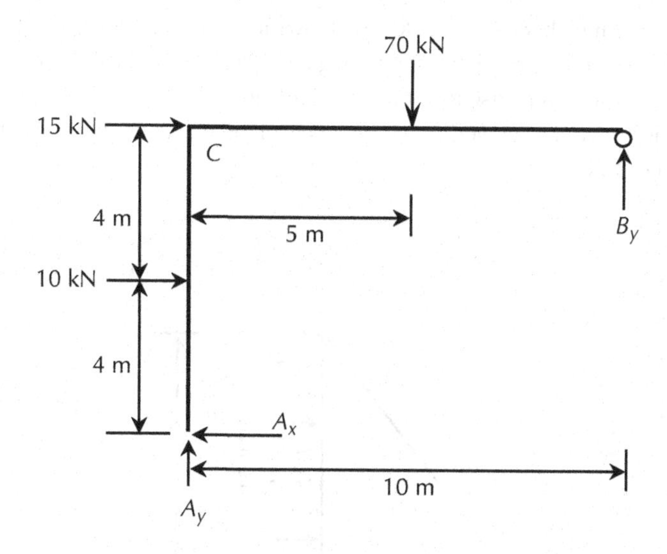

FIGURE 6.8 The free-body diagram of the frame structure for Example 6.1.

Apply equilibrium equations as appropriate, as follows.

$$\sum M \text{ at } A = 0 \;(\circlearrowright + ve)$$

$$(70 \text{ kN})(5 \text{ m}) + 10 \text{ kN } (4 \text{ m}) + 15 \text{ kN } (8 \text{ m}) - B_y \,(10 \text{ m}) = 0$$

$350 \text{ kN.m} + 40 \text{ kN.m} + 120 \text{ kN.m} - 10B_y \text{ m} = 0$
$510 \text{ kN.m} - 10B_y \text{ m} = 0$
$10B_y \text{ m} = 510 \text{ kN.m}$
Therefore, $B_y = 51 \text{ kN} (\uparrow)$

$$\sum F_y = 0 \ (\uparrow +ve)$$

$A_y + B_y - 70 \text{ kN} = 0$
$A_y + 51 \text{ kN} - 70 \text{ kN} = 0$
$A_y - 19 \text{ kN} = 0$
Therefore, $A_y = 19 \text{ kN} (\uparrow)$

$$\sum F_x = 0 \ (\leftarrow +ve)$$

$-A_x + 10 \text{ kN} + 15 \text{ kN} = 0$
$-A_x + 25 \text{ kN} = 0$
Therefore, $A_x = 25 \text{ kN} (\leftarrow)$

Answers:

$A_x = 25 \text{ kN} (\leftarrow)$
$A_y = 19 \text{ kN} (\uparrow)$
$B_y = 51 \text{ kN} (\uparrow)$

Example 6.2:

A structure has two members *AB* and *BC*, as shown in Figure 6.9. Joints *A*, *B* and *C* are pin-connected. Member *AB* is vertical and supporting a horizontal load of 12 lb per foot throughout the member. There is no transverse load on the *BC* member.

Determine the horizontal and vertical components of the reactions at the pins *A*, *B* and *C*, and the member force at *BC*.

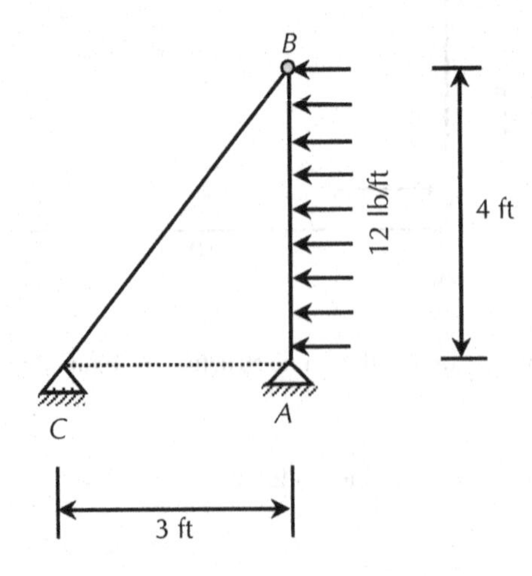

FIGURE 6.9 A frame structure with pin joints for Example 6.2.

SOLUTION

Member *AB* has a transverse load and, thus, *AB* is a frame member. Member *BC* is pin-connected at both ends and has no transverse load. Thus, *BC* is a two-force member. First, let us consider the whole structure as shown in Figure 6.10. Both supports (*A* and *C*) have two unknown reactions as they are pin supports.

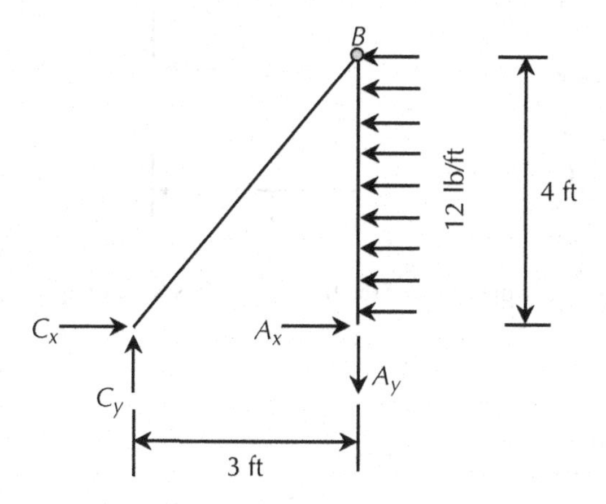

FIGURE 6.10 The free-body diagram of the whole frame for Example 6.2.

$$\sum M \text{ at } C = 0 \;(\lrcorner + ve)$$

$$(-12 \text{ lb/ft})(4 \text{ ft})\ (2 \text{ ft}) + A_y\ (3 \text{ ft}) = 0$$
$$-96 \text{ lb.ft} + 3A_y \text{ ft} = 0$$
$$3A_y \text{ ft} = 96 \text{ lb.ft}$$
$$\text{Therefore, } A_y = 32 \text{ lb } (\downarrow)$$

$$\sum F_y = 0 \;(\uparrow +ve)$$

$$C_y - A_y = 0$$
$$C_y = A_y$$
$$\text{Therefore, } C_y = 32 \text{ lb } (\uparrow)$$

$$\sum F_x = 0 \;(\rightarrow +ve)$$

$$A_x + C_x - (12 \text{ lb/ft})(4 \text{ ft}) = 0$$
$$A_x + C_x = 48 \text{ lb}$$
$$C_x = 48 \text{ lb} - A_x$$

Now consider the free-body diagram of each member as shown in Figure 6.11. Member *BC* is a two-force member, as the ends are pin-connected and there is no transverse load inside the member. It is reflected in the free-body diagrams. The hypotenuse of the slope triangle of member *BC* is $\sqrt{(3)^2 + (4)^2} = 5$.

In member *AB*,

$$\sum M \text{ at } B = 0 \;(\lrcorner + ve)$$

$$(12 \text{ lb/ft})(4 \text{ ft})\ (2 \text{ ft}) + A_y\ (0 \text{ ft}) - A_x(4 \text{ ft}) = 0$$
$$96 \text{ lb.ft} + 0 - A_x(4 \text{ ft}) = 0$$

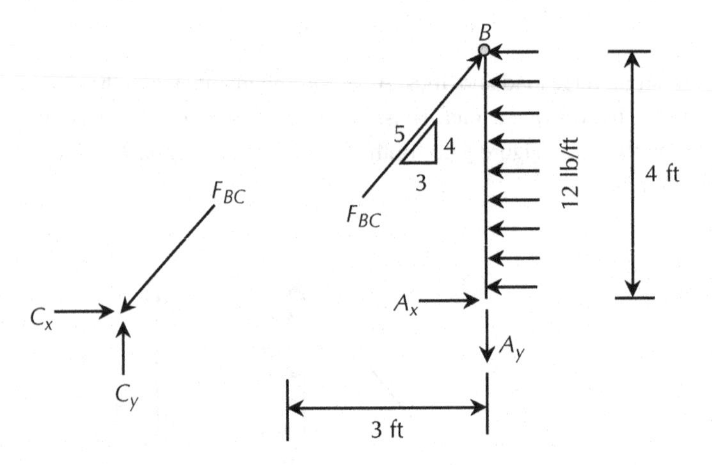

FIGURE 6.11 The free-body diagrams of the frame members for Example 6.2.

$A_x(4\text{ ft}) = 96\text{ lb.ft}$

Therefore, $A_x = 24\text{ lb }(\rightarrow)$

$C_x = 48\text{ lb} - A_x$

$C_x = 48\text{ lb} - 24\text{ lb}$

$C_x = 24\text{ lb }(\rightarrow)$

$$\sum F_y = 0\ (\uparrow +ve)$$

$$F_{BC}\left(\frac{4}{5}\right) - A_y = 0$$

$$F_{BC}\left(\frac{4}{5}\right) - 32\text{ lb} = 0$$

$$F_{BC} = 32\text{ lb}\left(\frac{5}{4}\right)$$

$F_{BC} = 40\text{ lb}$

Therefore, $F_{BC} = 40\text{ lb}$ (compressive)

The load at Pin *B* is also 40 lb along the member *BC*.

Let us verify the result. In member *AB*,

$$\sum F_x = 0\ (\leftarrow +ve)$$

$$-F_{BC}\left(\frac{3}{5}\right) - A_x + \left(12\frac{\text{lb}}{\text{ft}}\right)(4\text{ ft}) = 0$$

$$F_{BC}\left(\frac{3}{5}\right) = -24\text{ lb} + \left(12\frac{\text{lb}}{\text{ft}}\right)(4\text{ ft})$$

$$F_{BC}\left(\frac{3}{5}\right) = 24\text{ lb}$$

$$F_{BC} = 24\,\text{lb}\left(\frac{5}{3}\right)$$

$F_{BC} = 40$ lb (matches the result with the previous finding)

Answers:

$A_x = 24$ lb (→)
$A_y = 32$ lb (↓)
$C_x = 24$ lb (→)
$C_y = 32$ lb (↑)
$B = 40$ lb along the BC member
$F_{BC} = 40$ lb (compressive)

Note that the pin reactions at A and C can be presented in components (x and y). The components can be converted into resultant as $A = \sqrt{(A_x)^2 + (A_y)^2}$ and $B = \sqrt{(B_x)^2 + (B_y)^2}$.

Example 6.3:

A frame structure has three members *AC, CD* and *DB,* as shown in Figure 6.12. Joints *A, B* and *C* are pin-connected and Joint *D* is rigid. Member *BD* is vertical and supporting a horizontal load of 40 lb per foot throughout the member. There is no transverse load in members *AC* and *CD*.

Determine the horizontal and vertical components of the reactions at the pins *A, B* and *C*.

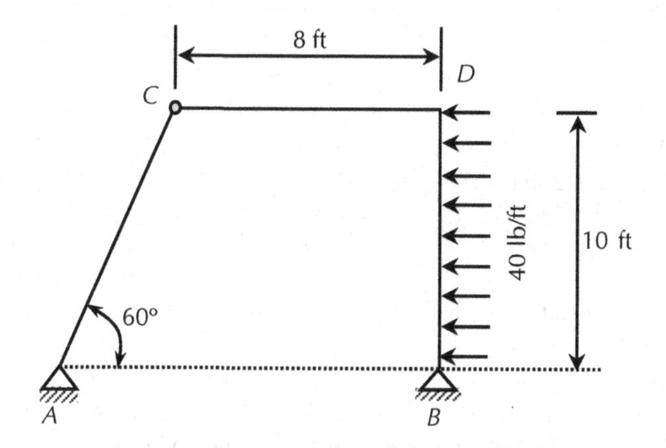

FIGURE 6.12 A frame structure with pin joints for Example 6.3.

SOLUTION

First let us consider the whole structure as shown in Figure 6.13. Both supports (*A* and *B*) have two unknown reactions as they are pin supports.

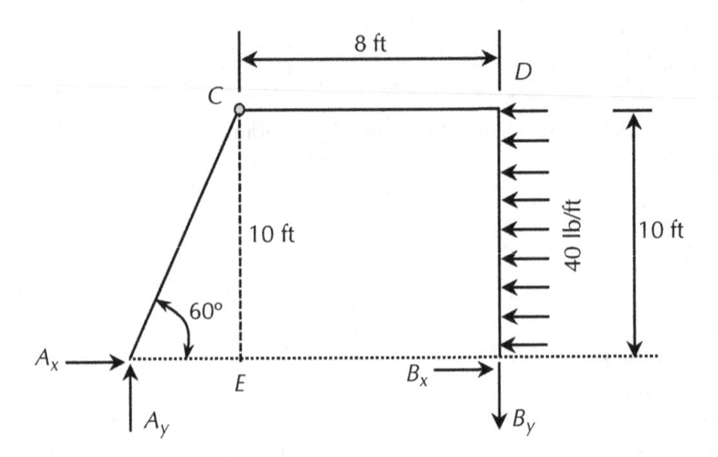

FIGURE 6.13 The free-body diagram of the whole frame structure for Example 6.3.

From ΔACE,

$$\tan 60 = \frac{10 \text{ ft}}{AE}$$

Therefore, $AE = 5.77$ ft

$$\sum M \text{ at } A = 0 \ (\lrcorner + ve)$$

$(-40 \text{ lb/ft})\ (10 \text{ ft})\ (5.0 \text{ ft}) + B_y\ (5.77 \text{ ft} + 8 \text{ ft}) = 0$

$-2{,}000 \text{ lb.ft} + 13.77\ B_y \text{ ft} = 0$

Therefore, $B_y = 145.2$ lb ($\downarrow$)

$$\sum F_y = 0 \ (\uparrow +ve)$$

$A_y - B_y = 0$

Therefore, $A_y = 145.2$ lb ($\uparrow$)

$$\sum F_x = 0 \ (\leftarrow +ve)$$

$A_x + B_x - (40 \text{ lb/ft})(10 \text{ ft}) = 0$

$A_x + B_x = 400 \text{ lb}$

$B_x = 400 \text{ lb} - A_x$

Now, consider the free-body diagram of each member as shown in Figure 6.14. *AC* is a two-force member, as it is pin-connected at both ends and has no transverse load. However, the member force *AC* is not shown, as it is not asked for in the problem. The problem asks for the support reactions in both the vertical and horizontal directions. This is why only the vertical and horizontal components of the pin reactions at *A* and *C* are shown.

In the *AC* member,

$$\sum M \text{ at } C = 0 \ (\lrcorner + ve)$$

$A_y\ (5.77 \text{ ft}) - A_x\ (10 \text{ ft}) = 0$

$(145.2 \text{ lb})(5.77 \text{ ft}) - A_x\ (10 \text{ ft}) = 0$

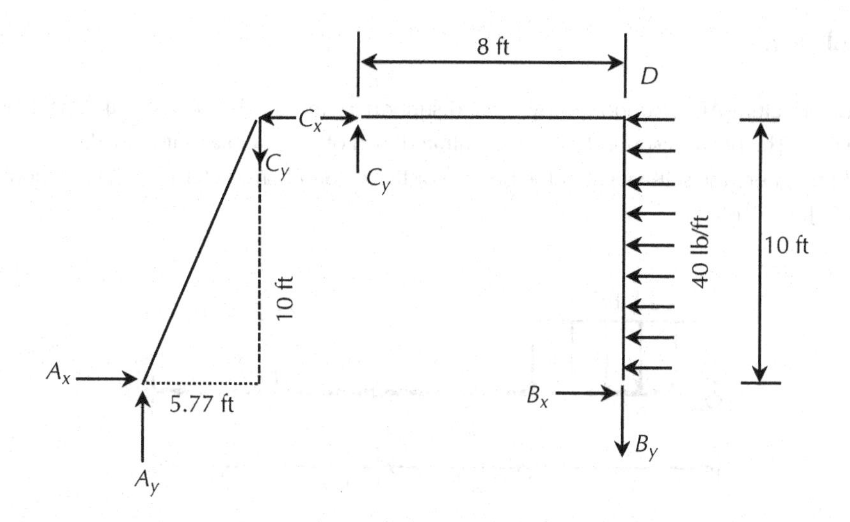

FIGURE 6.14 The free-body diagrams of the frame members for Example 6.3.

$837.8 \text{ lb.ft} - A_x (10 \text{ ft}) = 0$
Therefore, $A_x = 83.8$ lb (→)

Therefore, $B_x = 400 \text{ lb} - 83.8 \text{ lb} = 316.2$ lb (→)

$\sum F_x = 0$ (← +ve)

$A_x - C_x = 0$
$C_x = A_x$

Therefore, $C_x = 83.8$ lb (← at the *AC* member) (→ at the *CD* member)

$\sum F_y = 0$ (↑ +ve)

$A_y - C_y = 0$
$C_y = A_y$

Therefore, $C_y = 145.2$ lb (↓ at the *AC* member) (↑ at the *CD* member)

Let us verify the result. In the *BDC* member,

$\sum M \text{ at } C = 0$ (↲+ve)

$(40 \text{ lb/ft})(10 \text{ ft}) (5.0 \text{ ft}) + B_y (8 \text{ ft}) - B_x (10 \text{ ft}) = 0$
$2{,}000 \text{ lb.ft} + (145.2 \text{ lb}) (8 \text{ ft}) - B_x (10 \text{ ft}) = 0$
$2{,}000 \text{ lb.ft} + 1{,}161.6 \text{ lb.ft} - 10B_x \text{ ft} = 0$

Therefore, $B_x = 316.2$ lb (matches with the previous finding)

Answers:

$A_x = 83.8$ lb (→)
$A_y = 145.2$ lb (↑)
$B_x = 316.2$ lb (→)
$B_y = 145.2$ lb (↓)
$C_x = 83.8$ lb (← at the *AC* member)
$C_y = 145.2$ lb (↓ at the *AC* member)

Example 6.4:

A beam structure AB is supported by a fixed support B and a roller support at A as shown in Figure 6.15. The beam is supporting a concentrated load of 2 kip and a uniformly distributed load of 20 lb/ft. Ignore the self-weight of the beam. For these conditions, determine the reactions at A, B and C. Joint C is a pin.

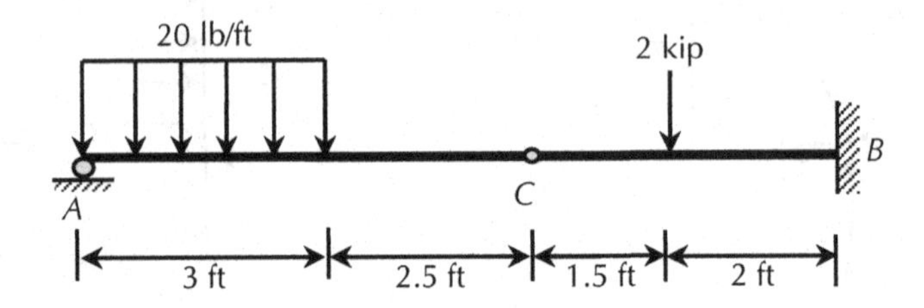

FIGURE 6.15 A beam with arbitrary loading for Example 6.4.

SOLUTION

The beam structure consists of two members, AC and CB, connected at C by a pin joint. Therefore, the moment at C is zero. Now, our next target is to find the horizontal and vertical force reactions at C. Let us draw the free-body diagrams of both members as shown in Figure 6.16.

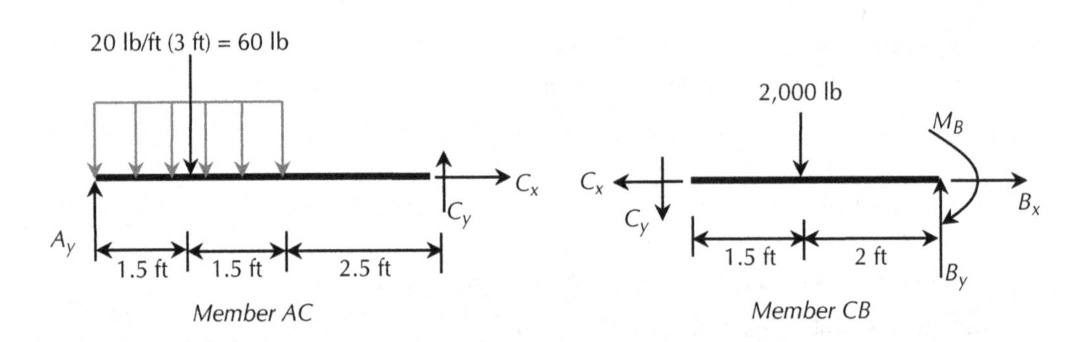

FIGURE 6.16 The free-body diagrams of the beam members for Example 6.4.

For member AC:

$\Sigma F_x = 0$ ($\rightarrow$ +ve)

$C_x = 0$

$\sum M$ at $A = 0$ ($\lrcorner$ + ve)

60 lb (1.5 ft) – C_y (5.5 ft) = 0

90 lb.ft – 5.5C_y ft = 0

5.5C_y ft = 90 lb.ft

C_y = 16.4 lb

$\Sigma F_y = 0$ (↑+ve)
$C_y + A_y - 60 \text{ lb} = 0$
$A_y = 60 - 16.4 \text{ lb} = 43.6 \text{ lb}$

For member *CB*:
$\Sigma F_x = 0$ (→ +ve)
$C_x = B_x = 0$

$\Sigma F_y = 0$ (↑+ve)
$-C_y + B_y - 2{,}000 \text{ lb} = 0$
$B_y = C_y + 2{,}000 \text{ lb}$
$B_y = 16.4 \text{ lb} + 2{,}000 \text{ lb}$
$B_y = 2{,}016.4 \text{ lb}$

$\sum M \text{ at } C = 0$ (↲+ve)
$2{,}000 \text{ lb } (1.5 \text{ ft}) - B_y (3.5 \text{ ft}) + M_B = 0$
$3{,}000 \text{ lb.ft} - 2{,}016.4 \text{ lb } (3.5 \text{ ft}) + M_B = 0$
$-4{,}057.3 \text{ lb.ft} + M_B = 0$
$M_B = 4{,}057.4 \text{ lb.ft}$

Answers:
$C_x = B_x = 0$
$C_y = 16.4 \text{ lb}$
$A_y = 43.6 \text{ lb}$
$B_y = 2{,}016.4 \text{ lb}$
$M_B = 4{,}057.4 \text{ lb.ft}$

Note that the frame structure can be analyzed considering the whole structure and each member in turn. Both procedures must yield the same answer. For example, in the current problem, let us check our answer considering the whole structure:

$\Sigma F_y = 0$ (upward positive)
$A_y + B_y - 60 \text{ lb} - 2{,}000 \text{ lb} = 0$ (C_y does not need to be included, as it is internal when considering the whole structure at a time)
$43.6 \text{ lb} + 2{,}016.4 \text{ lb} - 60 \text{ lb} - 2{,}000 \text{ lb} = 0$
$0 = 0$ (matches)

PRACTICE PROBLEMS

PROBLEM 6.1

A member, *ADB*, is supported by a pin support, *B*, and a roller support, *A*, as shown in Figure 6.17. It is supporting a concentrated load and a uniformly varying load. Ignore the weight of the beam. For these conditions, determine the internal reactions (axial, shear and moment) at *D*. Assume, the joint *D* is a rigid joint.

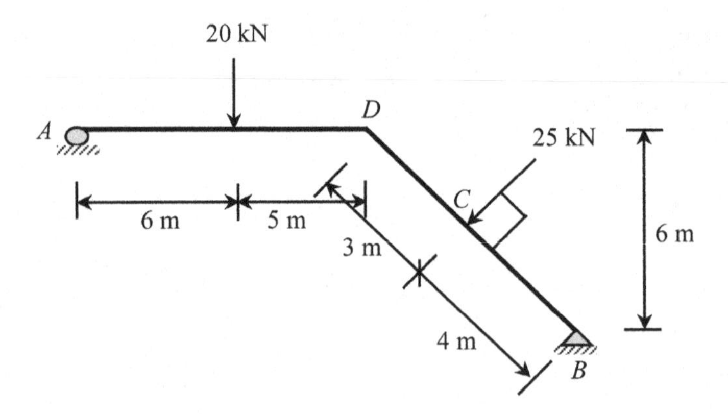

FIGURE 6.17 A frame structure with arbitrary loading for Problem 6.1.

PROBLEM 6.2

A beam structure, *AB*, is supported by a roller support, *A*, and a fixed support, *B*, as shown in Figure 6.18. It is supporting a concentrated load of 100 lb and a uniformly distributed load of 40 lb/ft. Ignore the weight of the beam. Determine the reactions at *A*, *B* and *C*. Joint *C* is pinned.

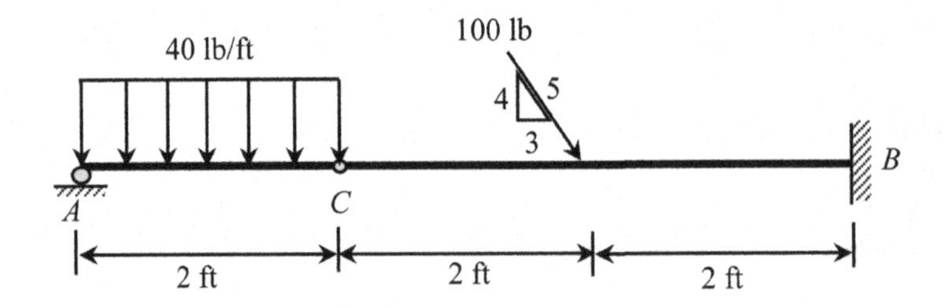

FIGURE 6.18 A frame structure with arbitrary loading for Problem 6.2.

PROBLEM 6.3

A frame structure is supported by pin supports, *A*, *B* and *C* as shown in Figure 6.19. It is supporting a concentrated load of 8 kip and a uniformly distributed load of 3 kip/ft. Ignore the weight of the beam. Determine the pin reactions at *A*, *B* and *C*. The joints *A*, *B* and *C* are pinned.

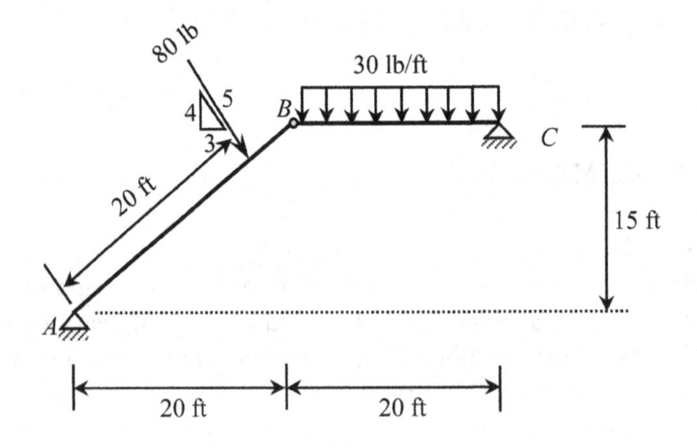

FIGURE 6.19 A frame structure with arbitrary loading for Problem 6.3.

PROBLEM 6.4

A frame structure is supporting two concentrated loads and a uniformly distributed load as shown in Figure 6.20. Neglect the weight of the beam. Determine the horizontal and vertical components of the reactions at *A*, *C*, and *D*. Assume the frame structure is pin connected at *A*, *C*, and *D*, and there is a fixed-connected joint at *B*.

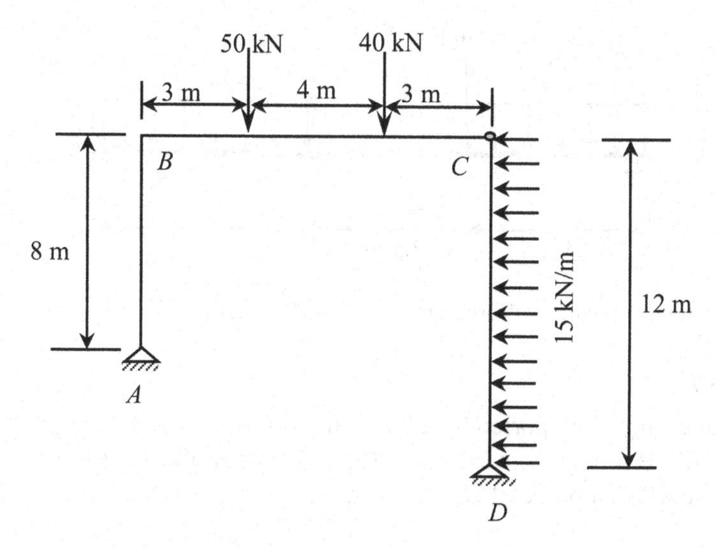

FIGURE 6.20 A frame structure with arbitrary loading for Problem 6.4.

PROBLEM 6.5

The frame structure in Figure 6.21 is loaded with a horizontal uniformly distributed load of 2 kip/ft on its vertical member *AE*. The top horizontal member *EF* is loaded with a concentrated load of 10 kip as shown. The structure is supported by two hinge supports (*A* and *B*). All other joints are hinged as well. Determine the reactions at *A, B, C, D, E*, and *F*.

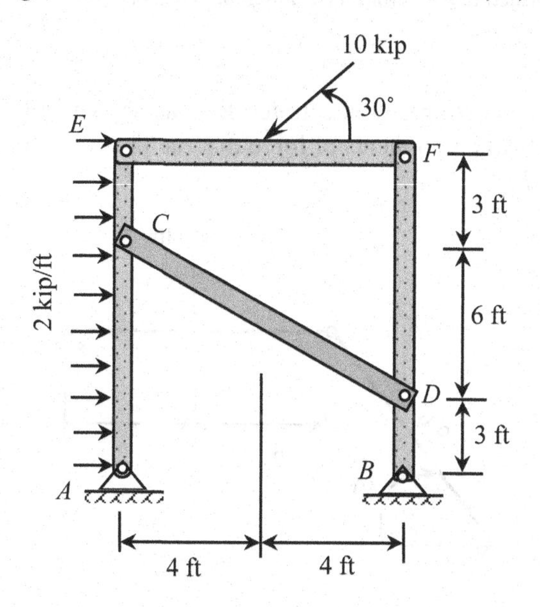

FIGURE 6.21 A frame structure with arbitrary loading for Problem 6.5.

PROBLEM 6.6

A beam structure, AB, is supported by a fixed support, B, and a roller support, A, as shown in Figure 6.22. It is supporting a concentrated load of 150 lb and two uniformly distributed loads of 40 lb/ft. Ignore the self-weight of the beam. For these conditions, determine the reactions at A, B and C. Joint C is a pin.

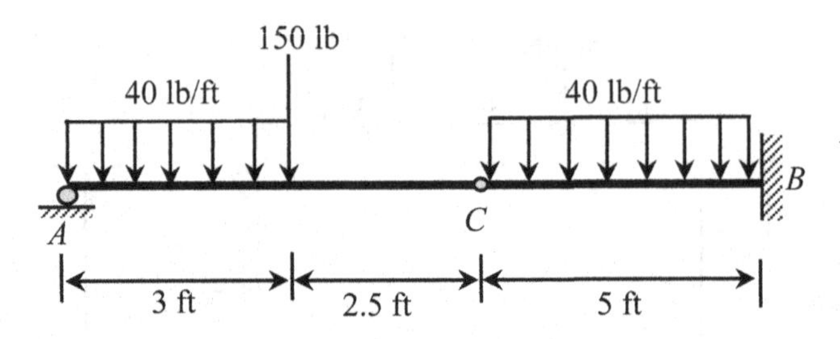

FIGURE 6.22 A frame structure with arbitrary loading for Problem 6.6.

PROBLEM 6.7

A beam structure, AC, is supported by a fixed support, A, and a roller support, C, as shown in Figure 6.23. Determine the reactions at the supports A and C of the compound beam. Assume A is fixed, B is a pin, and C is a roller.

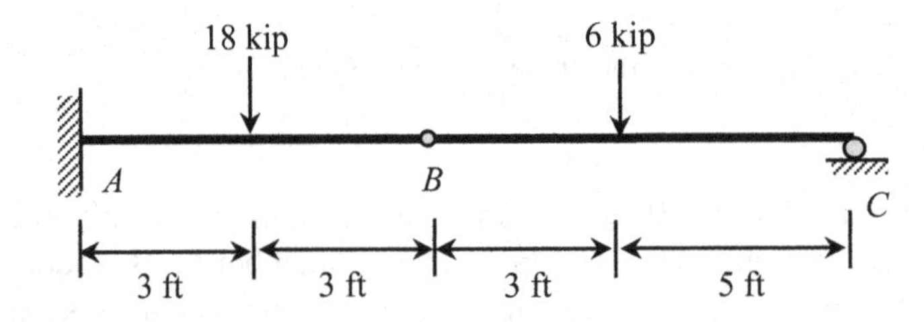

FIGURE 6.23 A frame structure with arbitrary loading for Problem 6.7.

PROBLEM 6.8

A frame structure is supporting a concentrated load as shown in Figure 6.24. Determine the horizontal and vertical components of force that the pin at C exerts on member BC and the member force at AB.

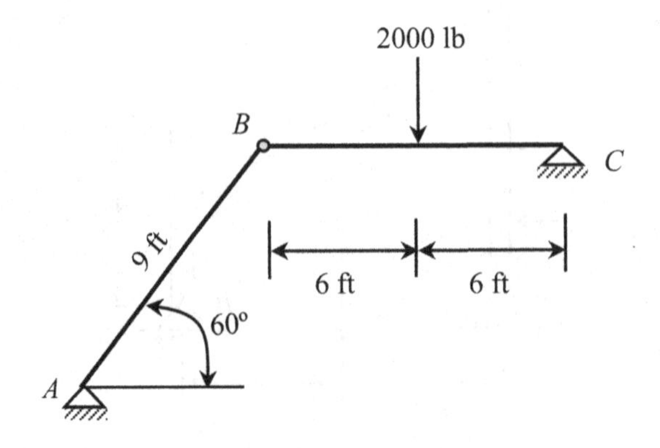

FIGURE 6.24 A frame structure with arbitrary loading for Problem 6.8.

7 Axial Force, Shear Force and Bending Moment in Beams

7.1 INTERNAL REACTIONS

Internal reactions develop inside the member when external loading is applied. The internal reactions are internal moment, shear force, and normal (axial) force. It is important to determine the internal moments, shear force and axial force at any point in a beam structure. These values can then be used to design the beam. In this section, the internal loads (moment, shear force and normal force) will be calculated for a specific location. The next section will expand this concept and provide diagrams showing the internal reactions at all locations.

Before the internal loads are determined, a few topics need to be reviewed, specifically, equilibrium equations, boundary conditions, and load types. Earlier, we discussed how there are three types of supports: roller, pin/hinge, and fixed. Roller supports can resist movement in one direction, pin/hinge supports can resist movement in two orthogonal directions and fixed supports can resist movement in two orthogonal directions and the rotation.

Let us consider a beam with any loading condition as shown in Figure 7.1. We are required to determine the internal reactions at location, C of the beam. To do so, we need to cut at C and consider the left part or the right part. Note that any structure or any member is in equilibrium condition at all points. Therefore, considering the left part or the right part yields the same result. However, whichever part is considered has a support (A or B). Therefore, before cutting at C, you must determine the support reaction at that part. Let us assume that we will consider the left part as the left of C has an easily manageable load. For this we need to determine the support reaction at A first, which can be done by drawing the free-body diagram of the whole member and applying $\sum M_B = 0$ and $\sum F_X = 0$.

Next a cut at C is made and the free-body diagram can be drawn as shown in Figure 7.2.

In the above free-body diagram, three unknown internal forces (N_C, V_C, and M_C) are shown. N_C means the reaction is a normal force at point C along the longitudinal direction of the beam. Away from the cut section, this value is considered positive (tension). V_C means the shear force perpendicular to the longitudinal direction of the beam. The downward direction in the left part of the cut section is assumed to be positive shear. This means the shown V_c in the above free-body diagram is positive. M_c means the internal moment developed at C of the beam. The counter-clockwise direction in the left part of the cut section is assumed to be positive.

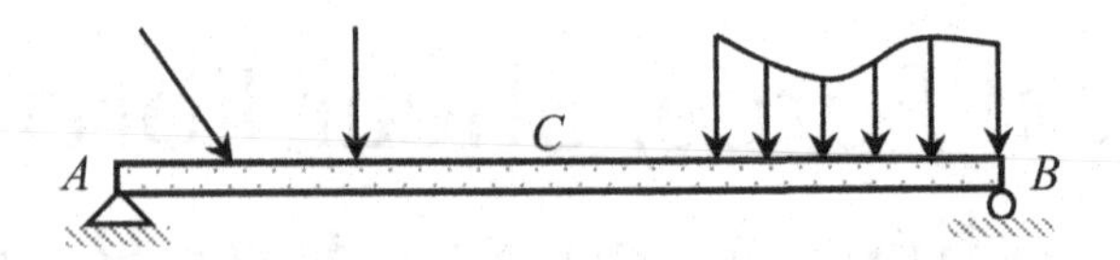

FIGURE 7.1 A beam with arbitrary loading.

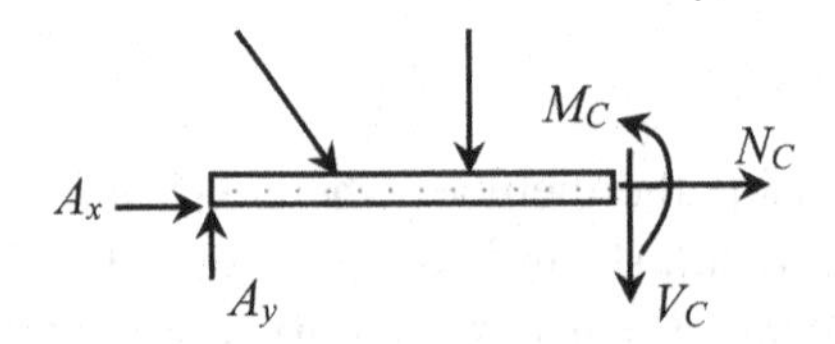

FIGURE 7.2 The free-body diagram of segment *AC* of the beam in Figure 7.1.

Then, applying the equilibrium equations to this cut section, the three unknown internal reactions at *C* (N_C, V_C, and M_C) can be determined. Regarding the support reactions A_x and A_y, these must be determined earlier before cutting the beam at *C*.

Instead of considering the left part, somebody may wish to consider the right part of the cut beam. In that case, the reaction at the right support, *B* is to be determined first. Then, a cut at *C* is to be made and the free-body diagram can be as shown in Figure 7.3.

In the above free-body diagram, N_C, V_C, and M_C are shown in opposite directions compared with what was done while considering the left part of the cut section. The directions shown in the above free-body diagram are positive for the right part of the cut section. The arrows in the above two free-body diagrams show the positive signs. For example, if the left part is considered, the counter-clockwise moment is considered positive. However, if the right part is considered, the clockwise moment is considered positive. Similarly, the downward *V* in the left part and the upward *V* in the right part are positive. Normal force is always assumed to be positive (tension) away from the member. The shearing force and bending moment sign conventions can also be described as follows (Figure 7.4):

- The bending moment is positive if it produces bending when the beam is concave upward (compression in the top fibers and tension in the bottom fibers).
- The shearing force is positive if the right portion of the beam moves downward with respect to the left.

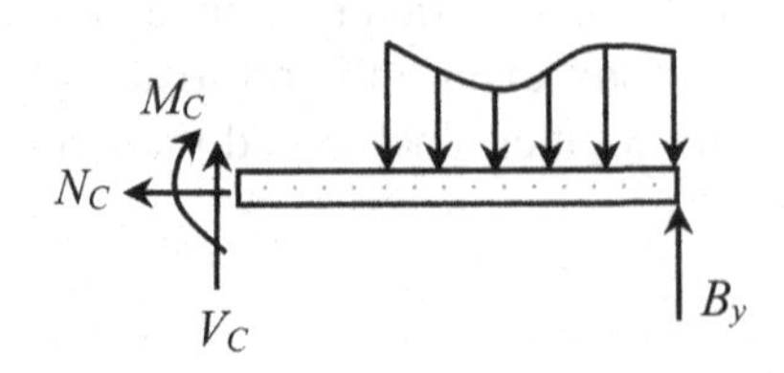

FIGURE 7.3 The free-body diagram of segment *CB* of the beam in Figure 7.1.

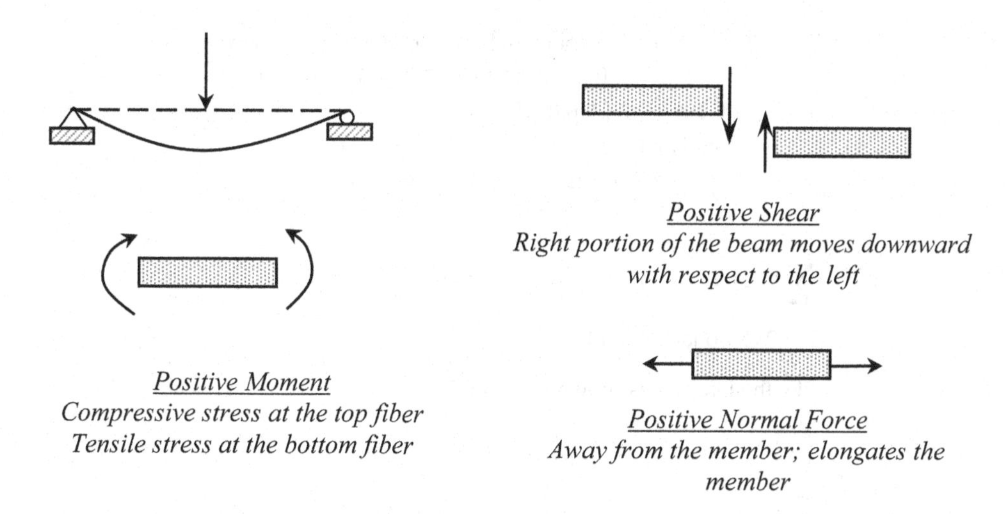

FIGURE 7.4 Sign convention for internal reactions (normal force, shear and moment).

Example 7.1:

For the simply supported beam shown in Figure 7.5, determine the internal normal force, shear force and moment at *C*.

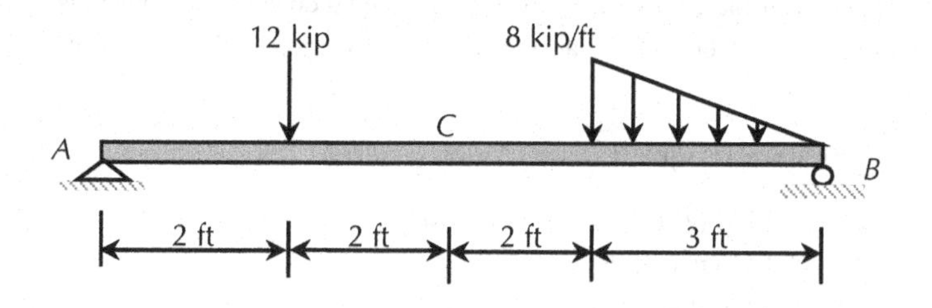

FIGURE 7.5 A beam with arbitrary loading for Example 7.1.

SOLUTION

If a cut at *C* is made, the left part of the structure is less loaded and expected to be easier to be analyzed. Therefore, let us find out the reaction at the left support considering the whole structure.

Consider the whole structure as shown in Figure 7.6:

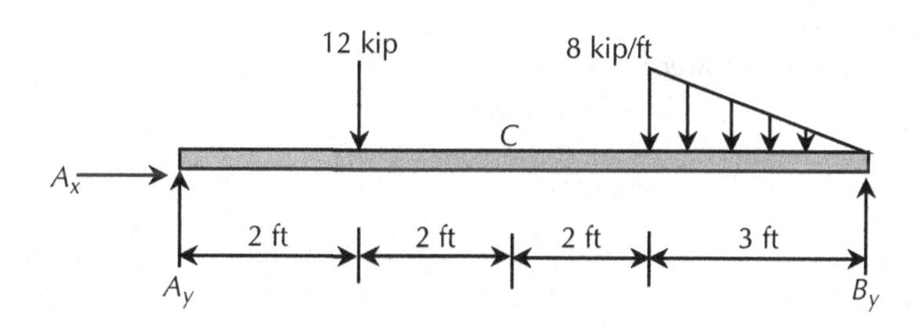

FIGURE 7.6 The free-body diagram of the beam for Example 7.1.

ΣM at $B = 0$ (assuming counter-clockwise moment is positive)

$$\tfrac{1}{2}\,(8\text{ kip/ft})(3\text{ ft})\,(2\text{ ft}) + 12\text{ kip}\,(7\text{ ft}) - A_y\,(9\text{ ft}) = 0$$
$$24\text{ kip.ft} + 84\text{ kip.ft} - 9A_y\text{ ft} = 0$$
$$108\text{ kip.ft} - 9A_y\text{ ft} = 0$$
$$9A_y\text{ ft} = 108\text{ kip.ft}$$
$$A_y = 12\text{ kip}$$

$$\sum F_x = 0\ (\rightarrow +ve)$$

$A_x = 0$ as no lateral load

Cut at C and consider the left part as shown in Figure 7.7:

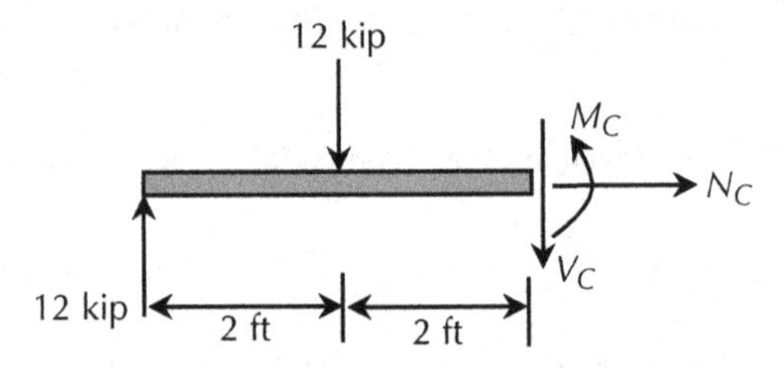

FIGURE 7.7 The free-body diagram of the segment AC of the beam for Example 7.1. (Note again that if the left part of the beam is considered, the shear force is considered downward; moment is considered counter-clockwise and normal force is considered away from the cut section. This is the positive sign convention.)

$$\sum M \text{ at } C = 0\ (\circlearrowleft + ve)$$

$$-M_C - 12\text{ kip}\,(2\text{ ft}) + 12\text{ kip}\,(4\text{ ft}) = 0$$
$$-M_C - 24\text{ kip.ft} + 48\text{ kip.ft} = 0$$
$$-M_C + 24\text{ kip.ft} = 0$$
$$M_C = 24\text{ kip.ft}$$

$$\sum F_x = 0\ (\rightarrow +ve)$$

$$N_C = 0$$

$$\sum F_y = 0\ (\uparrow +ve)$$

$$-V_C + 12\text{ kip} - 12\text{ kip} = 0$$
$$V_C = 0$$

Answers:

$M_C = 24$ kip.ft

$N_C = 0$

$V_C = 0$

Example 7.2:

A simply supported beam is supporting the loads as shown in Figure 7.8. Determine the internal normal force, shear force and moment at C. The point C is located 3 ft from the left support.

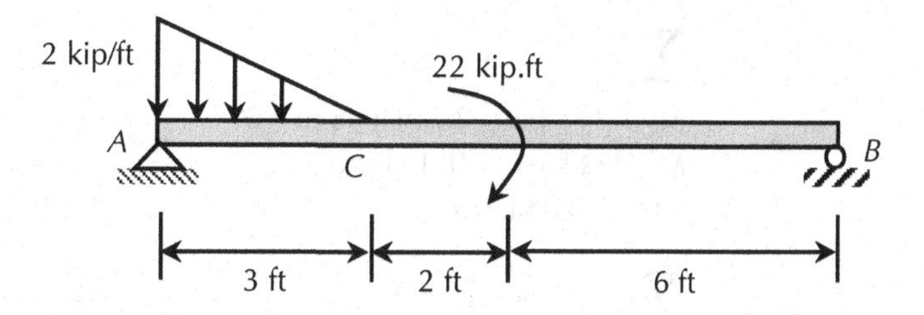

FIGURE 7.8 A beam with arbitrary loading for Example 7.2.

SOLUTION

If a cut at C is made, the right part of the structure is less loaded and expected to be easier to be analyzed. Therefore, let us find out the reaction at the right support considering the whole structure.

Consider the whole structure as shown in Figure 7.9:

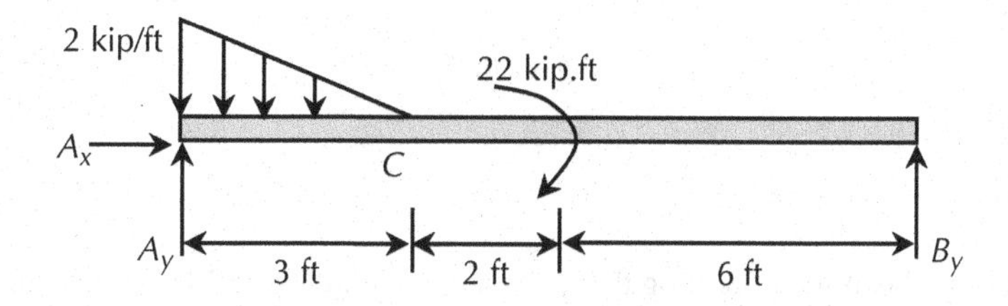

FIGURE 7.9 The free-body diagram of the beam for Example 7.2.

ΣM at $A = 0$ (assuming clockwise moment is positive)

$$\tfrac{1}{2}\,(2\text{ kip/ft})(3\text{ ft})\,(1\text{ ft}) + 22\text{ kip.ft} - B_y\,(11\text{ ft}) = 0$$
$$3\text{ kip.ft} + 22\text{ kip.ft} - 11B_y\text{ ft} = 0$$
$$25\text{ kip.ft} - 11B_y\text{ ft} = 0$$
$$11B_y\text{ ft} = 25\text{ kip.ft}$$
$$B_y = 2.27\text{ kip}$$

Cut at C and consider the right part as shown in Figure 7.10:

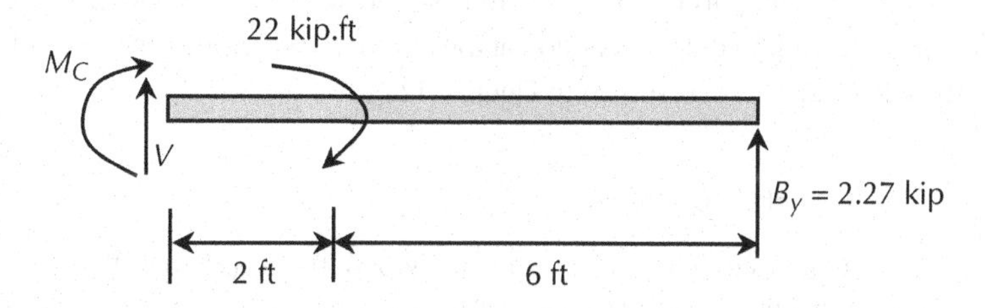

FIGURE 7.10 The free-body diagram of segment *CB* of the beam for Example 7.2. (Note again that if the right part of the beam is considered, the shear force is considered upward; the moment is considered clockwise and the normal force is considered away from the cut section. This is the positive sign convention.)

$$\sum M \text{ at } C = 0 \ (\circlearrowright + ve)$$

$M_C + 22$ kip.ft – 2.27 kip (8 ft) = 0
$M_C + 22$ kip.ft – 18.16 kip.ft = 0
$M_C = -3.84$ kip.ft

$$\sum F_x = 0 \ (\rightarrow +ve)$$

$N_C = 0$

$$\sum F_y = 0 \ (\uparrow +ve)$$

$V_C + 2.27$ kip = 0
$V_C = -2.27$ kip

Answers:

$M_C = -3.8$ kip.ft
$N_C = 0$
$V_C = -2.3$ kip

Example 7.3:

A simply supported beam is supporting the loads as shown in Figure 7.11. Determine the internal normal force, shear force and moment at *C*. Point *C* is located 5 ft from the left support *A*.

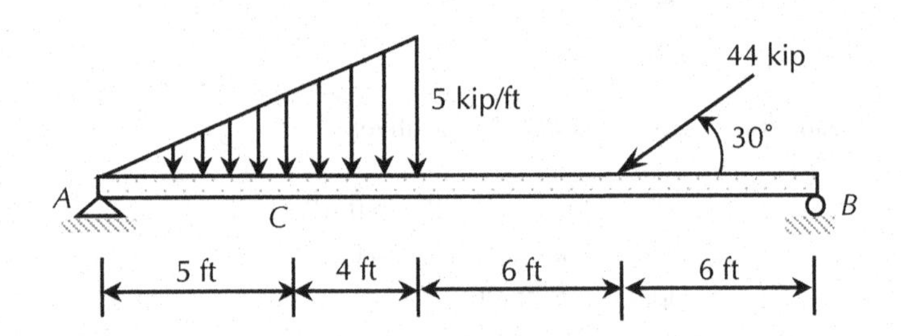

FIGURE 7.11 A beam with arbitrary loading for Example 7.3.

SOLUTION

If a cut at *C* is made, the left part of the structure is less loaded and expected to be easier to be analyzed. Therefore, let us find out the reaction at the left support considering the whole structure.

Consider the whole structure as shown in Figure 7.12:

$$\sum M \text{ at } B = 0 \ (\circlearrowright + ve)$$

– ½ (5 kip/ft)(9 ft) (15 ft) – 44 sin30 kip (6 ft) + A_y (21 ft) = 0
– 337.5 kip.ft – 132 kip.ft + $21A_y$ ft = 0
– 469.5 kip.ft + $21A_y$ ft = 0
$21A_y$ ft = 469.5 kip.ft
$A_y = 22.4$ kip

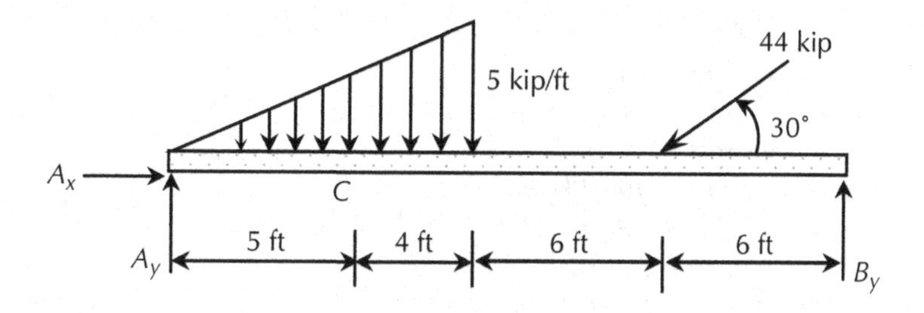

FIGURE 7.12 The free-body diagram of the beam for Example 7.3.

$$\sum F_x = 0 \ (\rightarrow +ve)$$

$$A_x - 44 \cos 30 \text{ kip} = 0$$
$$A_x - 38.1 \text{ kip} = 0$$
$$A_x = 38.1 \text{ kip}$$

Cut at C and consider the left part as shown in Figure 7.13:

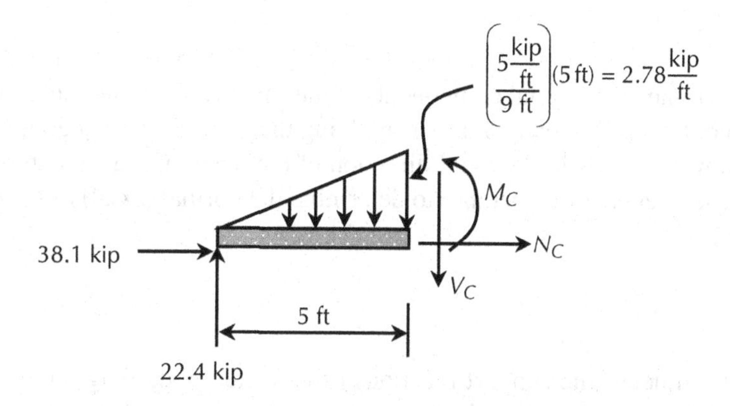

FIGURE 7.13 The free-body diagram of a small part of the beam for Example 7.3.

$$\sum M \text{ at } C = 0 \ (\lrcorner + ve)$$

$$-M_C + 22.4 \text{ kip } (5 \text{ ft}) - ½ \ (2.78 \text{ kip/ft})(5 \text{ ft}) \ (5/3 \text{ ft}) = 0$$
$$-M_C + 112 \text{ kip.ft} - 11.6 \text{ kip.ft} = 0$$
$$-M_C + 100.4 \text{ kip.ft} = 0$$
$$M_C = 100.4 \text{ kip.ft}$$

$$\sum F_x = 0 \ (\rightarrow +ve)$$

$$N_C + 38.1 \text{ kip} = 0$$
$$N_C = -38.1 \text{ kip}$$

$$\sum F_y = 0 \ (\uparrow +ve)$$

$$22.4 \text{ kip} - V_C - ½ \ (2.78 \text{ kip})(5 \text{ ft}) = 0$$
$$22.4 \text{ kip} - V_C - 6.95 \text{ kip} = 0$$

$$15.45 \text{ kip} - V_C = 0$$
$$V_C = 15.45 \text{ kip}$$

Answers:

$M_C = 100.4$ *kip.ft*
$N_C = -38.1$ *kip*
$V_C = 15.5$ *kip*

7.2 INTERNAL REACTIONS DIAGRAMS

In section 7.1, the procedure to determine the internal normal force, shear force and moment at a certain position has been discussed. However, in order to design the member, it is essential to calculate the maximum internal normal force, shear force and moment. For some easy loading cases, the point where the maximum internal reactions occur may be visualized. However, it is not always possible or may yield a wrong guess. To determine the maximum internal reactions exactly, the axial force, shear-force and moment diagrams are very helpful. Now, we will discuss the procedure to develop the diagrams.

Axial Force Diagram:

Step 1: Determine all the axial components of the support reactions using a free-body diagram of the whole beam and then applying the equilibrium equations.
Step 2: Follow the loads in the axial direction of the beam. Or, a cut can be made at different segments of the member to determine the normal (axial) force away from the member.

Shear Force Diagram:

Step 1: Determine all the support reactions using a free-body diagram of the whole beam and then applying equilibrium equations.
Step 2: Follow the loads in the transverse direction of the beam. For example, at the left support, if the left support reaction is upward then go upward, and then proceed accordingly.

Moment Diagram:

Step 1: Determine the area of the shear force; that is the moment.
Step 2: Keep on summing up the area of the force.

Three important items to remember are as follows:

- The moment is the maximum or minimum at the point where the shear force is zero.
- The slope of the moment diagram at any point equals the shear force at that point.
- The slope of the shear force diagram at any point equals the applied force at that point.
- The bending moments at the ends of a simply supported beam are zero.

Example 7.4:

Draw the shear force diagram and moment diagram for the following beam as shown in Figure 7.14.

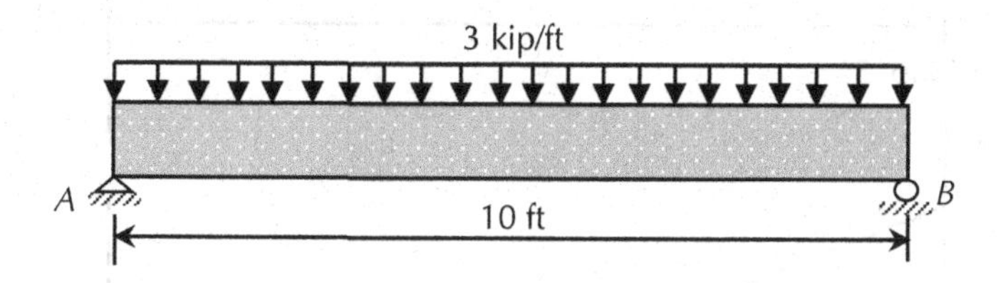

FIGURE 7.14 A beam with a uniform loading for Example 7.4.

SOLUTION

The first step of drawing the diagrams is to analyze the structure to determine all the support reactions. It is a customary practice to draw the diagram from the left side. Therefore, the reactions at the left support are determined first. The reactions at the right support may not be required but it is good to calculate these for checking the accuracy of the drawing. Let us calculate the reactions at the left support.

Consider the whole structure as shown in Figure 7.15:

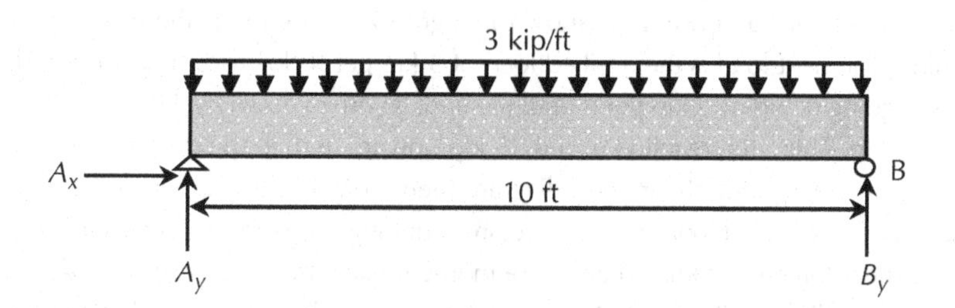

FIGURE 7.15 The free-body diagram of the beam for Example 7.4.

$$\sum M \text{ at } B = 0 \ (\circlearrowright +ve)$$

$$-\left(3\frac{\text{kip}}{\text{ft}}\right)(10\,\text{ft})\ (5\,\text{ft}) + A_y(10\,\text{ft}) = 0$$

$$-150 \text{ kip.ft} + A_y \ (10 \text{ ft}) = 0$$

$$A_y \ (10 \text{ ft}) = 150 \text{ kip.ft}$$

$$A_y = 15 \text{ kip } (\uparrow)$$

$$\sum F_y = 0 \ (\uparrow +ve)$$

$$A_y + B_y = 30$$

$$B_y = 15 \text{ kip } (\uparrow)$$

Now, let us draw the shear force diagram. After calculating the support reactions, the next step is to follow the vertical reactions and applied forces. From the left support, the first load is the

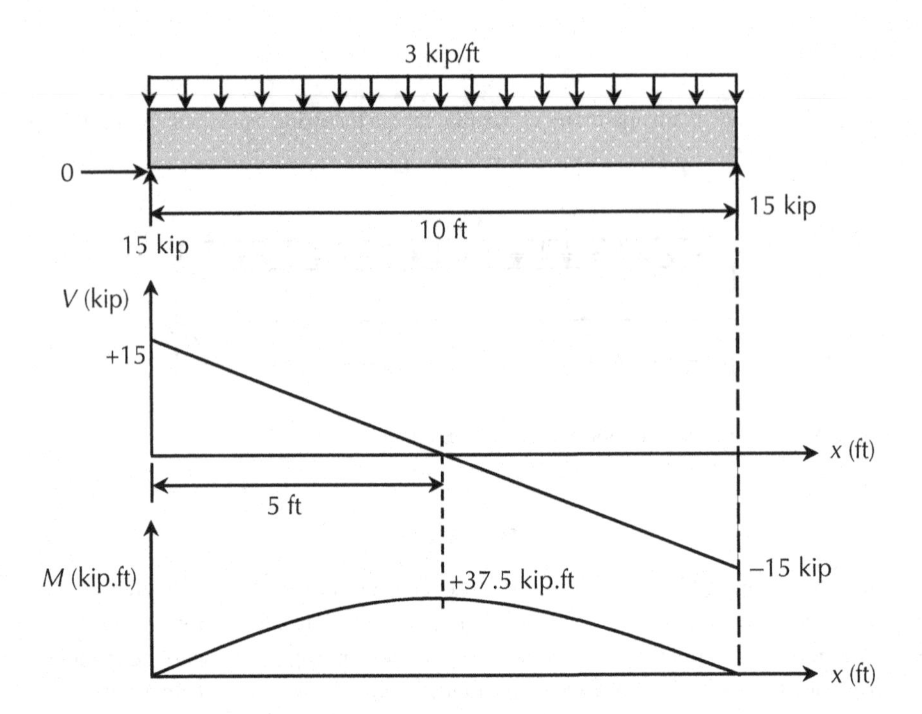

FIGURE 7.16 Shear force diagram and moment diagram for Example 7.4

vertical reaction of 15 kip (upward) as shown in Figure 7.16. So, insert the first point +15 kip at the beginning. The next load is the applied load of 3 kip per ft downward. So, in the shear force diagram, start going down by 3 kip per ft until the end. If you go down by 3 kip per ft for 10 ft, you will reach –15 kip. Then, the reaction at B is 15 kip upward and, in the diagram, go up by 15 kip and you will reach zero. The shear force diagram then looks like that in Figure 7.16. Remember, the diagram will be a closed loop unless there are rounding errors in the calculations.

Once the shear force diagram is drawn, the moment diagram can be drawn as the summation area of shear force diagram. Let us start drawing the moment diagram. At the left support, there is no moment; so, start with zero. Remember, at the external pin and roller supports and at free ends, moment is zero. Now, sum (algebraic sum as the sign matters) the area of the shear-force diagram to draw the moment diagram. The area of the first half of the shear force diagram is ½(15 kip)(5 ft) = 37.5 kip.ft; this is the moment at the middle of the beam. Continue summing up the shear force diagram to generate the moment diagram. The area of the second half of the shear force diagram is ½(–15 kip)(5 ft) = –37.5 kip.ft and reaches zero at the end of the beam. Once the shear force diagram is drawn, the moment diagram can be drawn as the summation area of the shear-force diagram.

Again, to remind you, the moment is the maximum/minimum at that point where the shear force is zero. The slope of the moment diagram at any point equals the shear force at that point. Also, the slope of the shear force diagram at any point equals the applied force at that point.

Rule: The moment is the maximum/minimum at the point where the shear force is zero. The moment at any point of a beam equals the summation of the shear force diagram from any end of the beam.

Example 7.5:

Draw the shear-force and moment diagrams for the following structure as shown in Figure 7.17.

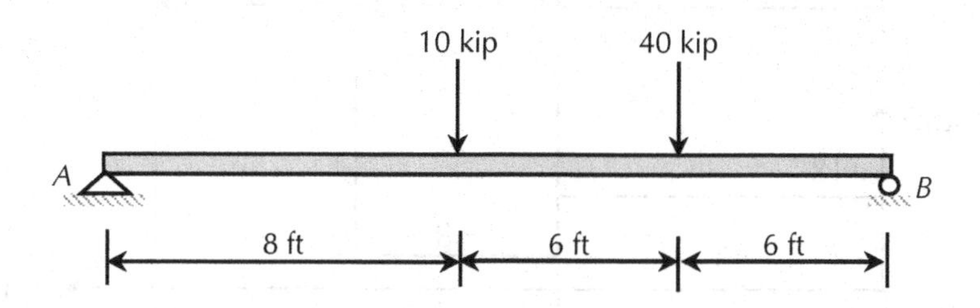

FIGURE 7.17 A beam with two concentrated loading for Example 7.5.

SOLUTION

The first step of drawing the free-body diagram is to analyze the structure to determine all the support reactions as shown in Figure 7.18. Let us analyze the structure first.

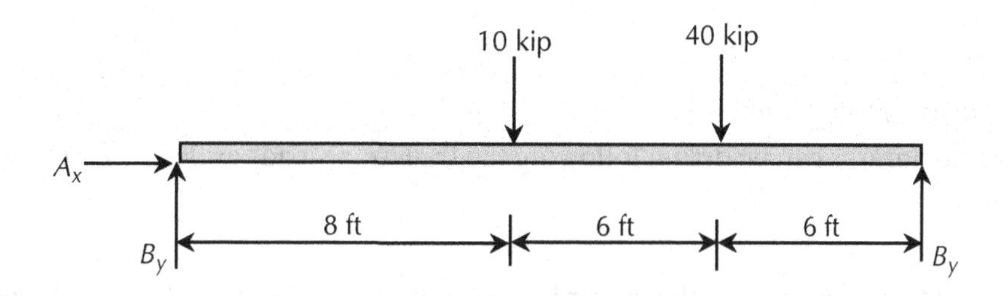

FIGURE 7.18 The free-body diagram of the beam for Example 7.5.

$$\sum M \text{ at } B = 0 \; (\circlearrowleft + ve)$$

$$40 \text{ kip } (6 \text{ ft}) + 10 \text{ kip } (12 \text{ ft}) - A_y \, (20 \text{ ft}) = 0$$

$$240 \text{ kip.ft} + 120 \text{ kip.ft} - A_y \, (20 \text{ ft}) = 0$$

$$A_y = 18 \text{ kip}$$

$$\sum F_y = 0 \; (\uparrow + ve)$$

$$A_y + B_y = 10 \text{ kip} + 40 \text{ kip}$$

$$B_y = 32 \text{ kip}$$

$$\sum F_x = 0 \; (\rightarrow + ve)$$

$$A_x = 0$$

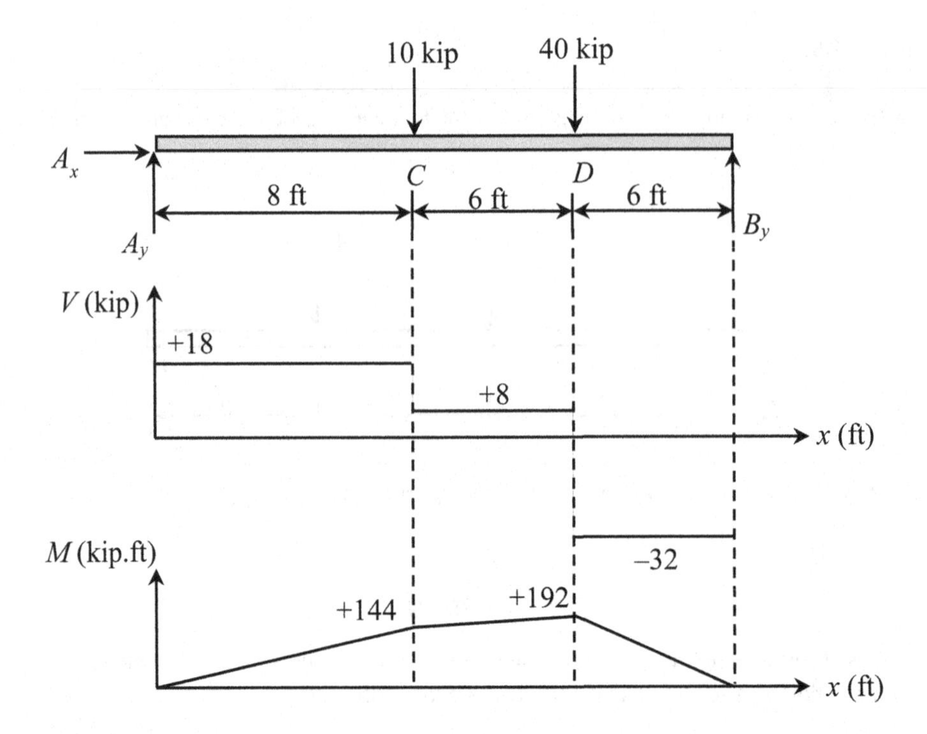

FIGURE 7.19 Shear-force and moment diagrams for Example 7.5

Note for Shear force diagram:

Shear force diagrams can be drawn following the transverse loads as described in the previous problem.

At the beginning, *A*:
At *A*, the first transverse load is the upward support reaction (18 kip) as shown in Figure 7.19. As this force is upward, we should draw the force upward in the shear force diagram.

Between *A* and *C*:
Between *A* and *C*, there is no load (zero load). Therefore, the slope of shear force is to be zero, meaning horizontal as shown. In other words, as there is no load applied between *A* and *C*, the shear force remains unchanged (18 kip).

At *C*:
At *C* ($x = 8$ ft), there is a downward concentrated load of 10 kip. Therefore, there will be a sharp drop in the shear-force diagram at point *C* by 10 kip. The shear force decreases to 8 kip.

Between *C* and *D*:
Between *C* and *D*, there is no load (zero load). Therefore, the shear force remains unchanged (8 kip).

At *D*:
At *D* ($x = 14$ ft), there is a downward concentrated load of 40 kip; so, there will be a sharp drop in the shear-force diagram at the point by 40 kip. The shear force decreases to –32 kip.

Between D and B:
Between D and B, there is no load (zero load). Therefore, the shear force remains unchanged (–32 kip).

At B:
At B (x = 20 ft), there is an upward vertical reaction of 32 kip. Therefore, there will be a vertical jump at the point by 32 kip. The shear force decreases to 0 kip.

Note for the Moment Diagram

Once a shear force diagram is drawn, the moment diagram can be drawn as the summation area of a shear force diagram. At the left support, there is no moment; so start with zero. Remember, at the external pin and roller supports and at free ends, moment is zero. Now, sum (algebraic sum as the sign matters) the area of the shear-force diagram to draw the moment diagram.

At the beginning, A:
The support is a pin and, thus, the moment is zero.

At C:
At C (x = 8 ft), the sum of the area of the shear-force diagram is (18 kip × 8 ft) 144 kip.ft. This means the moment at C is 144 kip.ft. The shear force between A and C is +18 kip. Therefore, the slope of the moment between A and C is +18 kip. This slope of the moment is positive constant, meaning a straight line with positive slope. In other words, between A and C, shear force is constant and positive; therefore, in this region, the slope of the moment diagram is constant and positive.

At D:
At D (x = 14 ft), the sum of the area of the shear-force diagram is (144 kip.ft + 8 kip × 6 ft) 192 kip.ft. This means the moment at D is 192 kip.ft. The shear force between C and D is +8 kip. Therefore, the slope of the moment between C and D is +8 kip. This slope of the moment is a positive constant, meaning a straight line with a positive slope.

At B:
At B (x = 20 ft), the sum of the area of the shear-force diagram is (192 kip.ft – 32 kip × 6 ft) 0 kip.ft. This means the moment at D is 0 kip.ft. The shear force between D and B is –32 kip. Therefore, the slope of the moment between D and B is –32 kip. This slope of the moment is negative constant, meaning a straight line with a negative slope.

Example 7.6:

Draw the shear force diagram and moment diagram for the following beam as shown in Figure 7.20.

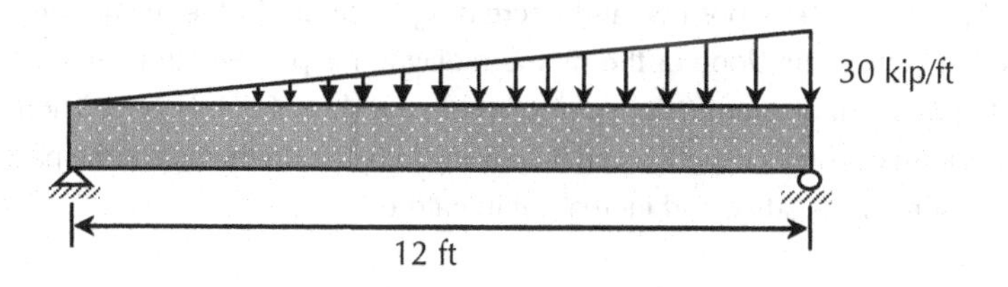

FIGURE 7.20 A beam with a uniformly varying load for Example 7.6.

SOLUTION

The first step of drawing the free-body diagram is to analyze the structure to determine all the support reactions as shown in Figure 7.21. Let us analyze the structure first.

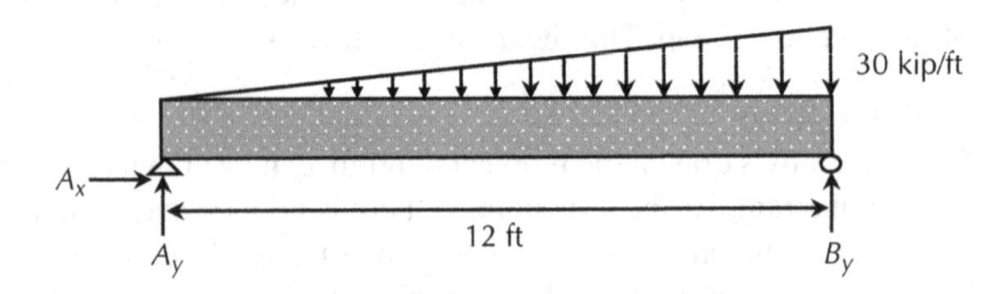

FIGURE 7.21 The free-body diagram of the beam for Example 7.6.

$$\sum M \text{ at } B = 0 \; (\circlearrowright + ve)$$

$$\tfrac{1}{2}(30\text{ kip/ft})(12\text{ ft})(4\text{ ft}) - A_y\,(12\text{ ft}) = 0$$

$$A_y = \frac{720\text{ kip.ft}}{12\text{ft}} = 60\text{ kip}$$

$$\sum F_y = 0 \; (\uparrow +ve)$$

$$A_y + B_y = \tfrac{1}{2}(30\text{ kip/ft})(12\text{ ft})$$
$$B_y = 120\text{ kip}$$

$$\sum F_x = 0 \; (\rightarrow +ve)$$

$$A_x = 0$$

Now, the shear force diagram (shown in Figure 7.22) can be drawn following the loads as described in Example 7.5. While drawing the diagrams, you may find it difficult to find out the shape of the curve. One more time to remind you that the moment is the maximum/minimum at that point where the shear force is zero. The slope of moment diagram at any point equals the shear force at that point. Also, the slope of shear force diagram at any point equals the applied force at that point. For example, while drawing a shear force diagram, you may be confused whether the shape is concave upward or downward. See that the applied triangular downward load (negative) is increasing from left to right. Therefore, the slope of the shear force diagram will be negative (zero at the beginning) and will increase from the left to the right.

Similarly, the shear force is positive and decreasing in nature in the initial part of the beam. Therefore, in that region, the slope of the moment diagram is positive and decreasing in nature. The slope of the moment diagram reaches zero where the shear force is zero. Then, the slope of the moment diagram becomes negative and increasing in nature in the rest of the part of the beam where the shear force is negative and increasing in nature.

Let us see how the zero point is calculated:

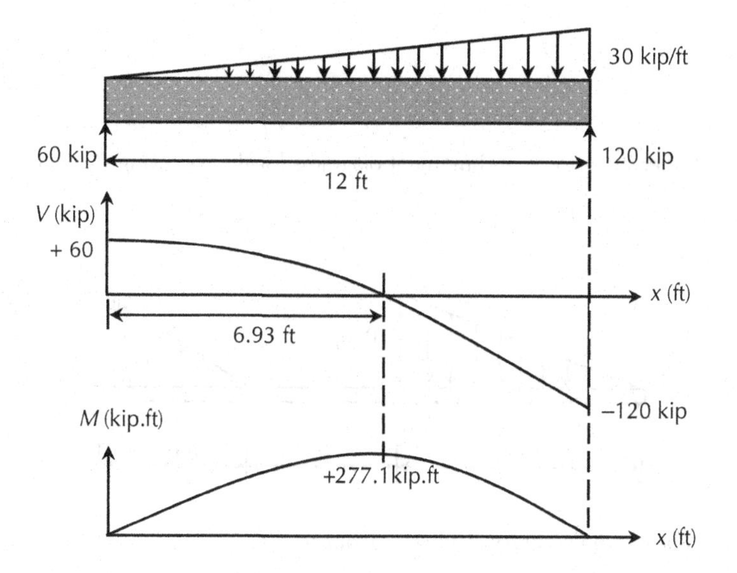

FIGURE 7.22 Shear-force and moment diagrams for Example 7.6.

Let the point of $V = 0$ be 'x' ft from the left.

$$A_y - \frac{1}{2}\left(\frac{30}{12}x\right)x = 0$$

$$60 - 1.25x^2 = 0$$

x = 6.93 ft (shown in Figure 7.23)

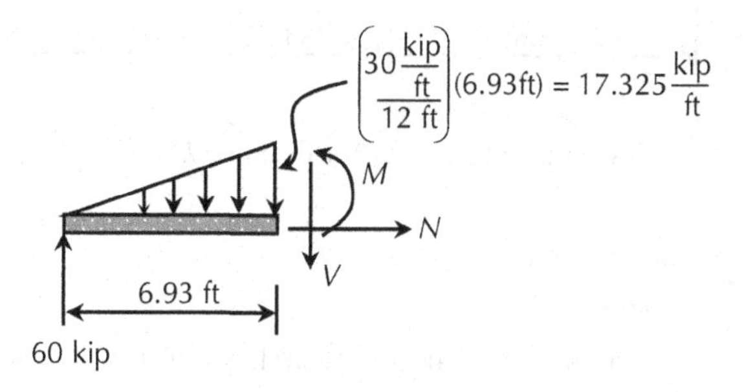

FIGURE 7.23 The free-body diagram of a small part of the beam for Example 7.6.

Now let us see how the maximum moment is calculated:

$$\sum M \text{ at Cut-point} = 0 \ (\lrcorner + ve)$$

$- M$ + 60 kip (6.93 ft) – ½ (17.325 kip/ft)(6.93 ft) (1/3 ft x 6.93 ft) = 0

$- M$ + 415.8 kip.ft – 138.67 kip.ft = 0

$- M$ + 277.1 kip.ft = 0

Alternatively, the area of the *V*-diagram up to 6.93 ft = 2/3 base x height = 2/3 (6.93 ft) (60 kip) = 277.2 kip.ft

Example 7.7:

Draw the axial force, shear-force and moment diagrams for the following structure as shown in Figure 7.24.

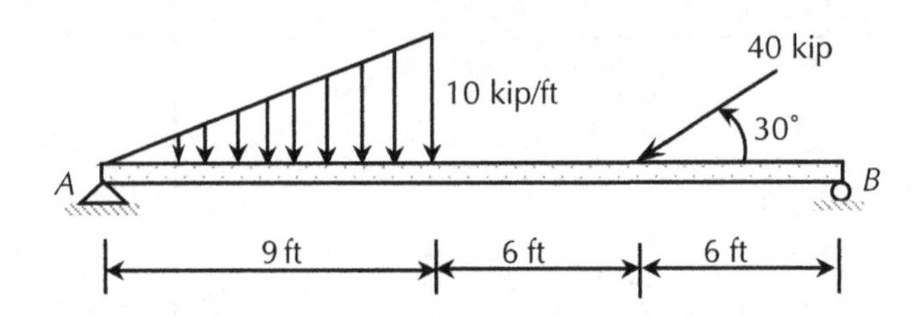

FIGURE 7.24 A beam with arbitrary loading for Example 7.7.

SOLUTION

Consider the whole structure as shown in Figure 7.25:

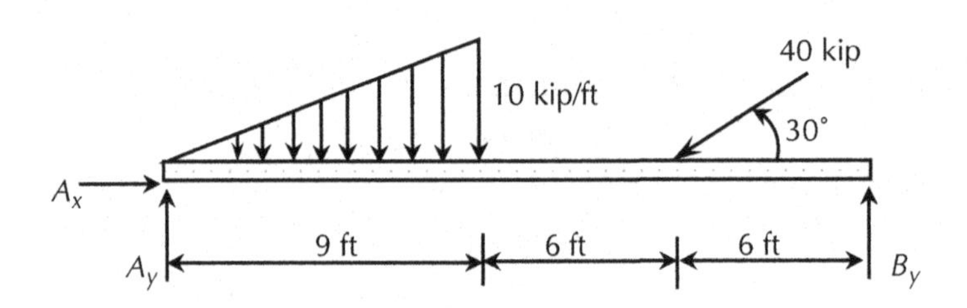

FIGURE 7.25 The free-body diagram of the beam for Example 7.7.

$\sum M \text{ at } B = 0$ (↲+ve)

$-\frac{1}{2}$ (10 kip/ft) (9 ft) (15 ft) – 40 sin30 kip (6 ft) + 21 $A_y = 0$

$A_y = 37.9$ kip

$\sum F_x = 0$ (→ +ve)

$A_x - 40 \cos 30$ kip = 0

$A_x = 34.64$ kip

$\sum F_y = 0$ (↑ +ve)

$A_y + B_y - \frac{1}{2}$ (10 kip/ft) (9 ft) – 40 sin30 kip = 0

$B_y = 27.1$ kip

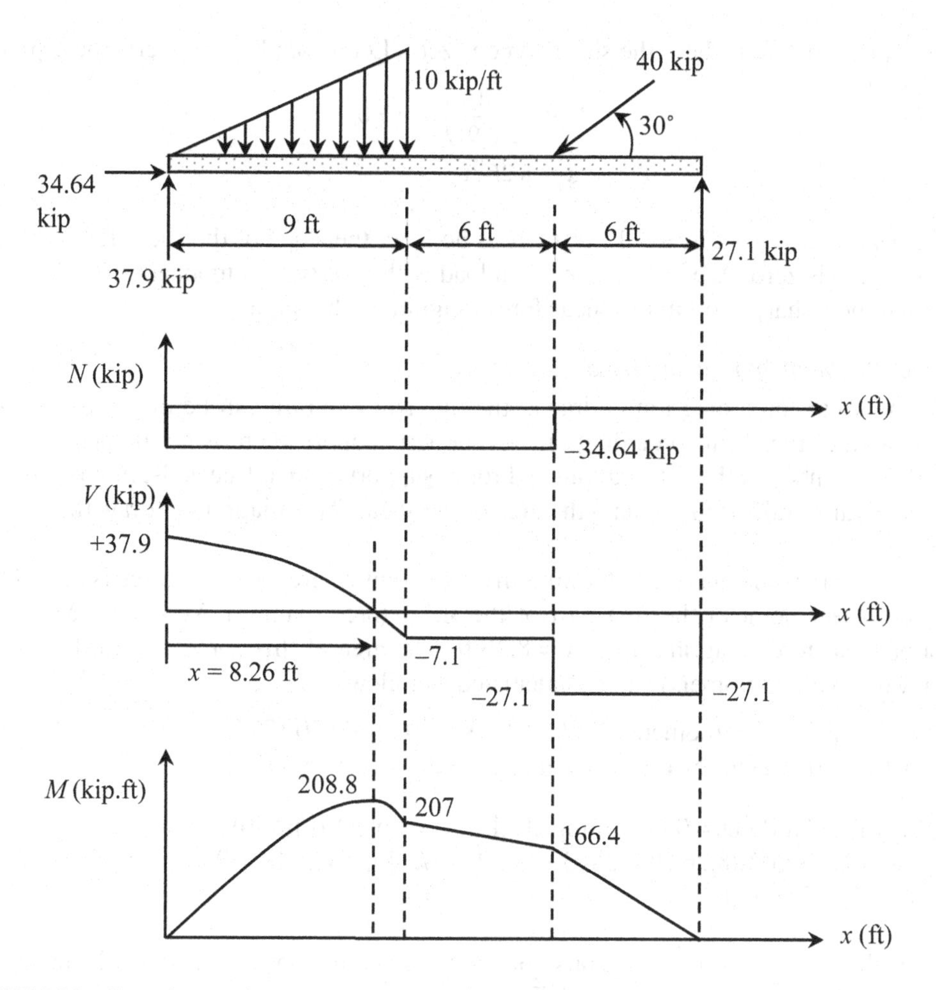

FIGURE 7.26 Shear-force and moment diagrams for Example 7.7.

Note for Axial Force Diagram (Figure 7.26):

We need to follow the loads in the longitudinal direction of the beam. The first axial load is the support reaction of 34.64 kip (negative or compressive as toward the beam; remember that an arrow shown away from the member is known tension for axial load). Continue this load as there is no other axial load. When you reach to $x = 15$ ft from the left 40cos30 will cancel it out; so the axial force becomes zero. After that, from $x = 15$ ft to 21 ft, there is no axial force.

Note for the Shear force diagram (Figure 7.26):

We need to follow the loads in the transverse direction of the beam. The first transverse load is the vertical component of the support reaction (37.9 kip) at *A*. As this force is upward, we should draw upward in the shear force diagram. Then, the curve should go down in triangular load format. The decrease in the shear curve at the beginning is very slow, as the load is very low at the beginning. In the later phase of the triangular loading area, the dropping down of the curve is very sharp. In other words, the load and the slope of the shear force diagram are similar. The triangular load is negative and increasing; therefore, the slope of of the shear-force diagram is negative and increasing. While decreasing the shear force, a point appears where the shear force is zero. Let us see how the zero-shear point is calculated:

$$37.9 - \frac{1}{2}\left(\frac{10}{9}\right)x(x) = 0$$

$$x = 8.26 \text{ ft}$$

In the next region, $x = 9$ ft to <15 ft, there is no load; the slope of the shear-force diagram at that region is zero. At $x = 15$ ft there is a load with a vertical component of – 20 kip; so, there will be a sharp drop in the shear-force diagram at that point.

Note for the Moment Diagram (Figure 7.26):

Once the shear force diagram is drawn, the moment diagram can be drawn as the summation area of the shear-force diagram. At the left support, there is no moment; so, start with 0. Remember, at the external pin and roller supports and at free ends, moment is zero. Now, sum (algebraic, sign matters) the area of the shear-force diagram to draw the moment diagram.

The moment is the maximum/minimum at the point where the shear force is zero. There is a zero-shear point in the first part of the shear force diagram. Maximum Moment = Area of shear force diagram up to $x = 8.26$ ft. The area of this convex-upward (concave-downward) region (moment) can be calculated as follows:

Approximate Area (moment): = 2/3(8.26 ft)(37.9 kip) = 209.75 kip.ft
Exact area (moment) by taking a cut at $x = 8.26$ ft is shown in Figure 7.27:

ΣM at the Cut Point = 0 (assuming clockwise moment is positive)
– ½ (9.18 kip/ft)(8.26 ft) (8.26/3 ft) – M + 37.9 kip (8.26 ft) = 0
M = 208.79 kip.ft

Both methods are almost equal. Thus, for convex-upward (concave-downward) shape, area = 2/3(base)(height) is very often used. For convex-downward (concave-upward) shape, area = 1/3(base)(height) is very often used.

Moment at x (9 ft) = 208.79 kip.ft – Area of shear force from 8.26 ft to 9.0 ft
= 208.79 kip.ft – 1/3(0.74 ft)(7.1 kip) = 1.75 kip
= 208.79 kip.ft – 1.75 kip.ft = 207 kip.ft

Then, M (x = 15 ft) = 207 kip.ft – 7.1 ft (6 ft) = 166.4 kip.ft
M (x = 21 ft) = 164.4 kip.ft – 27.1 ft (6 ft) = 1.8 kip.ft ≈ 0 kip.ft (rounding off error)

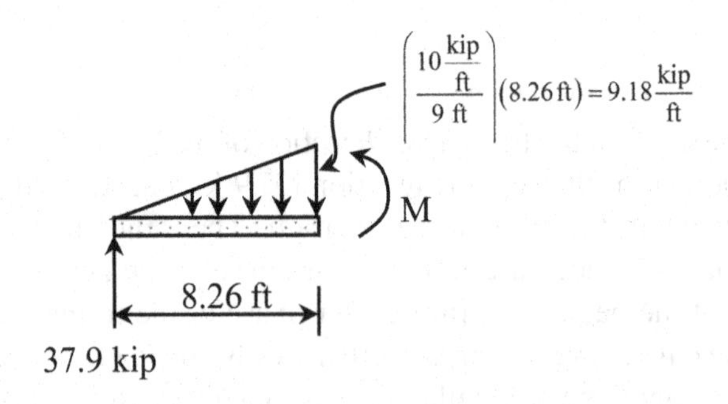

FIGURE 7.27 The free-body diagram of a small part of the beam for Example 7.7.

PRACTICE PROBLEMS

PROBLEM 7.1

Determine the internal normal force, shear force and moment at *C* for the simply supported beam as shown in Figure 7.28. Support *A* is a pin and support *B* is a roller.

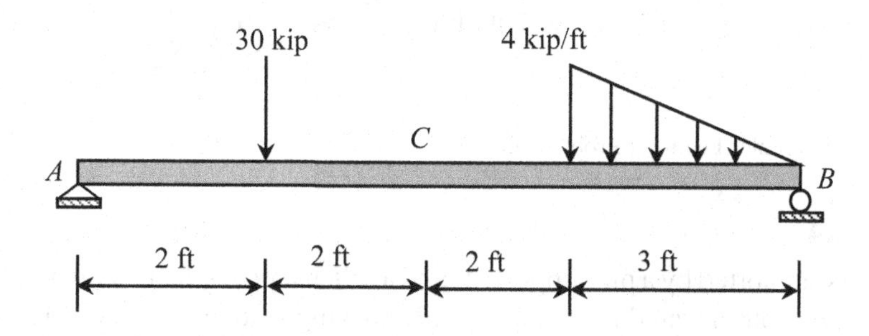

FIGURE 7.28 A beam with arbitrary loading for Problem 7.1.

PROBLEM 7.2

A beam, *AB,* is supported by a cable, *CD,* and a pin support, *A,* as shown in Figure 7.29. A concentrated load of 500 lb is applied at the end of the beam, *B*. The beam is made up with homogeneous material with a uniform self-weight of 20 lb per ft. The elongation and the self-weight of the cable are negligible. For these conditions, determine the internal reactions (axial, shear and moment) just to the left of *D* and at *E*. Point *E* is located 3 ft from the support *A*. Note that the symbol (') means foot and the symbol (") means inch.

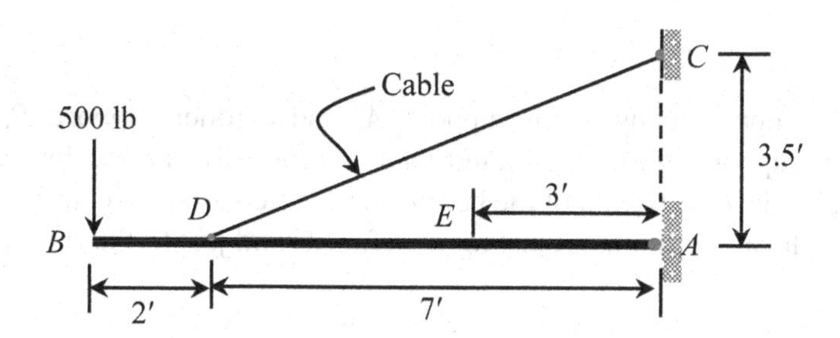

FIGURE 7.29 A complex beam for Problem 7.2.

PROBLEM 7.3

A beam, *AB,* is supported by a 7-ft strut, *CD,* and a pin support, *A,* as shown in Figure 7.30. A concentrated load of 120 lb is applied at the end of the beam, *B*. The beam is made up of homogeneous material with a uniform self-weight of 10 lb per ft. The elongation and the self-weight of the strut are negligible. For these conditions, determine the internal reactions (axial, shear and moment) just to the left of *D* and at *E*. Point *E* is located 3 ft from the support *A*. Note again that the symbol (') means foot and the symbol (") means inch.

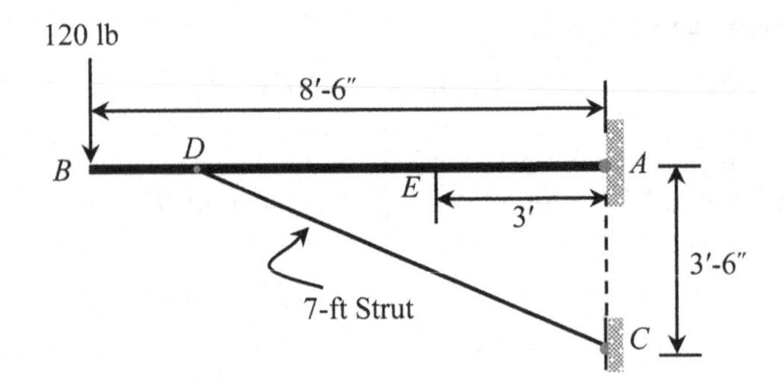

FIGURE 7.30 A complex beam for Problem 7.3.

PROBLEM 7.4

A beam, *AB*, is supported by a pin support, *A*, and a roller support, *B*, as shown in Figure 7.31. It is supporting a concentrated load, a uniformly varying load and a concentrated moment. Ignore the weight of the beam. For these conditions, determine the internal reactions (axial, shear and moment) just to the left of *C* and just to the right of *C*.

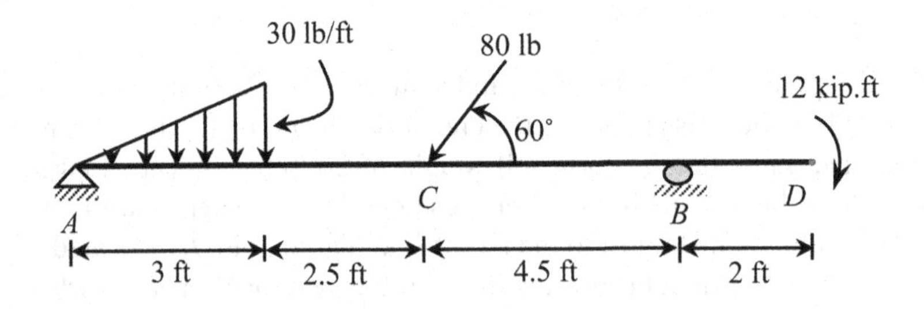

FIGURE 7.31 A beam with arbitrary loading for Problem 7.4.

PROBLEM 7.5

A beam, *AB*, is supported by a pin support, *A*, and a roller support, *B*, as shown in Figure 7.32. It is supporting a concentrated load, a uniformly varying load and a concentrated moment. Neglect the weight of the beam. For these conditions, determine the internal reactions (axial, shear and moment) just to the left of *C* and just to the right of *C*.

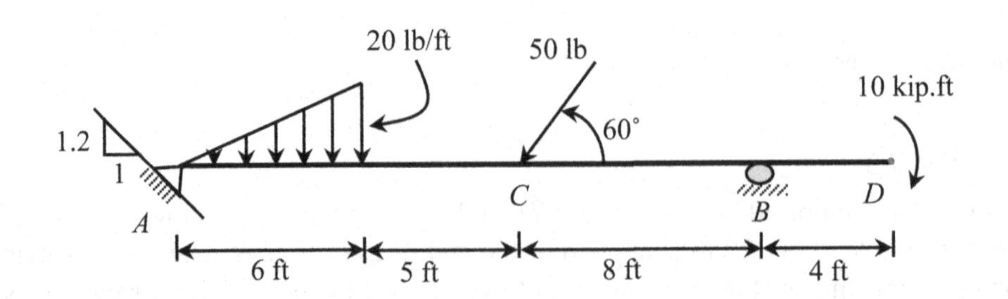

FIGURE 7.32 A beam with arbitrary loading for Problem 7.5.

PROBLEM 7.6

A member, *ADB*, is supported by a pin support, *B*, and a roller support, *A*, as shown in Figure 7.33. It is supporting a concentrated load and a uniformly varying load. Ignore the weight of the beam. For these conditions, determine the internal reactions (axial, shear and moment) at *D*. Joint *D* is a rigid joint.

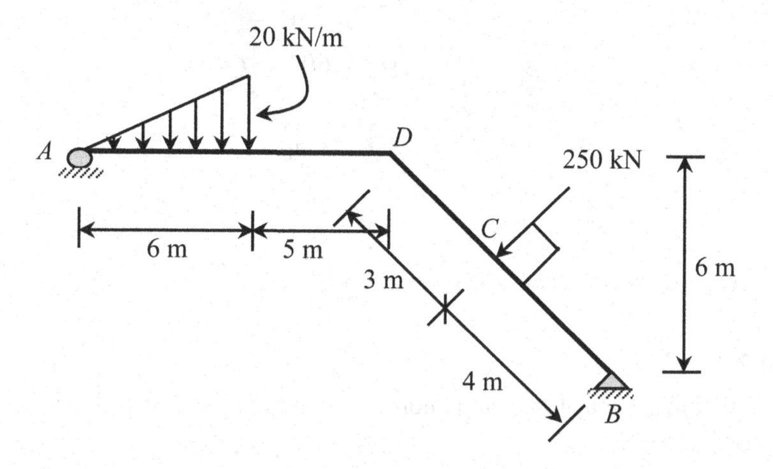

FIGURE 7.33 A complex beam for Problem 7.6.

PROBLEM 7.7

In Figure 7.34, assume the weight of the horse sign is 10 lb and the steel rods are weightless. Determine the internal reactions (axial, shear and moment) just to the left of *D* and just to the right of *D*.

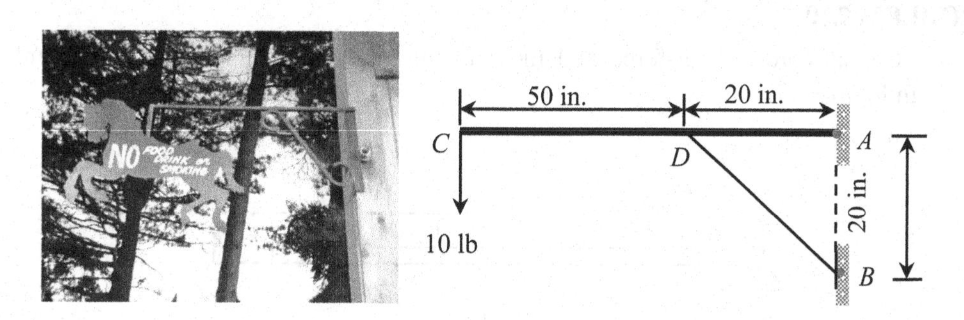

FIGURE 7.34 A complex beam for Problem 7.7.

PROBLEM 7.8

For the following structure as shown in Figure 7.35, determine the internal reactions (axial, shear and moment) just to the left of B and just to the right of B. The cable rolls over a frictionless pulley.

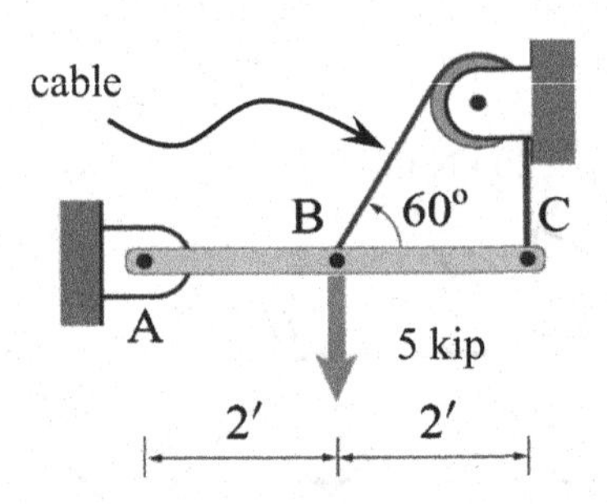

FIGURE 7.35 A beam with arbitrary loading for Problem 7.8.

PROBLEM 7.9

Draw the axial force, shear-force and moment diagrams for the following structure as shown in Figure 7.36.

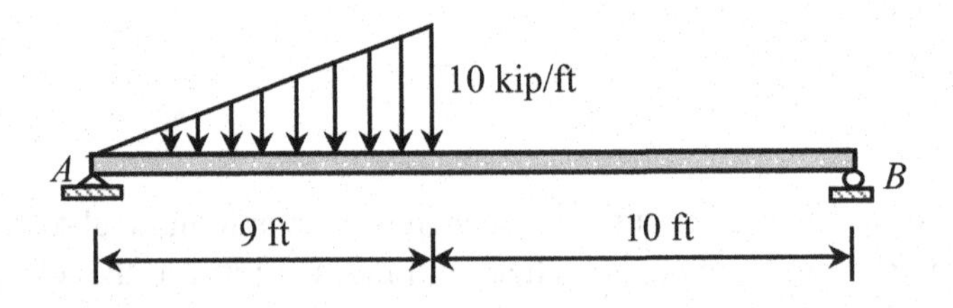

FIGURE 7.36 A beam with arbitrary loading for Problem 7.9.

PROBLEM 7.10

Draw the axial force, shear-force and moment diagrams for the following structure as shown in Figure 7.37.

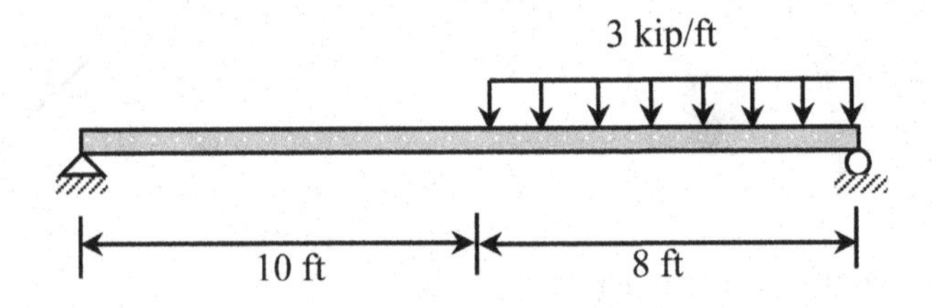

FIGURE 7.37 A beam with arbitrary loading for Problem 7.10.

PROBLEM 7.11

Draw the axial force, shear-force and moment diagrams for the following structure as shown in Figure 7.38.

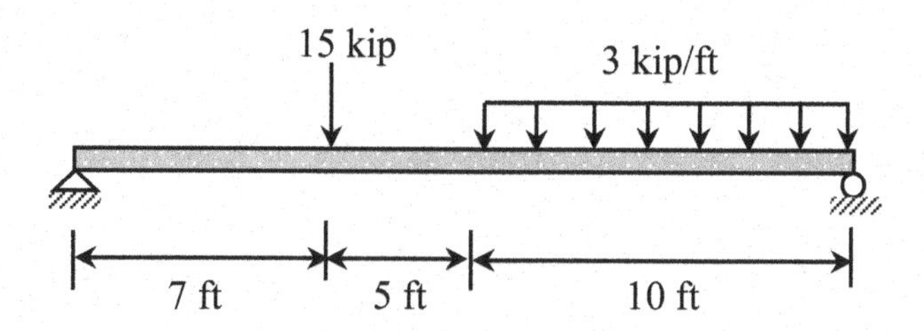

FIGURE 7.38 A beam with arbitrary loading for Problem 7.11.

PROBLEM 7.12

Draw the axial force, shear-force and moment diagrams for the following structure as shown in Figure 7.39.

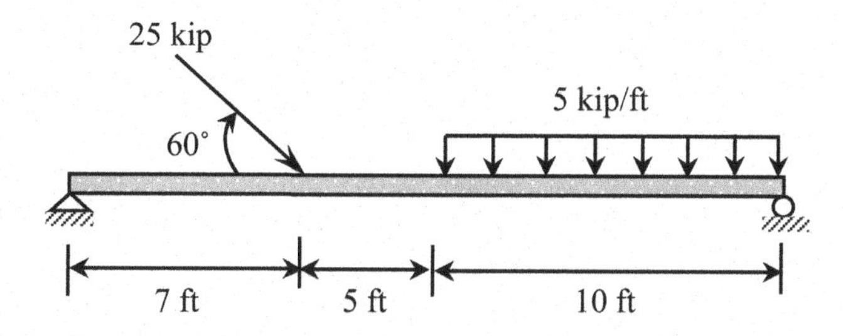

FIGURE 7.39 A beam with arbitrary loading for Problem 7.12.

PROBLEM 7.13

Draw the axial force, shear-force and moment diagrams for the following structure as shown in Figure 7.40.

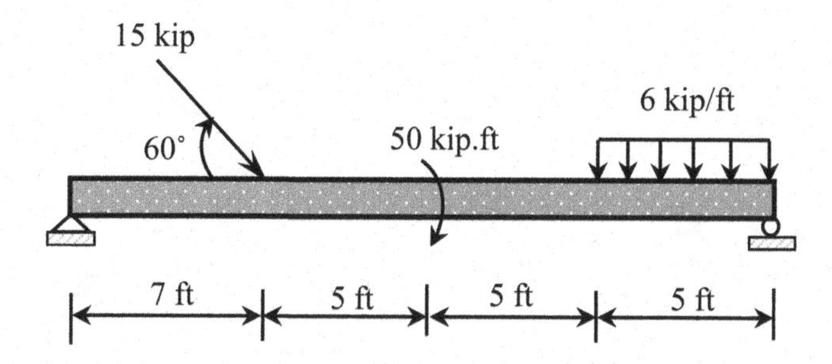

FIGURE 7.40 A beam with arbitrary loading for Problem 7.13.

8 Space Force Systems

8.1 GENERAL

In the previous chapters (Chapters 3 through 7), all the structures were considered planar or two- dimensional (2D) systems. The two-dimension approach was used even when analyzing the real space or three-dimensional (3D) structures, to simplify the problem analysis. However, in many structures, a planar assumption may not be appropriate and space (3D) consideration is more accurate. For example, in the Agora Tower in Taiwan shown in Figure 8.1, the load is distributed in a space system. The planar load distribution assumption is not appropriate. This chapter discusses how to analyze the space force system.

The vector method is the easiest way to analyze a 3D force system. The rectangular component method, which is geometric in nature and discussed in Chapter 2, can also be used. In this chapter, the vector method is discussed. The differences between these methods are also shown using examples. Now let us revise the vector analysis before applying it to analyze structures.

8.2 VECTOR METHOD

8.2.1 Scalars versus Vectors

Whatever we can measure is a called a quantity. For example, 2.5 lb or 10 m, 12 sec, etc. Some of these can be represented by magnitude only and some of these cannot be represented by magnitude only. For example, if a car moves along a straight path 7 m in the first second and 5 m in the following second in the same direction, then the displacement of the car is 7 m + 5 m = 12 m in two seconds. However, if the 2nd second the car comes back by 5 m, then the displacement of the car is 7 m – 5 m = 2 m in two seconds. This means without mentioning the direction, displacement cannot be fully expressed. The quantity that requires the direction along with its magnitude is called the vector quantity. A scalar is that quantity that can be represented by magnitude only.

Thus, any quantity is either a vector or a scalar. These two categories can be distinguished from one another by their distinct definitions: scalars are quantities that are fully described by a magnitude (or numerical value) alone. Vectors are quantities that are fully described by both a magnitude and a direction. Therefore, vectors have direction as well as magnitude. Some examples of scalar quantities include speed, volume, mass, temperature, power, energy, and time. Some examples of vector quantities include force, velocity, acceleration, displacement, and momentum. Vector representation is very often used in engineering dynamics.

8.2.2 Vector Representation

A vector ($\vec{F}$) can be determined as the product of the scalar quantity (magnitude of the force, F) times the unit vector ($\vec{u}$) along the direction as follows:

$$\vec{F} = F\vec{u}$$

FIGURE 8.1 Agora Tower in the Xinyi District of Taipei, Taiwan. *(Photo courtesy of Designboom)*

Now let us see a quantity can be expressed using vector. Let us consider that a car traveled 10 m from O to A in Figure 8.2. The travel path is in between x and y axes. Let us assume that the projection of 10 m along the x axis is 8 m, and along the y axis is 6 m. This means the particle moved 8 m along the x axis and 6 m along the y axis. Also, let us assume that $\vec{i}$ represents the unit vector along the x axis, meaning a magnitude of one (1.0) along the x axis. More clearly, $\vec{i}$ represents a quantity with a magnitude of one (1.0) and acts along the x axis. Similarly, let us assume that $\vec{j}$ represents the unit vector along the y axis.

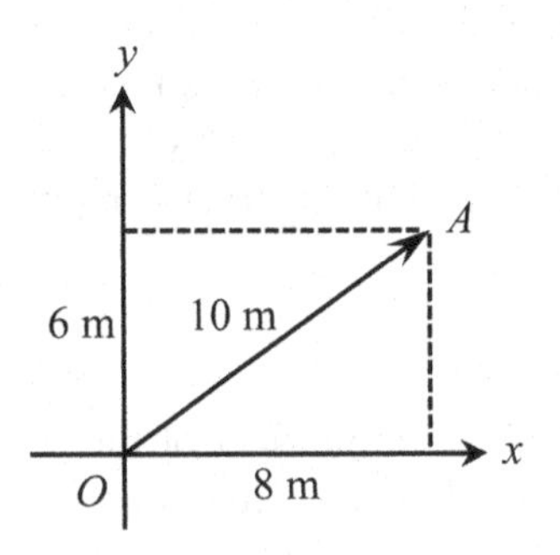

FIGURE 8.2 Representation of a vector.

Unit vector ($\vec{u}$) means a vector that shows the direction with a magnitude of 1.0. Therefore, multiplying a unit vector with a scalar force does not change the magnitude of the force; it just adds the direction with it.

Then, this displacement can be presented as $8\vec{i}+6\vec{j}$ m. Thus, any vector quantity can be represented as:

$$\vec{F} = F_x\vec{i} + F_y\vec{j}$$

The resultant direction (θ_x) with respect to the x axis is

$$\theta_x = \tan^{-1}\left(\frac{F_y}{F_x}\right)$$

Now let us discuss the scenario for a 3D system where there are three dimensions of the system (x, y, and z).

In a 3D system, $\vec{F} = F_x\vec{i} + F_y\vec{j} + F_z\vec{k}$

The magnitude of $\vec{F}$ can be determined as: $F = \sqrt{(F_x)^2 + (F_y)^2 + (F_z)^2}$

where:

$F_x = F\cos\theta_x =$ the x component of the quantity

$F_y = F\cos\theta_y =$ the y component of the quantity

$F_z = F\cos\theta_z =$ the z component of the quantity

$\cos\theta_x = \frac{F_y}{F}$ the angle between the x axis and the direction of the quantity

$\cos\theta_y = \frac{F_y}{F} =$ the angle between the y axis and the direction of the quantity

$\cos\theta_z = \frac{F_z}{F}$ the angle between the z axis and the direction of the quantity

$\vec{i}, \vec{j}$, and $\vec{k}$ are the unit vectors along the x, y and z axes, respectively. The unit vector has a magnitude of one (1.0) but shows the direction. Therefore, if it is multiplied with a quantity, then the magnitude does not change, but the magnitude and direction can be expressed simultaneously.

8.2.3 Position Vector

The position vector is determined first to determine the unit vector.

A position vector is defined as a fixed vector that locates a point in space relative to another point. Consider two points, A and B, in a 3D space as shown in Figure 8.3. Let their coordinates be $A(x_A, y_A, z_A)$ and $B(x_B, y_B, z_B)$, respectively. The Positive Vector ($\vec{P}_{AB}$) directed from A to B can be determined as follows:

$$\vec{P}_{AB} = (x_B - x_A)\hat{i} + (y_B - y_A)\hat{j} + (z_B - z_A)\hat{k}$$

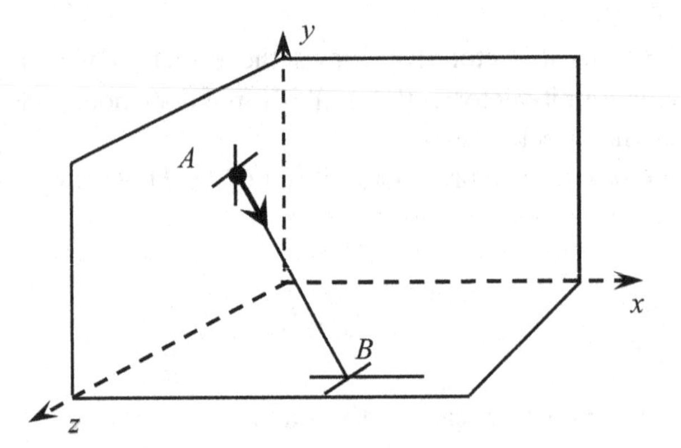

FIGURE 8.3 Vector method of expressing of forces.

Note that A is the starting point and B is the ending point. Always subtract the end coordinates from the start coordinates to determine the position vector. Then the position vector is divided by its magnitude to determine the unit vector ($\hat{u}_{AB}$) as follows.

$$\hat{u}_{AB} = \frac{\vec{P}_{AB}}{|P_{AB}|}$$

$$\hat{u}_{AB} = \frac{(x_B - x_A)\hat{i} + (y_B - y_A)\hat{j} + (z_B - z_A)\hat{k}}{\sqrt{(x_B - x_A)^2 + (y_B - y_A)^2 + (z_B - z_A)^2}}$$

Example 8.1:

A 100-N force can be represented by the diagonal of a rectangular cuboid as shown in Figure 8.4. Determine the x, y, and z components using the rectangular component method.

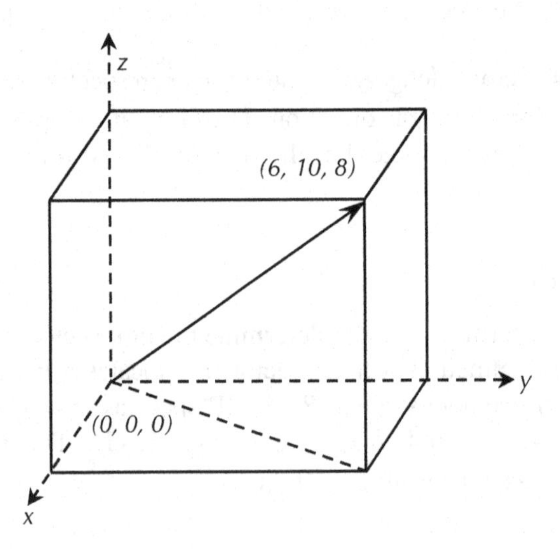

FIGURE 8.4 A rectangular cuboid for Example 8.1.

SOLUTION

Let us analyze the cube as shown in Figure 8.5.

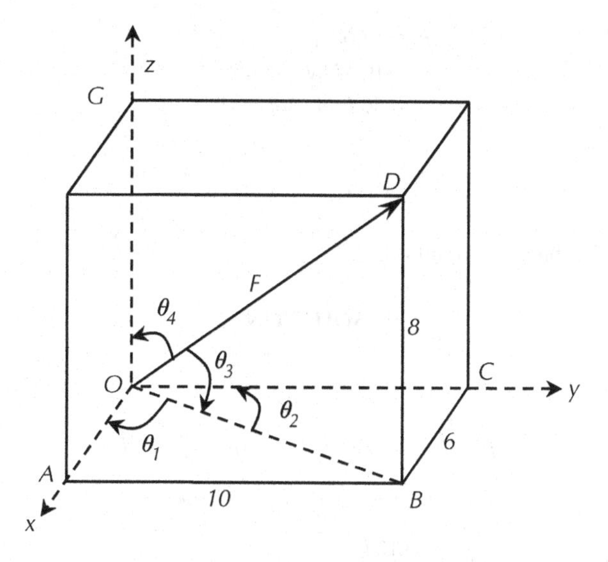

FIGURE 8.5 Analysis of the rectangular cuboid for Example 8.1.

$$\theta_1 = \tan^{-1}\left(\frac{AB}{OA}\right)$$
$$= \tan^{-1}\left(\frac{10}{6}\right)$$
$$= 59^{\circ}$$

$$\theta_2 = \tan^{-1}\left(\frac{BC}{OC}\right)$$
$$= \tan^{-1}\left(\frac{6}{10}\right)$$
$$= 31^{\circ}$$

$$\theta_3 = \tan^{-1}\left(\frac{BD}{OB}\right)$$
$$= \tan^{-1}\left(\frac{8}{\sqrt{(6)^2 + (10)^2}}\right)$$
$$= 34.5^{\circ}$$

$$\theta_4 = 90 - \theta_3$$
$$= 90^{\circ} - 34.5^{\circ}$$
$$= 55.5^{\circ}$$

$$F_x = F\cos\theta_3\cos\theta_1$$
$$= 100\text{ N }(\cos 34.5)(\cos 59)$$
$$= 42.4\text{ N } (\textit{Answer})$$

$$F_y = F\cos\theta_3\cos\theta_2$$
$$= 100\text{ N }(\cos 34.5)(\cos 31)$$
$$= 70.7\text{ N } (\textit{Answer})$$

$$F_z = F\cos\theta_4$$
$$= 100\text{ N }(\cos 55.5)$$
$$= 56.6\text{ N } (\textit{Answer})$$

Example 8.2:

Solve Example 8.1 by the vector method.

SOLUTION

Position Vector:

$$\vec{P}_{AB} = (x_B - x_A)\hat{i} + (y_B - y_A)\hat{j} + (z_B - z_A)\hat{k}$$
$$\vec{P}_{AB} = (6-0)\hat{i} + (10-0)\hat{j} + (8-0)\hat{k}$$
$$\vec{P}_{AB} = 6\vec{i} + 10\vec{j} + 8\vec{k}$$

Unit Vector:

$$\hat{u}_{AB} = \frac{\vec{P}_{AB}}{|P_{AB}|}$$
$$\hat{u}_{AB} = \frac{6\hat{i} + 10\hat{j} + 8\hat{k}}{\sqrt{(6)^2 + (10)^2 + (8)^2}}$$
$$\hat{u}_{AB} = 0.424\hat{i} + 0.707\hat{j} + 0.566\hat{k}$$

Force Vector:

$$\vec{F}_{AB} = F\hat{u}_{AB}$$
$$\vec{F}_{AB} = (100\text{ N})(0.424\hat{i} + 0.707\hat{j} + 0.566\hat{k})$$
$$\vec{F}_{AB} = 42.4\hat{i} + 70.7\hat{j} + 56.6\hat{k}$$

Therefore,

$F_x = 42.4$ N *(Answer)*
$F_y = 70.7$ N *(Answer)*
$F_z = 56.6$ N *(Answer)*

Note: Comparing Examples 8.1 and 8.2, the rectangular component method in space system is tedious and needs more attention compared to the vector method. However, some individuals may like the rectangular component method in a space system, as this method shows how a force is resolved into components and then sub-components. The vector method is more likely a mathematical calculation without visualizing the force components. It should be noted that the resolution of forces in a space system is truly more cumbersome when the analysis involves multiple forces.

Example 8.3:

A force of 10 kN is acting along *AB* as shown in Figure 8.6. Determine the vector representation of the force.

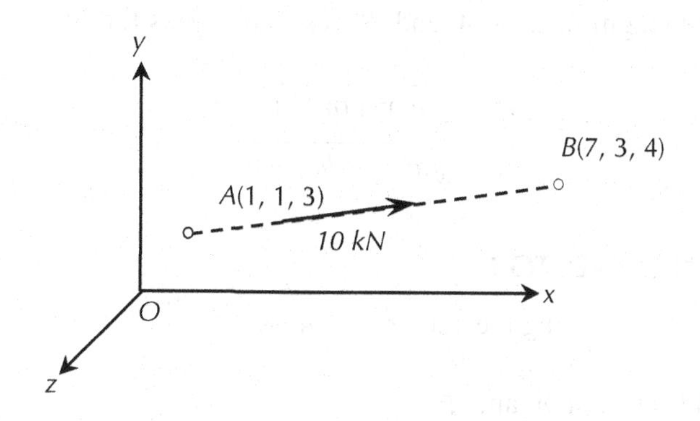

FIGURE 8.6 A force vector for Example 8.3.

SOLUTION

Position Vector:

$$\vec{P}_{AB} = (x_B - x_A)\hat{i} + (y_B - y_A)\hat{j} + (z_B - z_A)\hat{k}$$

$$\vec{P}_{AB} = (7-1)\hat{i} + (3-1)\hat{j} + (4-3)\hat{k}$$

$$\vec{P}_{AB} = 6\hat{i} + 2\hat{j} + \hat{k}$$

Unit Vector:

$$\hat{u}_{AB} = \frac{\vec{P}_{AB}}{|P_{AB}|}$$

$$\hat{u}_{AB} = \frac{6\hat{i} + 2\hat{j} + \hat{k}}{\sqrt{(6)^2 + (2)^2 + (1)^2}}$$

$$\hat{u}_{AB} = \frac{1}{\sqrt{41}}(6\hat{i} + 2\hat{j} + \hat{k})$$

Force Vector:

$$\vec{F}_{AB} = F\hat{u}_{AB}$$

$$\vec{F}_{AB} = (10\text{ kN})\frac{1}{\sqrt{41}}(6\hat{i} + 2\hat{j} + \hat{k})$$

$$\vec{F}_{AB} = 9.375\hat{i} + 3.125\hat{j} + 1.5625\hat{k} \text{ } (Answer)$$

8.3 VECTOR PRODUCTS

Let us assume two vectors: $\vec{A} = a_x\hat{i} + a_y\hat{j} + a_z\hat{k}$ and $\vec{B} = b_x\hat{i} + b_y\hat{j} + b_z\hat{k}$

The dot-product of two vectors is the scalar quantity and can be determined as:

$$\vec{A}.\vec{B} = (a_x b_x) + (a_y b_y) + (a_z b_z) = |A|\ |B| \cos\theta = \vec{B}.\vec{A}$$

where:

$|A|$ and $|B|$ are the magnitudes of $\vec{A}$ and $\vec{B}$, respectively, as follows:

$$|A| = \sqrt{(a_x)^2 + (a_y)^2 + (a_z)^2}$$
$$|B| = \sqrt{(b_x)^2 + (b_y)^2 + (b_z)^2}$$

8.4 PROJECTION OF VECTOR

The projection of vector $\vec{B}$ along the vector $\vec{A} = |B| \cos\theta = \dfrac{|A||B|\cos\theta}{|A|} = \dfrac{\vec{A}.\vec{B}}{|A|}$

where θ is the angle between $\vec{A}$ and $\vec{B}$

Example 8.4:

Two force vectors, $\vec{A} = \hat{i} + 2\hat{j} + 2\hat{k}$ kN and $\vec{B} = 4\hat{i} + 8\hat{j} - \hat{k}$ kN, are acting as shown in Figure 8.7. Determine the total force along the force $\vec{A}$.

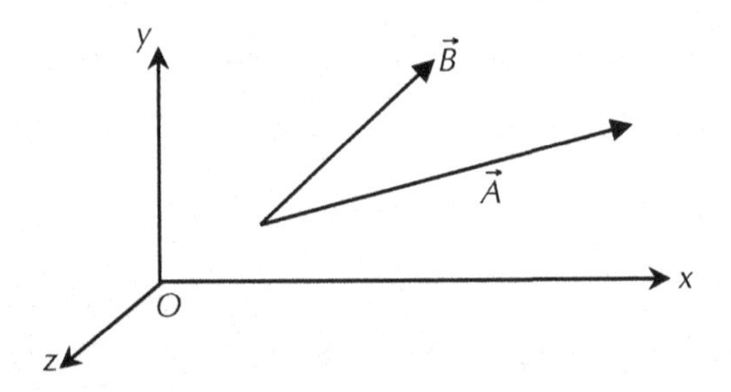

FIGURE 8.7 Two vectors for Example 8.4.

SOLUTION

The total force along the force $\vec{A}$ is the summation of the magnitude of force $\vec{A}$ and the projection of force $\vec{B}$ along the force $\vec{A}$.

$$\vec{A} = \hat{i} + 2\hat{j} + 2\hat{k}$$
$$\vec{B} = 4\hat{i} + 8\hat{j} - \hat{k}$$

$$\vec{A}.\vec{B} = (a_x b_x) + (a_y b_y) + (a_z b_z)$$

$$\vec{A}.\vec{B} = (1)(4) + (2)(8) + (2)(-1)$$

$$\vec{A}.\vec{B} = 4 + 16 - 2$$

$$\vec{A}.\vec{B} = 18$$

$$|A| = \sqrt{(a_x)^2 + (a_y)^2 + (a_z)^2}$$

$$|A| = \sqrt{(1)^2 + (2)^2 + (2)^2} = 3$$

The projection of vector $\vec{B}$ along vector, $\vec{A} = |B| \cos\theta$

$$\vec{A} = \frac{\vec{A}.\vec{B}}{|A|}$$

$$\vec{A} = \frac{18}{3} \text{ kN}$$

$$\vec{A} = 6 \text{ kN}$$

The total force along the direction of force $\vec{A} = 3 \text{ kN} + 6 \text{ kN}$

$\vec{A} = 9 \text{ kN}$

Answer: 9.0 kN.

8.5 RESULTANT OF VECTORS

If there is more than one force, say $\vec{A}$ and $\vec{B}$, respectively, as follows:

$$\vec{A} = a_x\hat{i} + a_y\hat{j} + a_z\hat{k} \text{ and } \vec{B} = b_x\hat{i} + b_y\hat{j} + b_z\hat{k}$$

The resultant (R) of these two vectors can be determined as follows:

$$\vec{R} = \vec{A} + \vec{B} = (a_x + b_x)\hat{i} + (a_y + b_y)\hat{j} + (a_z + b_z)\hat{k}$$

$$|R| = \sqrt{(a_x + b_x)^2 + (a_y + b_y)^2 + (a_z + b_z)^2}$$

If there are more than two forces, then the corresponding components can be added as follows:

$$\vec{R} = \vec{A} + \vec{B} + \ldots = (a_x + b_x + \ldots)\hat{i} + (a_y + b_y + \ldots)\hat{j} + (a_z + b_z + \ldots)\hat{k}$$

$$|R| = \sqrt{(a_x + b_x + \ldots)^2 + (a_y + b_y + \ldots)^2 + (a_z + b_z + \ldots)^2}$$

Example 8.5:

Three coplanar forces are acting on a point as shown in Figure 8.8. Determine the magnitude and direction (with respect to the 900-N force) of the resultant. The 900-N and the 200-N forces are acting along the horizontal and the vertical directions, respectively.

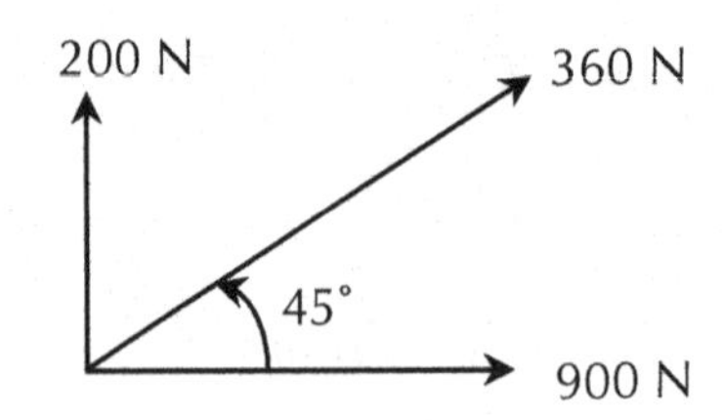

FIGURE 8.8 Three concurrent forces for Example 8.5.

SOLUTION

The vector form of 900-N force:

$$\vec{F}_1 = F_1 \cos\theta\, \hat{i} + F_1 \sin\theta\, \hat{j}$$
$$= 900 \cos 0 \hat{i} + 900 \sin 0 \hat{j}$$
$$= 900\hat{i}$$

The vector form of 360-N force:

$$\vec{F}_2 = F_2 \cos\theta\, \hat{i} + F_2 \sin\theta\, \hat{j}$$
$$= 360 \cos 45 \hat{i} + 360 \sin 45 \hat{j}$$
$$= 255\hat{i} + 255\hat{j}$$

The vector form of 200-N force:

$$\vec{F}_3 = F_3 \cos\theta\, \hat{i} + F_3 \sin\theta\, \hat{j}$$
$$= 200 \cos 90 \hat{i} + 200 \sin 90 \hat{j}$$
$$= 200\hat{j}$$

The vector form of the resultant force:

$$\vec{R} = \vec{F}_1 + \vec{F}_2 + \vec{F}_3$$
$$= (900 + 255 + 0)\hat{i} + (0 + 255 + 200)\hat{j}$$
$$= 1{,}155\hat{i} + 455\hat{j}$$

The magnitude of the resultant force:

$$R = \sqrt{(1{,}155\text{ N})^2 + (455\text{ N})^2}$$
$$R = 1{,}241\text{ N}$$

The direction of R with respect to the 900-N force, $\theta = \tan^{-1}\left|\frac{455\text{ N}}{1{,}155\text{ N}}\right| = 21.5^\circ$ (shown in Figure 8.9)

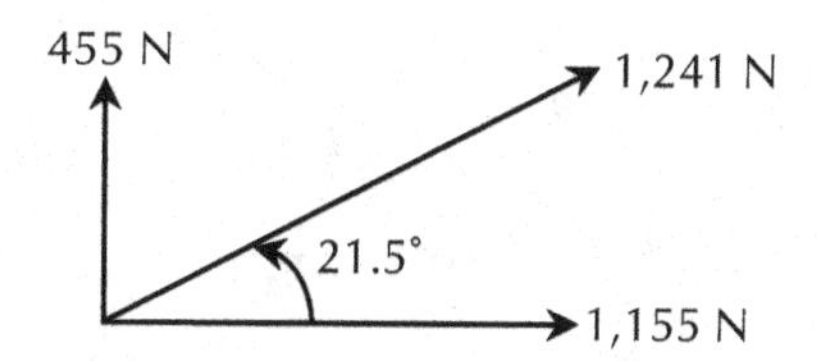

FIGURE 8.9 The resultant of the forces for Example 8.5.

Answers: 1,241 N, 21.5° with respect to 900 N force.

Example 8.6:

Three forces $\vec{P}, \vec{Q}$, and $\vec{R}$ (kN) are acting on a point, O, as shown in Figure 8.10. Determine the magnitude of the resultant and its direction.

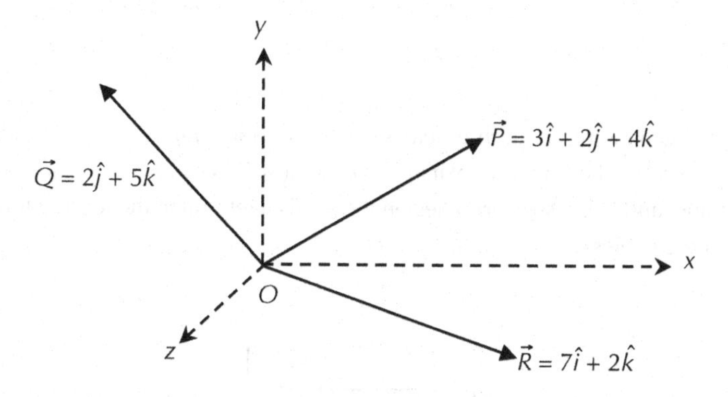

FIGURE 8.10 Three concurrent forces system for Example 8.6.

SOLUTION

Resultant vector:

$$\begin{aligned}\vec{R} &= \vec{P} + \vec{Q} + \vec{R}\\ &= (3\hat{i} + 2\hat{j} + 4\hat{k}) + (0\hat{i} + 2\hat{j} + 5\hat{k}) + (7\hat{i} + 0\hat{j} + 2\hat{k})\\ &= 10\hat{i} + 4\hat{j} + 11\hat{k}\end{aligned}$$

Resultant:

$$R = \sqrt{(F_x)^2 + (F_y)^2 + (F_z)^2}$$
$$R = \sqrt{(10\text{ kN})^2 + (4\text{ kN})^2 + (11\text{ kN})^2}$$
$$R = 15.4\text{ kN}$$

Direction:

$$\begin{aligned}\theta_x &= \cos^{-1}\left(\frac{F_x}{F}\right)\\ &= \cos^{-1}\left(\frac{10}{15.4}\right)\\ &= 49.5^{\circ}\end{aligned}$$

$$\theta_y = \cos^{-1}\left(\frac{F_y}{F}\right)$$
$$= \cos^{-1}\left(\frac{4}{15.4}\right)$$
$$= 74.95^{\circ}$$

$$\theta_z = \cos^{-1}\left(\frac{F_z}{F}\right)$$
$$= \cos^{-1}\left(\frac{11}{15.4}\right)$$
$$= 44.42^{\circ}$$

Answers: 15.4 kN, 49.5°, 74.95°, and 44.42° with respect to x, y, and z axes, respectively.

Example 8.7:

A weight of 5 kN is handing using three cables (*AB*, *AC*, and *AD*) as shown in Figure 8.11. The intersection of the cables lies in the *xy* plane; cable support *D* lies in the *xz* plane; cable support *C* lies in the *yz* plane; and cable support *B* lies on the *y* axis. Determine the tension force developed in each of the three cables.

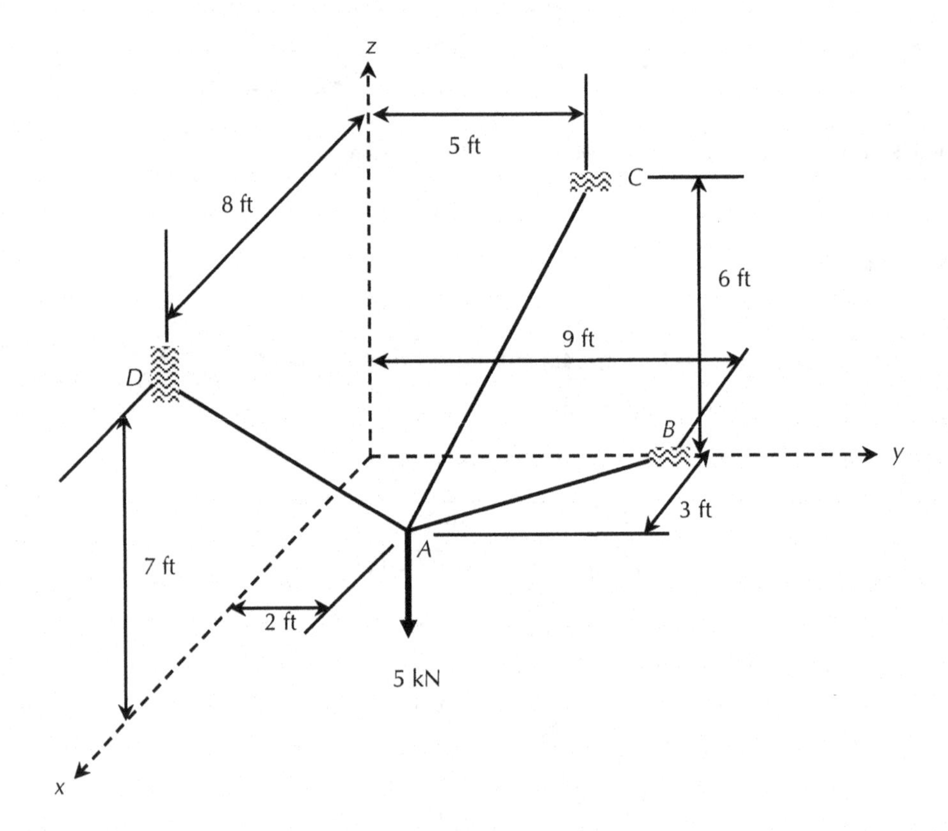

FIGURE 8.11 A space force system for Example 8.7.

SOLUTION

Let us find the coordinates first to determine the unit vectors along each cable.

$A \equiv (3, 2, 0)$ ft
$B \equiv (0, 9, 0)$ ft
$C \equiv (0, 5, 6)$ ft
$D \equiv (8, 0, 7)$ ft

The unit vectors along x, y, and z axes are $\hat{i}$, $\hat{j}$, and $\hat{k}$, respectively.

Position Vectors (P):

$$\begin{aligned}\vec{P}_{AB} &= (x_B - x_A)\hat{i} + (y_B - y_A)\hat{j} + (z_B - z_A)\hat{k} \\ &= (0-3)\hat{i} + (9-2)\hat{j} + (0-0)\hat{k} \\ &= -3\hat{i} + 7\hat{j}\end{aligned}$$

$$\begin{aligned}\vec{P}_{AC} &= (x_C - x_A)\hat{i} + (y_C - y_A)\hat{j} + (z_C - z_A)\hat{k} \\ &= (0-3)\hat{i} + (5-2)\hat{j} + (6-0)\hat{k} \\ &= -3\hat{i} + 3\hat{j} + 6\hat{k}\end{aligned}$$

$$\begin{aligned}\vec{P}_{AD} &= (x_D - x_A)\hat{i} + (y_D - y_A)\hat{j} + (z_D - z_A)\hat{k} \\ &= (8-3)\hat{i} + (0-2)\hat{j} + (7-0)\hat{k} \\ &= 5\hat{i} - 2\hat{j} + 7\hat{k}\end{aligned}$$

Vector Forces ($\vec{F} = F\hat{u}$):

$$\begin{aligned}\vec{F}_{AB} &= F_{AB}\hat{u}_{AB} \\ &= F_{AB}\frac{-3\hat{i} + 7\hat{j}}{\sqrt{(-3)^2 + (7)^2}} \\ &= F_{AB}(-0.394\hat{i} + 0.919\hat{j})\end{aligned}$$

$$\begin{aligned}\vec{F}_{AC} &= F_{AC}\hat{u}_{AC} \\ &= F_{AC}\frac{-3\hat{i} + 3\hat{j} + 6\hat{k}}{\sqrt{(-3)^2 + (3)^2 + (6)^2}} \\ &= F_{AC}(-0.408\hat{i} + 0.408\hat{j} + 0.817\hat{k})\end{aligned}$$

$$\vec{F}_{AD} = F_{AD}\hat{u}_{AD}$$
$$= F_{AD}\frac{5\hat{i} - 2\hat{j} + 7\hat{k}}{\sqrt{(5)^2 + (-2)^2 + (7)^2}}$$
$$= F_{AD}(0.566\hat{i} - 0.226\hat{j} + 0.793\hat{k})$$

Load, $\vec{W} = -5\hat{k}$

Resultant Equation:

$$\vec{F}_{AB} + \vec{F}_{AC} + \vec{F}_{AD} + \vec{W} = 0$$

$$-0.394F_{AB}\hat{i} + 0.919F_{AB}\hat{j} - 0.408F_{AC}\hat{i} + 0.408F_{AC}\hat{j} + 0.817F_{AC}\hat{k}$$
$$+ 0.566F_{AD}\hat{i} - 0.226F_{AD}\hat{j} + 0.793F_{AD}\hat{k} = -5\hat{k}$$

$$\text{or,}(-0.394F_{AB} - 0.408F_{AC} + 0.566F_{AD})\hat{i} + (0.919F_{AB} + 0.408F_{AC} - 0.226F_{AD})\hat{j}$$
$$+ (0.817F_{AC} + 0.793F_{AD})\hat{k} = 5\hat{k}$$

Equating the $\hat{i}$, $\hat{j}$, and $\hat{k}$ components yields

$$-3.94F_{AB} - 0.408F_{AC} + 0.566F_{AD} = 0$$
$$0.919F_{AB} + 0.408F_{AC} - 0.226F_{AD} = 0$$
$$0F_{AB} + 0.817F_{AC} + 0.793_{AD} = 5$$

Solving the above three equations:

$F_{AB} = -1.33$ kN *(Answer)*
$F_{AC} = 4.13$ kN *(Answer)*
$F_{AD} = 2.05$ kN *(Answer)*

Example 8.8:

The coordinates (feet) of a three-leg water tower in Kenney, Illinois are shown in Figure 8.12. On the top of the tower (100 feet above the ground), there is a mass of 120 kip. If the masses of tower legs are negligible, how much will each leg carry?

a) A water tower in Kenney, Illinois *(Photo courtesy of Flickr)*

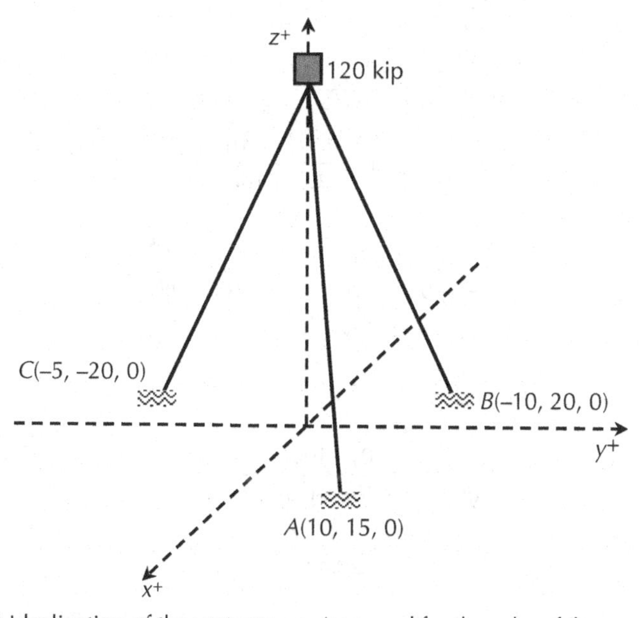

b) Idealization of the water tower (assumed for the sake of the example)

FIGURE 8.12 A space force system for Example 8.8.

SOLUTION

Let the top of the tower be the origin, *O(0, 0, 0)*. Then, the coordinates of *A*, *B* and *C* will be

A(10, 15, –100)
B(–10, 20, –100)
C(–5, –20, –100)

Unit vectors assuming *O* is the origin:

$$\vec{P}_{OA} = (x_A - x_O)\hat{i} + (y_A - y_O)\hat{j} + (z_A - z_O)\hat{k}$$
$$= (10-0)\hat{i} + (15-0)\hat{j} + (-100-0)\hat{k}$$
$$= 10\hat{i} + 15\hat{j} - 100\hat{k}$$
$$\hat{u}_{OA} = \frac{\vec{P}_{OA}}{|\vec{P}_{OA}|}$$
$$= \frac{10\hat{i} + 15\hat{j} - 100\hat{k}}{\sqrt{(10)^2 + (15)^2 + (-100)^2}}$$
$$= \left(\frac{1}{10.16}\hat{i} + \frac{1.5}{10.16}\hat{j} - \frac{10}{10.16}\hat{k}\right)$$

$$\vec{P}_{OB} = (x_B - x_O)\hat{i} + (y_B - y_O)\hat{j} + (z_B - z_O)\hat{k}$$
$$= (-10-0)\hat{i} + (20-0)\hat{j} + (-100-0)\hat{k}$$
$$= -10\hat{i} + 20\hat{j} - 100\hat{k}$$
$$\hat{u}_{OB} = \frac{\vec{P}_{OB}}{|\vec{P}_{OB}|}$$
$$= \frac{-10\hat{i} + 20\hat{j} - 100\hat{k}}{\sqrt{(-10)^2 + (20)^2 + (-100)^2}}$$
$$= \left(-\frac{1}{10.25}\hat{i} + \frac{2}{10.25}\hat{j} - \frac{10}{10.25}\hat{k}\right)$$

$$\vec{P}_{OC} = (x_C - x_O)\hat{i} + (y_C - y_O)\hat{j} + (z_C - z_O)\hat{k}$$
$$= (-5-0)\hat{i} + (-20-0)\hat{j} + (-100-0)\hat{k}$$
$$= -5\hat{i} - 20\hat{j} - 100\hat{k}$$
$$\hat{u}_{OC} = \frac{\vec{P}_{OC}}{|\vec{P}_{OC}|}$$
$$= \frac{-5\hat{i} - 20\hat{j} - 100\hat{k}}{\sqrt{(-5)^2 + (-20)^2 + (-100)^2}}$$
$$= \left(-\frac{1}{20.42}\hat{i} - \frac{4}{20.42}\hat{j} - \frac{20}{20.42}\hat{k}\right)$$

Force vectors:

$$\vec{F}_{OA} = F_{OA}\left(\frac{1}{10.16}\hat{i} + \frac{1.5}{10.16}\hat{j} - \frac{10}{10.16}\hat{k}\right)$$
$$\vec{F}_{OB} = F_{OB}\left(-\frac{1}{10.25}\hat{i} + \frac{2}{10.25}\hat{j} - \frac{10}{10.25}\hat{k}\right)$$
$$\vec{F}_{OC} = F_{OC}\left(-\frac{1}{20.42}\hat{i} - \frac{4}{20.42}\hat{j} - \frac{20}{20.42}\hat{k}\right)$$

$\vec{W} = 120\vec{k}$ (The unit vectors along the x, y, and z axes are $\hat{i}$, $\hat{j}$, and $\hat{k}$, respectively)

Now,

$$\vec{F}_{OA} + \vec{F}_{OB} + \vec{F}_{OC} + \vec{W} = 0$$
$$\vec{F}_{OA} + \vec{F}_{OB} + \vec{F}_{OC} = -\vec{W}$$

$$F_{OA}\left(\frac{1}{10.16}\hat{i} + \frac{1.5}{10.16}\hat{j} - \frac{10}{10.16}\hat{k}\right) + F_{OB}\left(-\frac{1}{10.25}\hat{i} + \frac{2}{10.25}\hat{j} - \frac{10}{10.25}\hat{k}\right) +$$
$$F_{OC}\left(-\frac{1}{20.42}\hat{i} - \frac{4}{20.42}\hat{j} - \frac{20}{20.42}\hat{k}\right) = -120\hat{k}$$

$$\left(\frac{F_{OA}}{10.16} - \frac{F_{OB}}{10.25} - \frac{F_{OC}}{20.42}\right)\hat{i} + \left(\frac{1.5F_{OA}}{10.16} + \frac{2F_{OB}}{10.25} - \frac{4F_{OC}}{20.42}\right)\hat{j} -$$
$$\left(\frac{10F_{OA}}{10.16} + \frac{10F_{OB}}{10.25} + \frac{20F_{OC}}{20.42}\right)\hat{k} = -120\hat{k}$$

Equating the $\hat{i}$, $\hat{j}$, and $\hat{k}$ components yields

$$0.0984\ F_{OA} - 0.0976\ F_{OB} - 0.0489\ F_{OC} = 0$$
$$0.148\ F_{OA} + 0.195\ F_{OB} - 0.196\ F_{OC} = 0$$
$$0.984\ F_{OA} + 0.976\ F_{OB} + 0.98\ F_{OC} = 120$$

Solving,

$F_{OA} = 47.2$ kip *(Answer)*
$F_{OB} = 19.8$ kip *(Answer)*
$F_{OC} = 55.3$ kip *(Answer)*

PRACTICE PROBLEMS

PROBLEM 8.1

The magnitude (in kip) and direction of a force can be represented by the *AB* line as shown in Figure 8.13. The coordinates of *A* and *B* are also shown. Determine the vector form of this force. Determine also its magnitude, scalar components along three orthogonal directions, and orientation with respect to the three orthogonal directions.

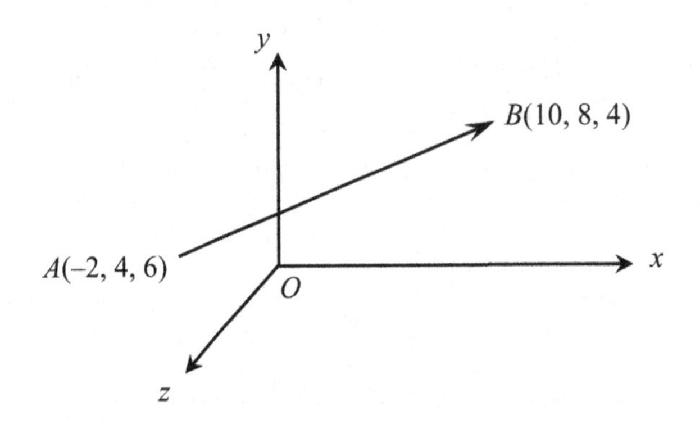

FIGURE 8.13 A force vector for Problem 8.1.

PROBLEM 8.2

The magnitude (in kip) and direction of a force can be represented by the *AB* line as shown in Figure 8.14. The coordinates of *A* and *B* are also shown. Determine the vector form of this force. Determine also its magnitude, scalar components along three orthogonal directions, and orientation with respect to the three orthogonal directions.

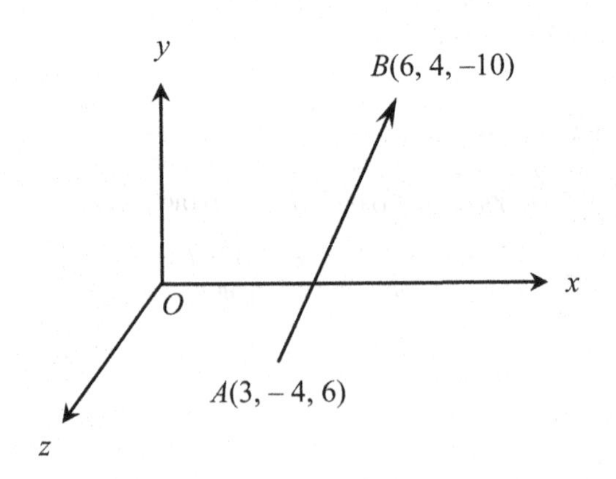

FIGURE 8.14 A force vector for Problem 8.2.

PROBLEM 8.3

The magnitude (in kip) and direction of two forces can be represented by lines *OA* and *OB* as shown in Figure 8.15. The coordinates of *A* and *B* are also shown. Determine the

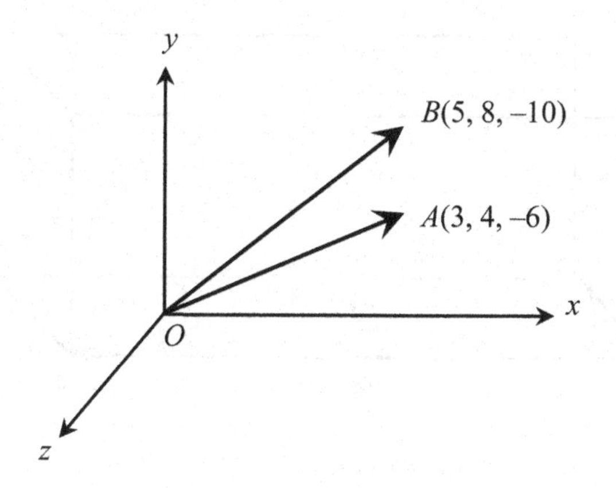

FIGURE 8.15 Two force vectors for Problem 8.3.

magnitude of the resultant, its scalar components along three orthogonal directions, and orientation with respect to the three orthogonal directions.

PROBLEM 8.4

The magnitude (in kip) and direction of three forces can be represented by lines *OA*, *OB* and *OC* as shown in Figure 8.16. The coordinates of *A*, *B* and *C* are also shown. Determine the resultant force vector, its magnitude of the resultant, its scalar components along three orthogonal directions, and orientation with respect to the three orthogonal directions.

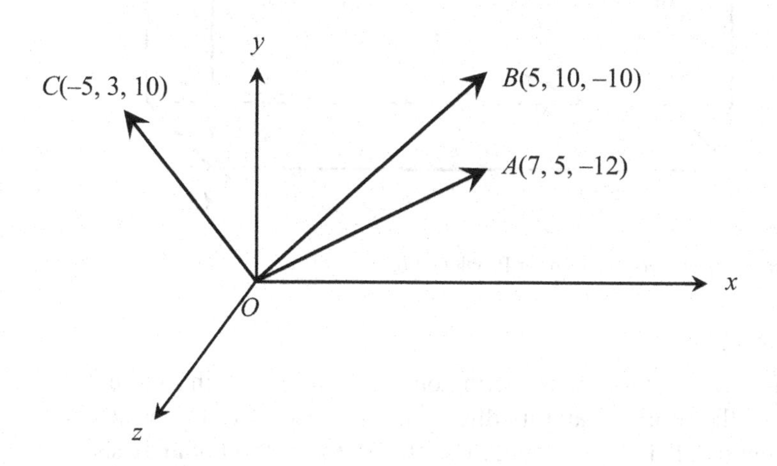

FIGURE 8.16 A concurrent force system for Problem 8.4.

PROBLEM 8.5

Concurrent forces are acting at a corner of a rectangular prism (cuboid) as shown in Figure 8.17. Determine the magnitude of the resultant and its direction.

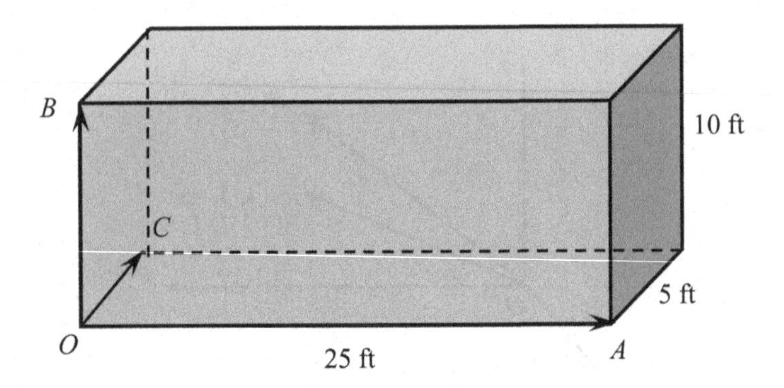

FIGURE 8.17 A space force system for Problem 8.5.

PROBLEM 8.6

In Figure 8.18, *OA*, *OB* and *OC* are three sides of a rectangular prism (cuboid). The coordinates of the origin *O* and the diagonal *D* are (0, 0, 0) and (15, 8, -5), respectively. The magnitude of the force acting along *OD* is 400 N, which is the resultant of these three forces. Determine the force vectors along *OA*, *OB* and *OC*.

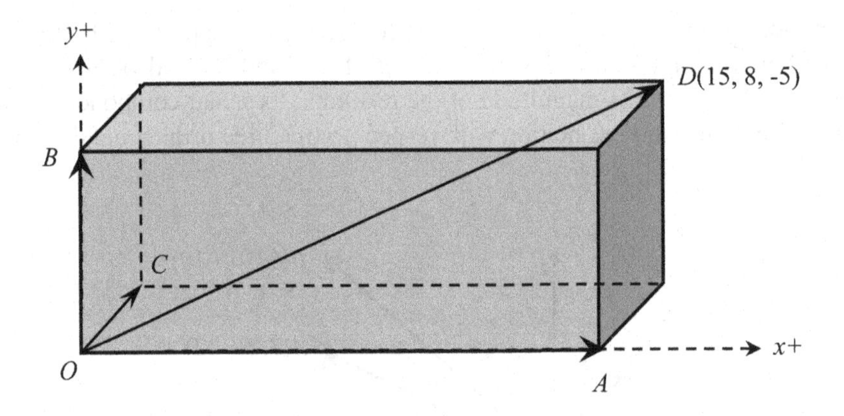

FIGURE 8.18 A space force system for Problem 8.6.

PROBLEM 8.7

Two planar-concurrent forces are acting on a hook as shown in Figure 8.19. Determine the magnitude of the resultant and its direction with respect to the 50-kN force. The 50-kN force is acting parallel to the ground. Use the vector method of analysis.

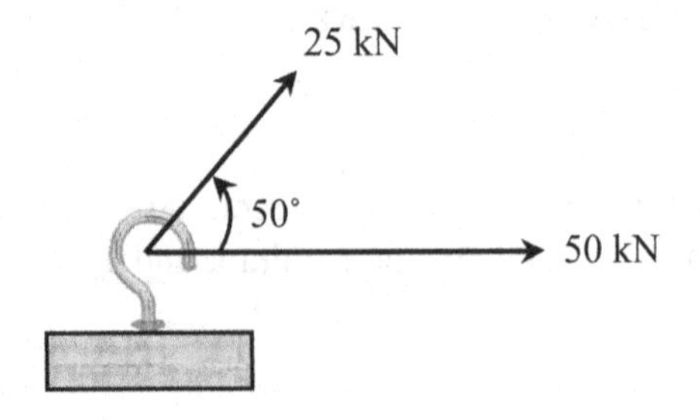

FIGURE 8.19 A concurrent force System for Problem 8.7.

PROBLEM 8.8

Three concurrent forces are acting on a hook in three orthogonal directions as shown in Figure 8.20. Determine the magnitude of the resultant and its direction. Use the vector method of analysis.

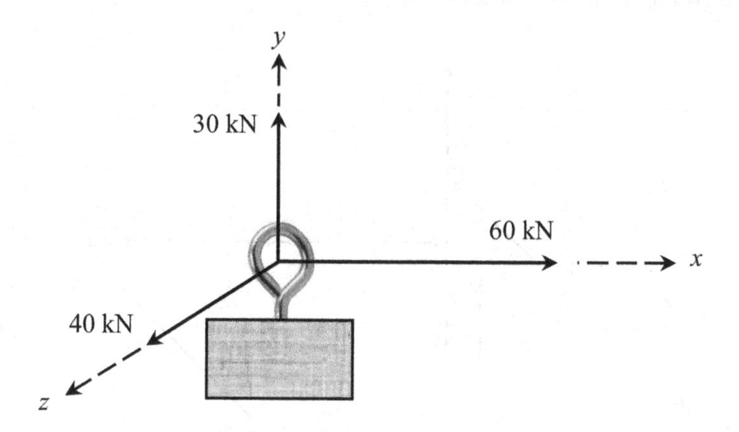

FIGURE 8.20 A concurrent force system for Problem 8.8.

PROBLEM 8.9

A weight of 200 lb is being hung using a cable (AC) and a strut (AB) as shown in Figure 8.21. The supports B and C lie on the xz plane. Determine the forces developed in cable and strut.

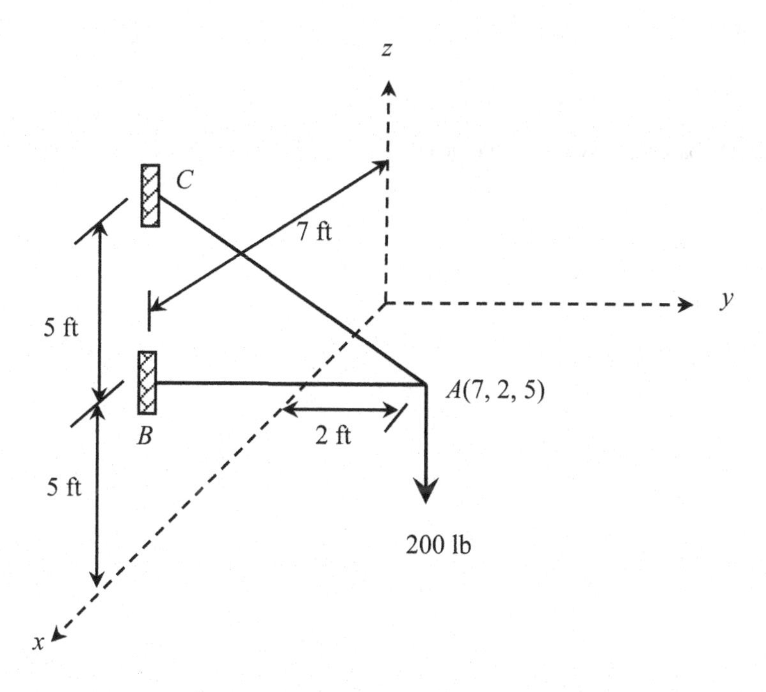

FIGURE 8.21 A concurrent force system for Problem 8.9.

PROBLEM 8.10

A weight of 5 kip is being hung using three cables (*AB*, *AC*, and *AD*) as shown in Figure 8.22. The intersection of the cables lies on the *xy* plane; cable support *D* lies on the *xz* plane; cable support *C* lies on the *yz* plane; and cable support *B* lies on the *z* axis. Determine the tension force developed in each of the three cables.

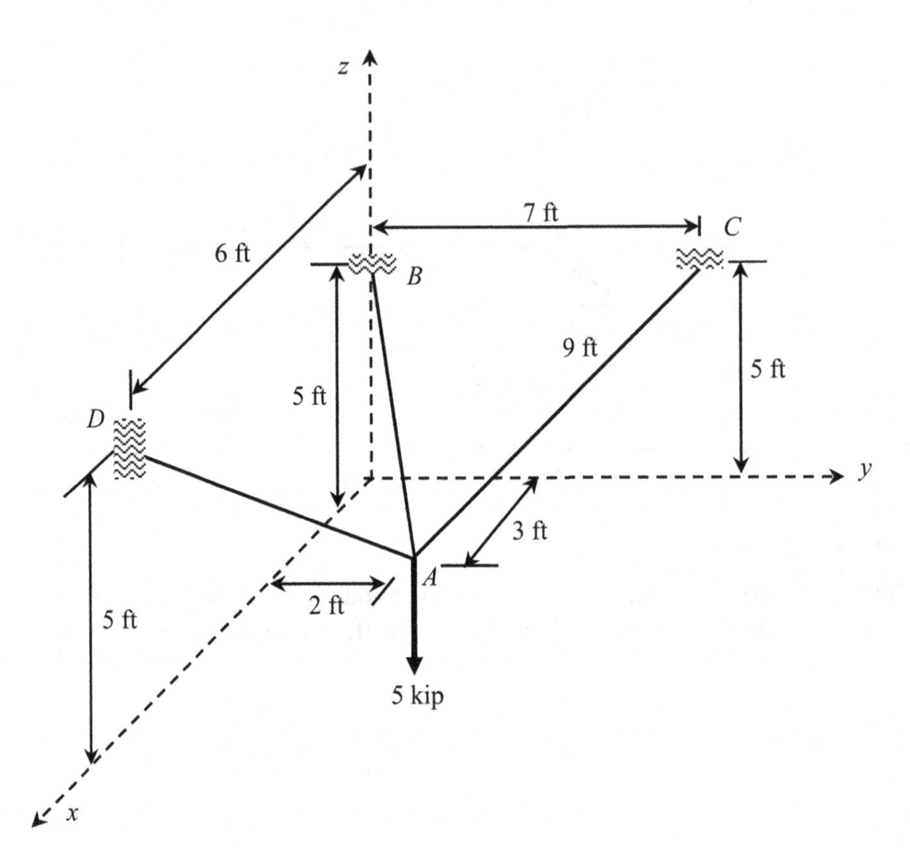

FIGURE 8.22 A space force system for Problem 8.10.

9 Centroids of Area

9.1 CENTROIDS

The centroid (or the center of gravity) of an object is the point where the masses of all grains of the object orients. It is the point about which the area could be balanced if it were supported by that point. The term 'centroid' is derived from the word 'center'. It can be considered the geometric center of an area. For example, the centroid of a circular disk acts at the center; the centroid of a square/rectangle acts at the intersection of the diagonals. In engineering, the determination of the centroid is essential for determining the centroid of distributed loads, and also to determine the moment of inertia for a composite section. Figure 9.1 shows the importance of determining the centroid in civil engineering.

The study of centroid is important in engineering to locate the point of load application in a structure, determining the neutral plane in a structural member that does not deform, analyzing structure in an efficient way, etc. For example, designing supporting structures, carrying large objects, calculating theoverturning potential, foundation design, etc. are impossible without the knowledge of the centroid. The procedures to determine the centroids of regular sections, composite sections and irregular sections are discussed in this chapter. Composite sections (rectangular, circular, etc.) can be divided into several regular sections whose centroids are known. For irregular-shaped sections, which cannot be divided into regular sections, calculus is the only option. However, most of the engineering structure can be divided into pieces of regular sections.

9.2 CENTROID OF REGULAR AREAS

The centroids of some common regular areas are listed in Figure 9.2. These centroids can be determined using calculus. Some examples are given later in this chapter.

9.3 CENTROID OF COMPOSITE AREAS

Centroids of common regular sections such as square, rectangle, triangle, circle, semicircle, etc. are readily known from Figure 9.2. For composite section, the centroid is to be calculated using the composite section rule. Composite areas are composed of several regular areas whose centroids are known. At the beginning, the composite areas are divided into several regular sections. Let us assume, a composite area has been divided into n numbers of regular areas such as $A_1, A_2,...A_n$ and the x-coordinates are $x_1, x_2, ... x_n$, respectively. The number of divisions should be as small as possible to make the computation easy. Then, the x-coordinate of the centroid of the composite section can be written as:

$$\bar{x} = \frac{\sum x\Delta A}{\sum \Delta A} = \frac{x_1A_1 + x_2A_2 + \ldots + A_n}{A_1 + A_2 + \ldots + A_n}$$

FIGURE 9.1 Maintaining the center of gravity. *(Courtesy of Civil Engineering Discoveries)*

Similarly, the y-coordinate of the centroid of the composite section can be written as:

$$\bar{y} = \frac{\sum y\Delta A}{\sum \Delta A} = \frac{y_1A_1 + y_2A_2 + \ldots + A_n}{A_1 + A_2 + \ldots + A_n}$$

If the composite section has an axis of symmetry, the centroid lies on the axis of symmetry. If the axis of symmetry is horizontal, $\bar{y}$ is on the axis of symmetry and $\bar{x}$ is to be calculated. If the axis of symmetry is vertical, $\bar{x}$ lies on the axis of symmetry and $\bar{y}$ is to be calculated. If an area has two axes of symmetry, the centroid is located at the intersection of these two axes of symmetry. Figure 9.3 shows a few examples of sections having two axes of symmetry. The centroid is denoted as C. No calculation is necessary to locate the centroid of these sections at it can be located visually.

If two axes of symmetry do not exist, the composite area can be divided into a number of regular sections and the above equations are used. Let us work on a few examples to

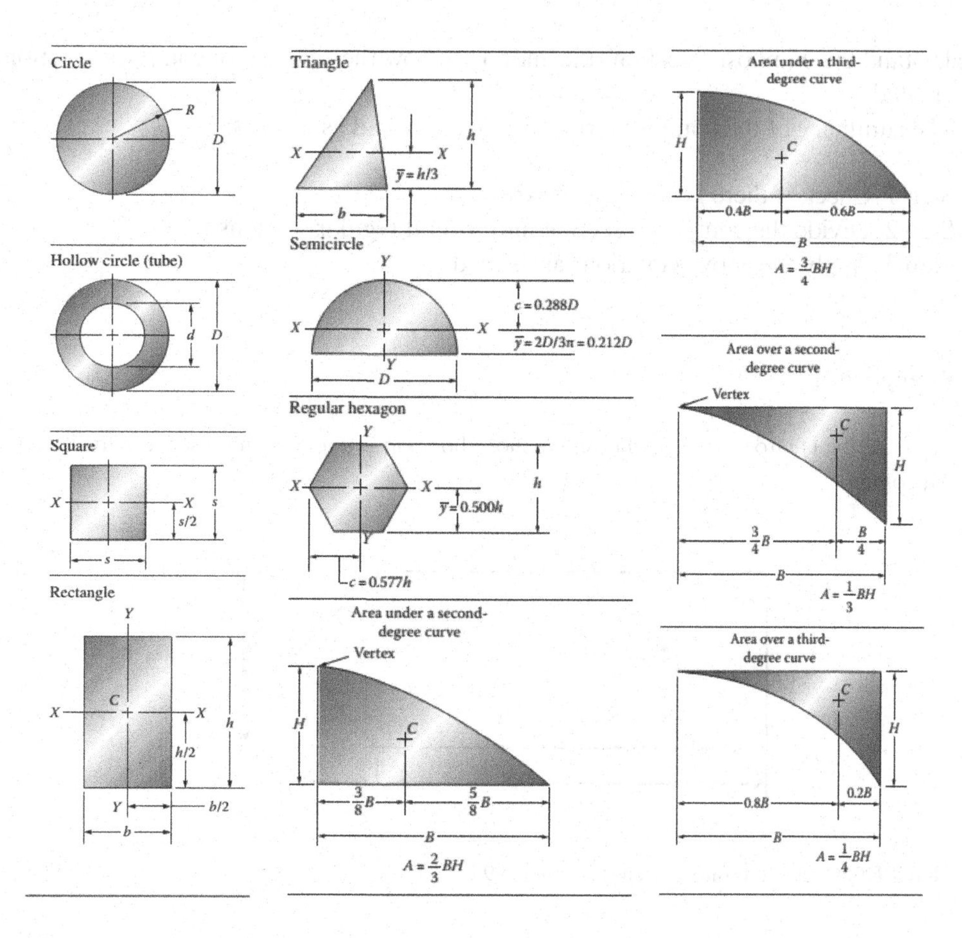

FIGURE 9.2 Centroids of some common areas. *(From Mott, R. L. and Untener, J. A. (2017). Applied Strength of Materials, 6th edition, CRC Press, Boca Raton, FL.)*

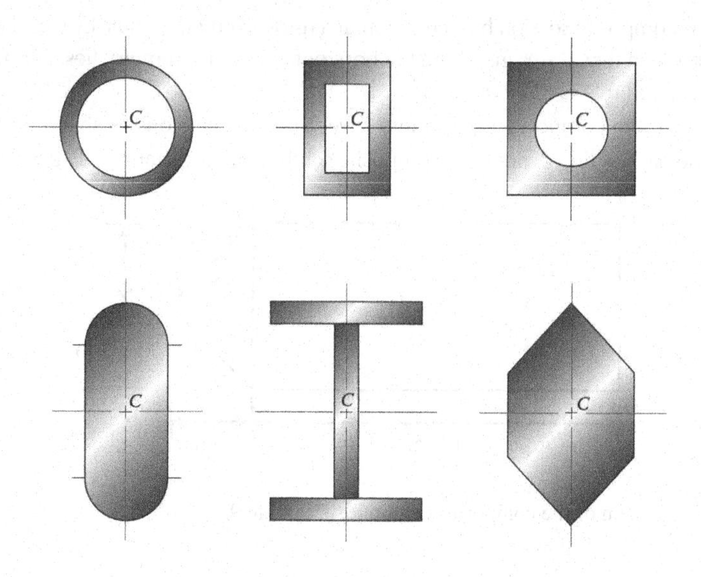

FIGURE 9.3 Centroids of some common areas. *(From Mott, R. L. and Untener, J. A. (2017). Applied Strength of Materials, 6th edition, CRC Press, Boca Raton, FL, figure 6-3.)*

understand the composite section rule and to see how the centroid of composite sections is calculated.

The summary of the composite rule can be described as follows:

Step 1. Check if there is any axis of symmetry
Step 2. Divide the composite section into several regular sections
Step 3. Apply the above equations as needed

Example 9.1:

Calculate the centroid of the composite section shown in Figure 9.4 with respect to the bottom-left corner.

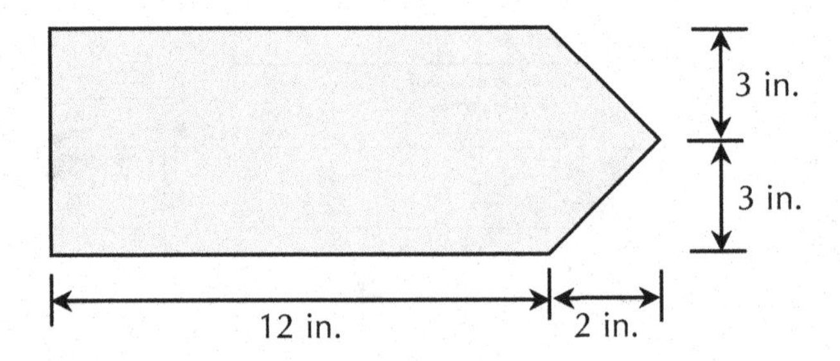

FIGURE 9.4 A composite section for Example 9.1.

SOLUTION

Step 1. This composite section has a horizontal axis of symmetry. Thus, by visual inspection, it can be seen that $\bar{y} = 3$ in. where the horizontal axis of symmetry lies. For $\bar{x}$, the composite section rule is to be applied.

Step 2. Divide the composite section into several regular sections as shown in Figure 9.5. Here the dashed line divides the composite section into a rectangle of 12 in. by 6 in. and

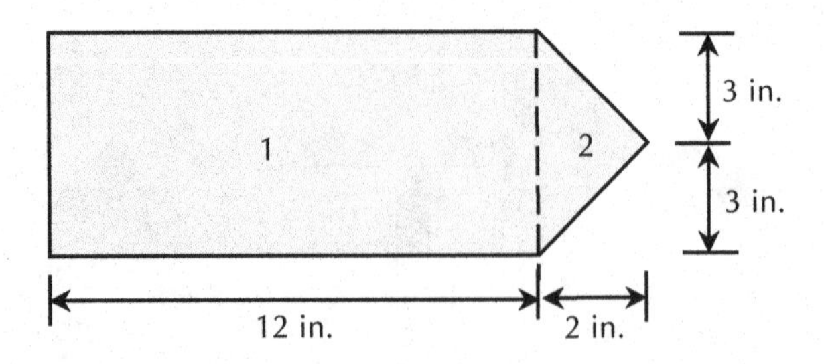

FIGURE 9.5 Division of the composite section for Example 9.1.

a triangle of 6 in. base and 2 in. height. Note that there may be different options to divide the shape. Let the rectangle be Section 1 and the triangle be Section 2.

Step 3. Apply the equation to calculate $\bar{x}$

x_1 = x-coordinate of the centroid of section 1 = 12 in./2 = 6 in.
A_1 = Area of section 1 = 12 x 6 in.^{2}

x_2 = x-coordinate of the centroid of section 2 = 12 in. + (1/3)(2) in.
A_2 = Area of section 2 = ½ (6 x 2) in.^{2}

$$\bar{x} = \frac{\sum x\Delta A}{\sum \Delta A}$$

$$\bar{x} = \frac{x_1 A_1 + x_2 A_2}{A_1 + A_2}$$

$$\bar{x} = \frac{(6\text{ in.})(12\text{ in.}\times 6\text{ in.}) + \left(12\text{ in.} + \frac{1}{3}(2\text{ in.})\right)\left(\frac{1}{2}\times 6\times 2\text{ in.}^2\right)}{12\times 6\text{ in.}^2 + \frac{1}{2}(6)(2)\text{ in.}^2}$$

$$\bar{x} = 6.51\text{ in.}$$

Answer:

(6.5, 3.0) in.

Note that there are many different ways a composite section can be divided. Whatever division you use, the result must be the same. For example, the composite section discussed in Example 9.1 can be extended into a rectangular section and two triangular sections may be deducted from the extended section as shown in Figure 9.6. After the calculation, you must obtain the same answer of $\bar{x} = 6.51$ in. from the bottom-left corner.

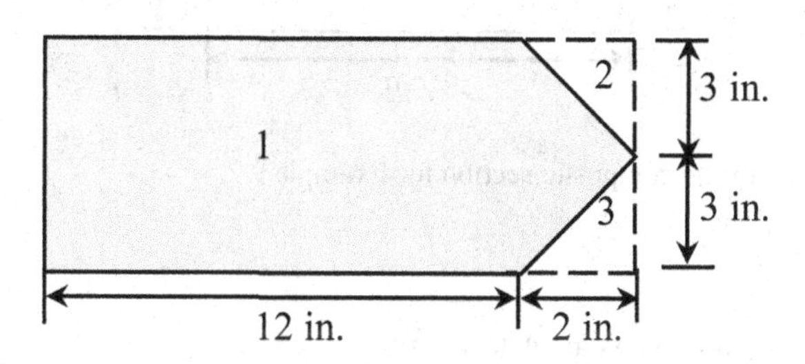

FIGURE 9.6 Alternative division of the composite section for Example 9.1.

Example 9.2:

Calculate the centroid of the section shown in Figure 9.7 with respect to the bottom-left corner.

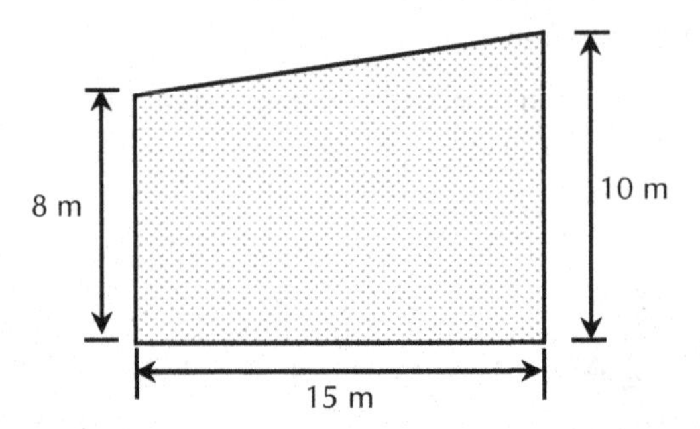

FIGURE 9.7 A composite section for Example 9.2.

SOLUTION

Step 1. This composite section has no axis of symmetry. Therefore, we will have to use the composite section rule to find out both $\bar{x}$ and $\bar{y}$.

Step 2. Divide the composite section into several regular sections as shown in Figure 9.8. Here the dashed line divides the composite section into a rectangle of 15 m by 8 m and a right-angled triangle of 15 m base and 2 m height. Let the rectangle be Section 1 and the triangle be Section 2.

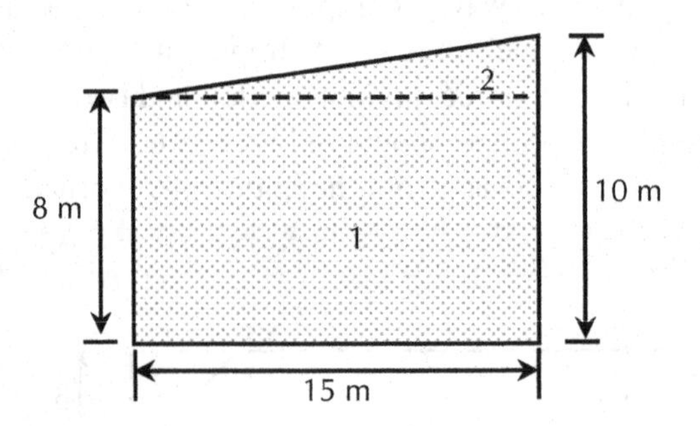

FIGURE 9.8 Division of the composite section for Example 9.2.

Step 3. Apply the equations to calculate $(\bar{x}, \bar{y})$

x_1 = x-coordinate of the centroid of section 1 = 15 m/2 = 7.5 m
y_1 = y-coordinate of the centroid of section 1 = 8 m/2 = 4 m
A_1 = Area of section 1 = 8 x 15 m^2

x_2 = x-coordinate of the centroid of section 2 = (2/3)(15) m = 10 m
y_2 = y-coordinate of the centroid of section 2 = 8 m + 2/3 (2) = 8 + 0.67 m

A_2 = Area of section 2 = ½ (15 x 2) m²

$$\bar{x} = \frac{x_1A_1 + x_2A_2}{A_1 + A_2}$$

$$\bar{x} = \frac{7.5\,\text{m}(8\,\text{m}\times 15\,\text{m}) + 10\,\text{m}\left(\frac{1}{2}\times 15\,\text{m}\times 2\,\text{m}\right)}{(8\,\text{m}\times 15\,\text{m}) + \left(\frac{1}{2}\times 15\,\text{m}\times 2\,\text{m}\right)}$$

$$\bar{x} = 7.78\,\text{m}$$

$$\bar{y} = \frac{y_1A_1 + y_2A_2}{A_1 + A_2}$$

$$\bar{y} = \frac{4\,\text{m}(8\,\text{m}\times 15\,\text{m}) + (8\,\text{m} + 0.67\,\text{m})\left(\frac{1}{2}\times 15\,\text{m}\times 2\,\text{m}\right)}{(8\,\text{m}\times 15\,\text{m}) + \left(\frac{1}{2}\times 15\,\text{m}\times 2\,\text{m}\right)}$$

$$\bar{y} = 4.52\,\text{m}$$

Answer:

(7.8, 4.5) m

Example 9.3:

Calculate the centroid of the T-shape as shown in Figure 9.9 with respect to the horizontal reference line. The units are in inches.

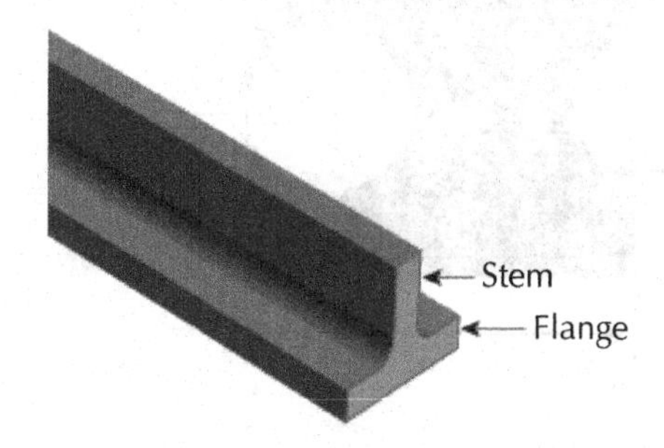

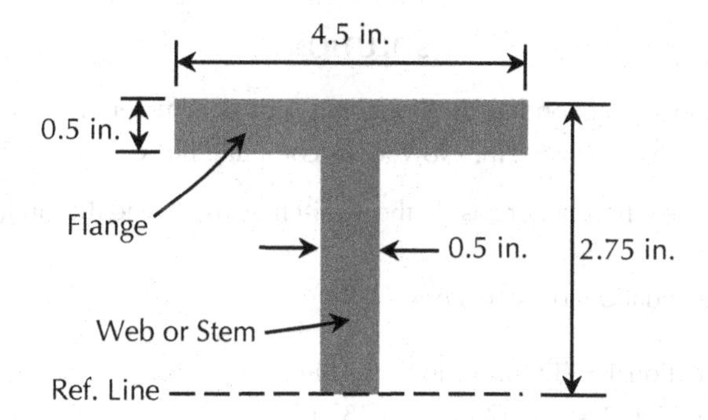

FIGURE 9.9 A T-section for Example 9.3.

SOLUTION

Step 1. This composite section has a vertical axis of symmetry. Therefore, $\bar{x}$ can be located visually. However, the problem is asking you to find out the $\bar{y}$ only.

Step 2. Divide the composite section into regular sections. Let the flange be '1' and the web be '2'.

Step 3. Apply the equation to calculate $\bar{y}$

y_1 = y-coordinate of the centroid of section 1 = 2.25 + 0.5/2 in. = 2.5 in.

y_2 = y-coordinate of the centroid of section 2 = 2.25 in./2 = 1.125 in.

A_1 = Area of section 1 = (4.5)(0.5) in.^{2} = 2.25 in.^{2}

A_2 = Area of section 2 = (0.5)(2.25) in.^{2} = 1.125 in.^{2}

$$\bar{y} = \frac{y_1 A_1 + y_2 A_2}{A}$$

$$\bar{y} = \frac{(2.5\text{ in.})(2.25\text{ in.}^2) + (1.125\text{ in.})(1.125\text{ in.}^2)}{2.25\text{ in.}^2 + 1.125\text{ in.}^2}$$

$$\bar{y} = 2.04\text{ in.}$$

Answer:

2.0 in.

Example 9.4:

Calculate the centroid of the square with a semi-circular, and a circular hollow as shown in Figure 9.10 with respect to the bottom-left corner.

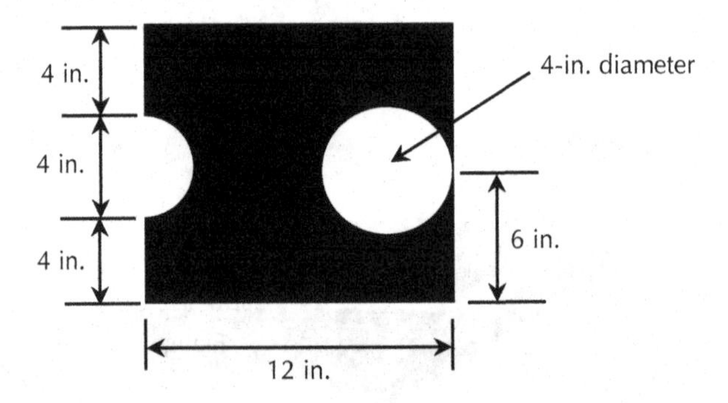

FIGURE 9.10 A composite section for Example 9.4.

SOLUTION

Step 1. This composite section has the horizontal axis of symmetry. Therefore, the $\bar{y}$ can be located visually as $\bar{y} = 6.0$ in. Now let us compute the $\bar{x}$.

Step 2. Consider, the whole sqaure as '1', the semi-circle as '2', and the circle as '3' as shown in Figure 9.11.

Step 3. Apply the equation to calculate $\bar{x}$

A_1 = Area of section 1 = (12 in.)(12 in.) = 144 in.^{2}

A_2 = Area of section 2 = ½ (π)(2 in.)^2 = 6.28 in.^{2}

A_3 = Area of section 3 = (π)(2 in.)^2 = 12.56 in.^{2}

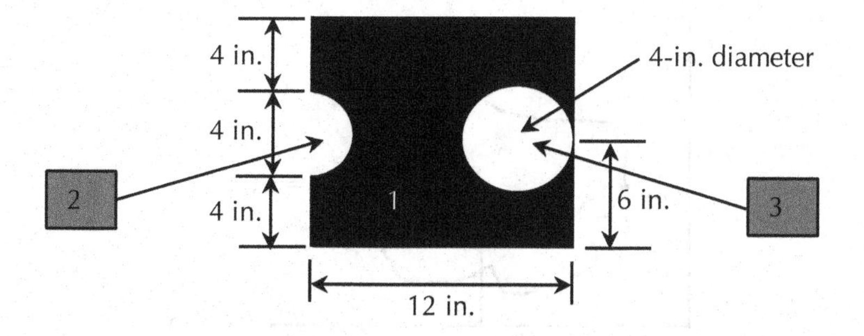

FIGURE 9.11 Division of the composite section for Example 9.4.

x_1 = x-coordinate of the centroid of section 1= 6.0 in.

x_2 = x-coordinate of the centroid of section 2 = $\frac{2D}{3\pi} = \frac{2(4.0\text{ in.})}{3\pi} = 0.85\text{ in.}$

x_3 = x-coordinate of the centroid of section 3 = 12 in. – 2 in. = 10 in.

$$\bar{x} = \frac{\sum x\Delta A}{\sum \Delta A}$$

$$\bar{x} = \frac{x_1A_1 - x_2A_2 - x_3A_3}{A_1 - A_2 - A_3}$$

$$\bar{x} = \frac{(6.0\text{ in.})(144\text{ in.}^2) - (0.85\text{ in.})(6.28\text{ in.}^2) - (10\text{ in.})(12.56\text{ in.}^2)}{144\text{ in.}^2 - 6.28\text{ in.}^2 - 12.56\text{ in.}^2}$$

$$\bar{x} = 5.86\text{ in.}$$

Let us check $\bar{y}$ to be sure.

y_1 = y-coordinate of the centroid of section 1 = 6.0 in.
y_2 = y-coordinate of the centroid of section 2 = 6.0 in.
y_3 = y-coordinate of the centroid of section 3 = 6.0 in.

$$\bar{y} = \frac{\sum y\Delta A}{\sum \Delta A}$$

$$\bar{y} = \frac{y_1A_1 - y_2A_2 - y_3A_3}{A_1 - A_2 - A_3}$$

$$\bar{y} = \frac{(6.0\text{ in.})(144\text{ in.}^2) - (6.0\text{ in.})(6.28\text{ in.}^2) - (6.0\text{ in.})(12.56\text{ in.}^2)}{144\text{ in.}^2 - 6.28\text{ in.}^2 - 12.56\text{ in.}^2}$$

$$\bar{y} = 6.0\text{ in.}$$

Answer:

(5.9, 6.0) in.

9.4 CENTROID OF IRREGULAR AREAS

The procedure to calculate the centroid of composite areas that can be divided into several regular sections such as squares, rectangles, triangles, circles, etc. are discussed in the previous section. However, the methods described in those sections are not good to determine

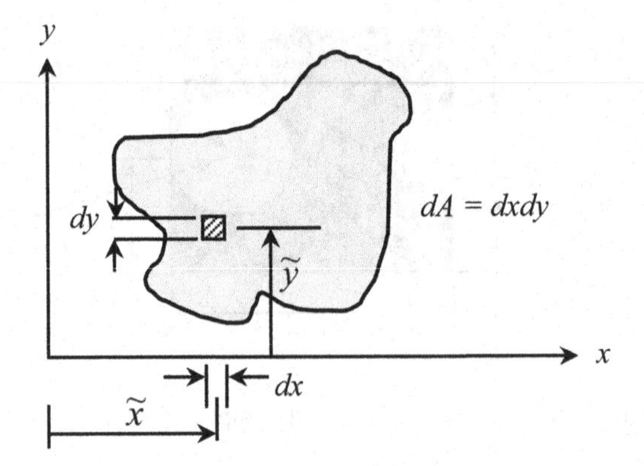

FIGURE 9.12 Definition of centroid.

the centroid of irregular sections. This section describes how the centroid and moment of inertia can be determined for irregular sections using calculus.

Let us assume any arbitrary shape shown in Figure 9.12.

An infinitesimal small area, dA, inside the object is shown. Let $(\tilde{x}, \tilde{y})$ be the centroid of the small area dA. Then, the centroid $(\overline{x}, \overline{y})$ of the whole section can be determined as:

$$\overline{x} = \frac{\int \tilde{x} dA}{\int dA} = \frac{\int \tilde{x} dA}{A}$$

$$\overline{y} = \frac{\int \tilde{y} dA}{\int dA} = \frac{\int \tilde{y} dA}{A}$$

Let us see some examples of determining the centroid using calculus.

9.4.1 Centroid of Rectangle

Let us assume a rectangle with base b and height h on the xy plane as shown in Figure 9.13. An infinitesimal segment dx is assumed at x-distance from the origin; the ordinate is h at that infinitesimal segment.

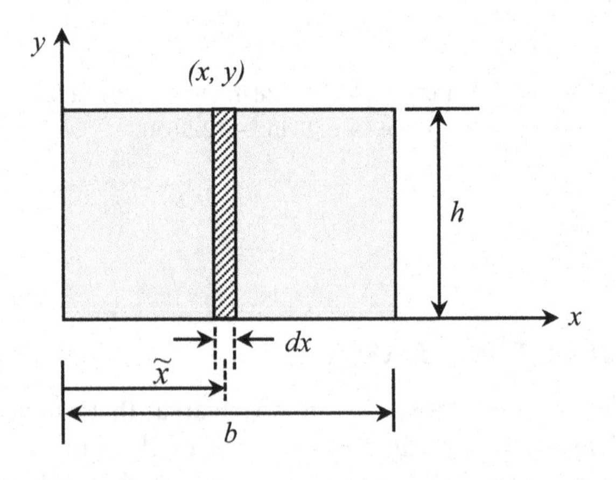

FIGURE 9.13 Determining the centroid of a rectangle.

The infinitesimal segment can be assumed to be a rectangle and its area can be written as hdx. Therefore, the x-coordinate of the centroid can be found as:

$$\bar{x} = \frac{\int \tilde{x} dA}{\int dA}$$

$$\bar{x} = \frac{\int (x)(hdx)}{\int hdx}$$

$$\bar{x} = \frac{h}{h}\frac{\int xdx}{\int dx}$$

$$\bar{x} = \frac{\left[\frac{x^2}{2}\right]_{x=0}^{b}}{[x]_{x=0}^{b}}$$

$$\bar{x} = \frac{\left[\frac{b^2}{2} - \frac{0^2}{2}\right]}{[b-0]}$$

$$\bar{x} = \frac{b^2}{2b}$$

$$\bar{x} = \frac{b}{2}$$

Similarly, $\bar{y} = \frac{\int \tilde{y} dA}{\int dA} = \frac{h}{2}$

9.4.2 Centroid of Triangle

Let us assume a triangle with base b and height h in xy plane as shown in Figure 9.14. An infinitesimal segment dx is assumed at x-distance from the origin; the ordinate is y at that infinitesimal segment.

From the similar angle triangle rules, $\frac{x}{y} = \frac{b}{h}$

Therefore, $x = \frac{by}{h}$

The infinitesimal segment can be assumed to be rectangle and its area can be written as ydx. Therefore, the x-coordinate of the centroid can be found as:

$$\bar{x} = \frac{\int \tilde{x} dA}{\int dA}$$

$$\bar{x} = \frac{\int \left(\frac{by}{h}\right)(xdy)}{\int xdy}$$

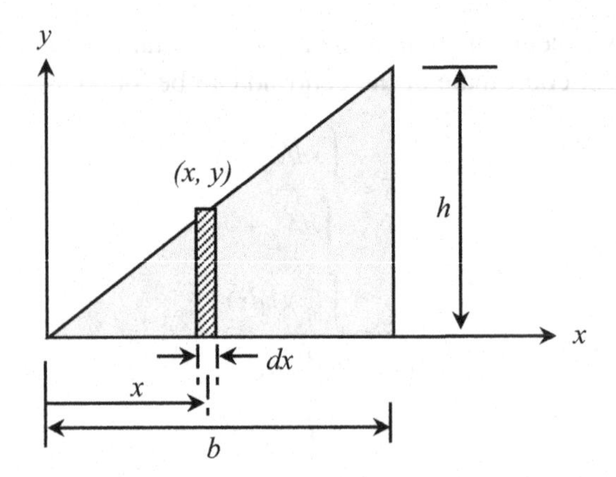

FIGURE 9.14 Determining the centroid of a triangle.

$$\bar{x} = \frac{\int \left(\frac{by}{h}\right)\left(\frac{by}{h}\right) dy}{\int \left(\frac{by}{h}\right) dy}$$

$$\bar{x} = \frac{b}{h} \frac{\int y^2 dy}{\int y dy}$$

$$\bar{x} = \frac{b}{h} \left(\frac{\left[\frac{y^3}{3}\right]_{y=0}^{h}}{\left[\frac{y^2}{2}\right]_{y=0}^{h}} \right)$$

$$\bar{x} = \frac{b}{h} \left(\frac{h^3}{3} x \frac{2}{h^2} \right)$$

$$\bar{x} = \frac{2}{3} b$$

Similarly, $\bar{y} = \frac{\int \tilde{y} dA}{\int dA} = \frac{2}{3} h$

9.4.3 Centroid of Quarter Circle

Let us consider a quarter circle with radius R as shown in Figure 9.15. A small element of the angle of $d\theta$ in the quarter circle is assumed which is located at an angle of θ from the x axis. The small element takes the shape of a triangle as the bounded angle $d\theta$ is very small. The arc of the small element can be determined as follows:

$$S = R\, d\theta$$

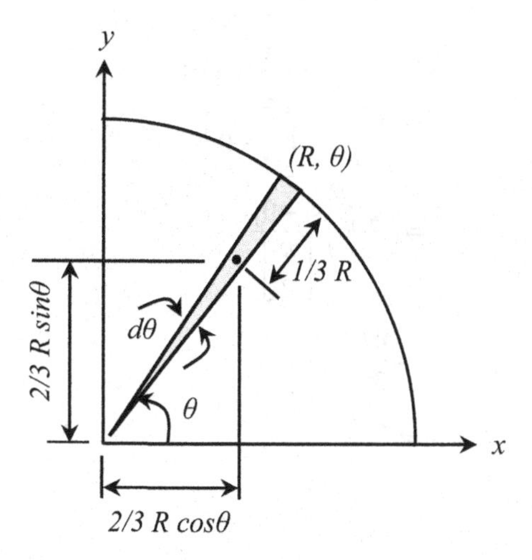

FIGURE 9.15 Determining the centroid of a quarter circle.

As the subtended angle $d\theta$ is very small, the slice can be considered a triangle. The area of the small triangular element can be calculated as follows:

$$dA = \frac{1}{2}(R)(R\,d\theta) = \frac{R^2}{2}d\theta$$

The centroid of the small triangular element can be written as:

$$\tilde{x} = \frac{2}{3}R\cos\theta$$

$$\tilde{y} = \frac{2}{3}R\sin\theta$$

$$\overline{x} = \frac{\int \tilde{x}\,dA}{\int dA} = \frac{\int_0^{\pi/2}\left(\frac{2}{3}R\cos\theta\right)\left(\frac{R^2}{2}\right)d\theta}{\int_0^{\pi/2}\left(\frac{R^2}{2}\right)d\theta}$$

$$= \frac{\left(\frac{2}{3}R\right)\left(\frac{R^2}{2}\right)\int_0^{\pi/2}\cos\theta d\theta}{\left(\frac{R^2}{2}\right)\int_0^{\pi/2}d\theta}$$

$$= \left(\frac{2}{3}R\right)\frac{[\sin\theta]_0^{\pi/2}}{[\theta]_0^{\pi/2}}$$

$$=\left(\frac{2}{3}R\right)\frac{\left[\sin\frac{\pi}{2}-\sin 0\right]}{\left[\frac{\pi}{2}-0\right]}$$

$$=\left(\frac{2}{3}R\right)\frac{[1-0]}{\left[\frac{\pi}{2}\right]}$$

$$=\frac{4R}{3\pi}$$

$$=\frac{2D}{3\pi}$$

$$\bar{y}=\frac{\int \tilde{y}dA}{\int dA}$$

$$=\frac{\int_0^{\pi/2}\left(\frac{2}{3}R\sin\theta\right)\left(\frac{R^2}{2}\right)d\theta}{\int_0^{\pi/2}\left(\frac{R^2}{2}\right)d\theta}$$

$$=\frac{\left(\frac{2}{3}R\right)\left(\frac{R^2}{2}\right)\int_0^{\pi/2}\sin\theta d\theta}{\left(\frac{R^2}{2}\right)\int_0^{\pi/2}d\theta}$$

$$=\left(\frac{2}{3}R\right)\frac{[-\cos\theta]_0^{\pi/2}}{[\theta]_0^{\pi/2}}$$

$$=\left(\frac{2}{3}R\right)\frac{\left[-\cos\frac{\pi}{2}-(-\cos 0)\right]}{\left[\frac{\pi}{2}-0\right]}$$

$$=\left(\frac{2}{3}R\right)\frac{[-0+1]}{\left[\frac{\pi}{2}\right]}$$

$$=\frac{4R}{3\pi}$$

$$=\frac{2D}{3\pi}$$

9.4.4 Centroid of an Irregular Section

Let us determine the centroid of a section whose shape can be represented by $y = x^2 + x$ as shown in Figure 9.16. The boundary of the section is $x = 4$ in.

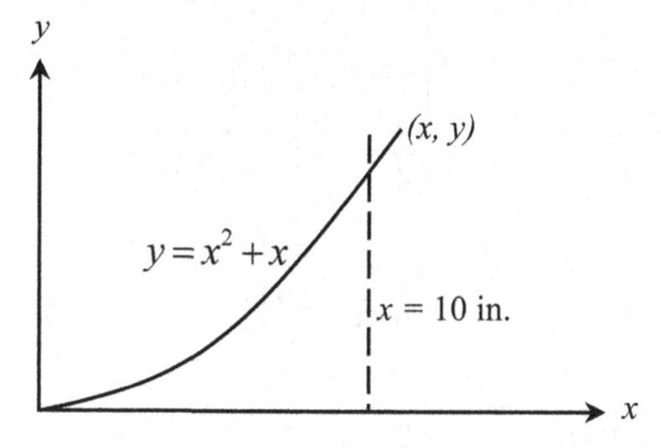

FIGURE 9.16 An irregular section.

Assume an infinitesimal small area dA as shown in Figure 9.17.

$$\bar{x} = \frac{\int \tilde{x}\,dA}{\int dA} = \frac{\int x(ydx)}{\int ydx}$$

$$= \frac{\int_0^4 x(x^2 + x)dx}{\int_0^4 (x^2 + x)dx}$$

$$= \frac{\int_0^4 (x^3 + x^2)dx}{\int_0^4 (x^2 + x)dx}$$

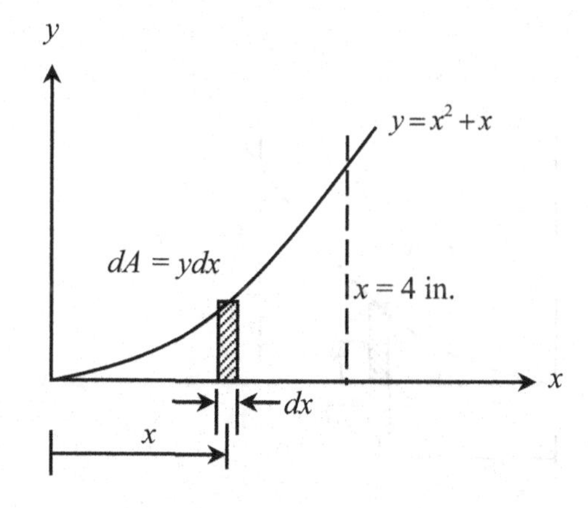

FIGURE 9.17 Determining the x-coordinate of the centroid of an irregular section.

$$= \frac{\left[\frac{x^4}{4} + \frac{x^3}{3}\right]_0^4}{\left[\frac{x^3}{3} + \frac{x^2}{2}\right]_0^4}$$

$$= \frac{\left[\frac{4^4}{4} + \frac{4^3}{3}\right] - \left[\frac{0^4}{4} + \frac{0^3}{3}\right]}{\left[\frac{4^3}{3} + \frac{4^2}{2}\right] - \left[\frac{0^3}{3} + \frac{0^2}{2}\right]}$$

$$= \frac{85.33}{29.33} = 2.9 \text{ in.}$$

From Figure 9.18:

$$\bar{y} = \frac{\int \tilde{y} dA}{\int dA} = \frac{\int \frac{y}{2}(ydx)}{29.33}$$

$$= \frac{1}{2} \frac{\int_0^4 (x^2 + x)^2 dx}{29.33}$$

$$= \frac{\int_0^4 (x^4 + 2x^3 + x^2) dx}{58.66}$$

$$= \frac{\left[\frac{x^5}{5} + \frac{2x^4}{4} + \frac{x^3}{3}\right]_0^4}{58.66}$$

$$= \frac{\left[\frac{(4)^5}{5} + \frac{2(4)^4}{4} + \frac{(4)^3}{3}\right] - \left[\frac{(0)^5}{5} + \frac{2(0)^4}{4} + \frac{(0)^3}{3}\right]}{58.66}$$

$$= \frac{354.133}{58.66}$$

$$= 6 \text{ in.}$$

Answer: (2.9 in., 6.0 in.)

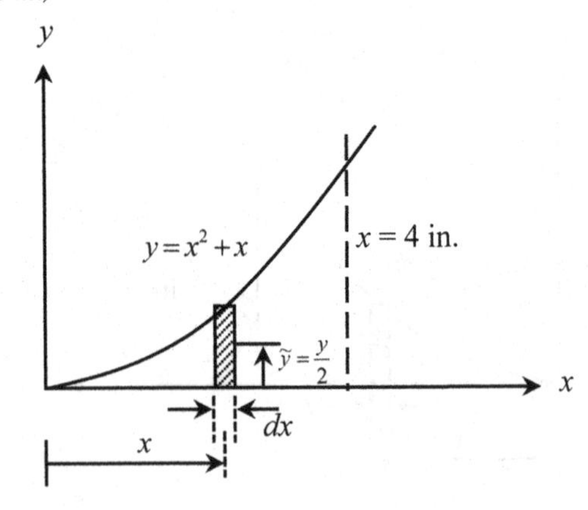

FIGURE 9.18 Determining the y-coordinate of the centroid of an irregular section.

PRACTICE PROBLEMS

PROBLEM 9.1

Determine the centroid of the angle section shown in Figure 9.19 with respect to the bottom-left corner.

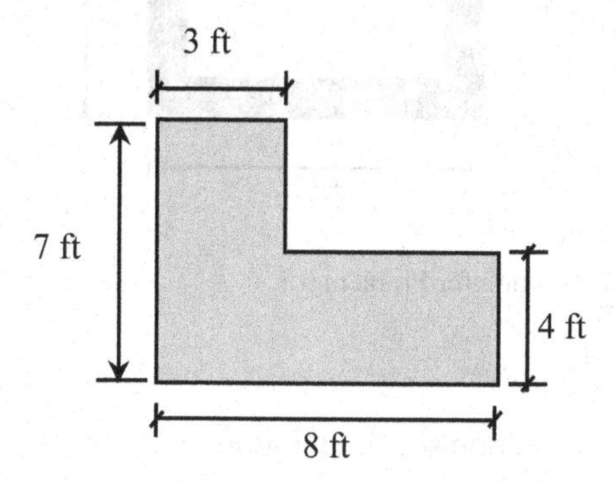

FIGURE 9.19 An L-section for Problem 9.1.

PROBLEM 9.2

Determine the centroid of the composite section shown in Figure 9.20 with respect to the bottom-left corner.

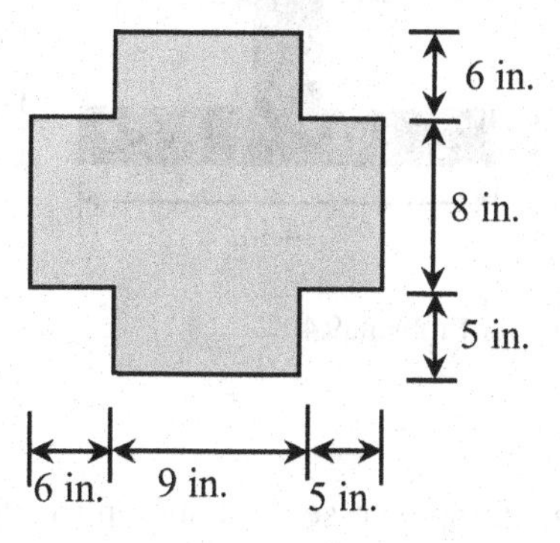

FIGURE 9.20 A composite section for Problem 9.2.

PROBLEM 9.3

Determine the centroid of the hollow-square section shown in Figure 9.21 with respect to the bottom-left corner.

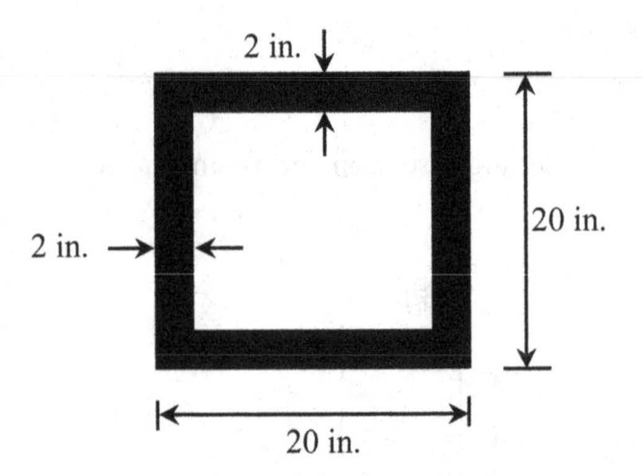

FIGURE 9.21 A hollow-square section for Problem 9.3.

PROBLEM 9.4

Determine the centroid of the section shown in Figure 9.22 with respect to the bottom-left corner. The section is symmetrical about the vertical axis passing through the centroid as shown.

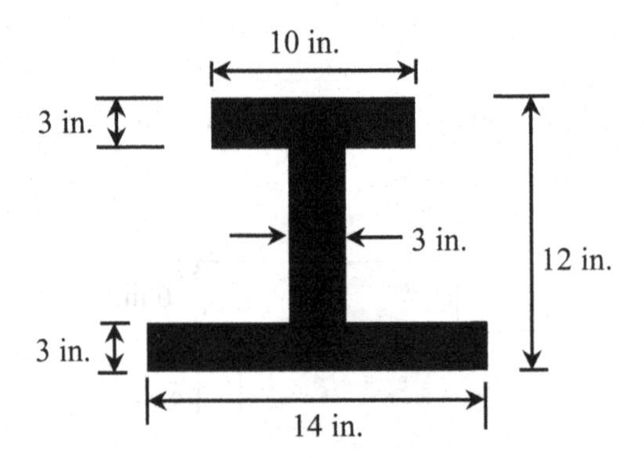

FIGURE 9.22 A composite section for Problem 9.4.

PROBLEM 9.5

Determine the centroid of the composite section shown in Figure 9.23 with respect to the bottom-left corner. The diameter of the circular hollow is 8 in. as shown.

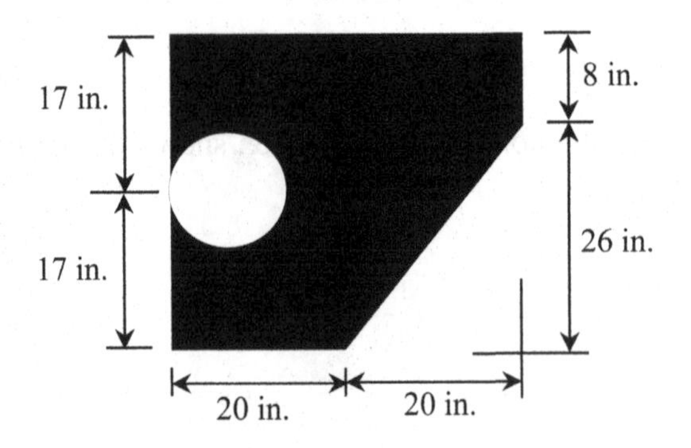

FIGURE 9.23 A composite section for Problem 9.5.

PROBLEM 9.6

Determine the centroid of the composite section shown in Figure 9.24 with respect to the bottom-left corner.

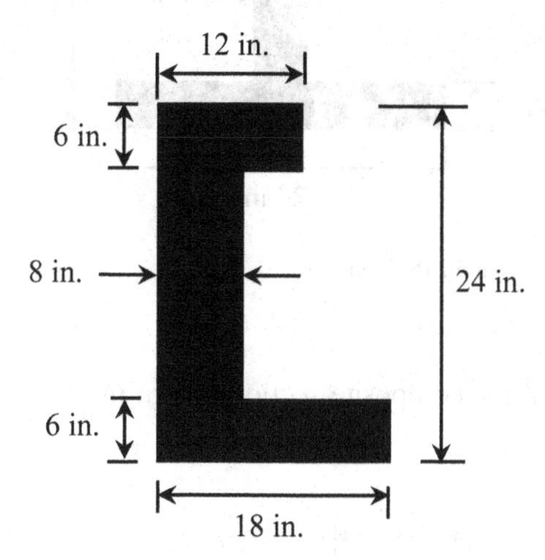

FIGURE 9.24 A composite section for Problem 9.6.

PROBLEM 9.7

Determine the centroid of the T-section shown in Figure 9.25 with respect to the bottom-left corner. The section is symmetrical about the vertical axis passing through the centroid.

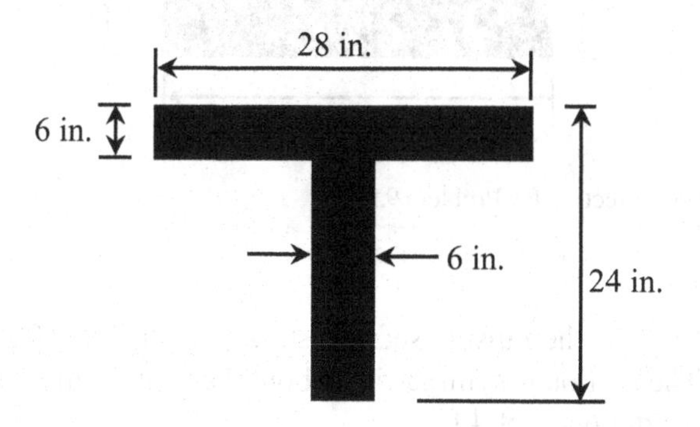

FIGURE 9.25 A T-section for Problem 9.7.

PROBLEM 9.8

Determine the centroid of the invert T-section shown in Figure 9.26 with respect to the bottom-left corner. The section is symmetrical about the vertical axis passing through the centroid.

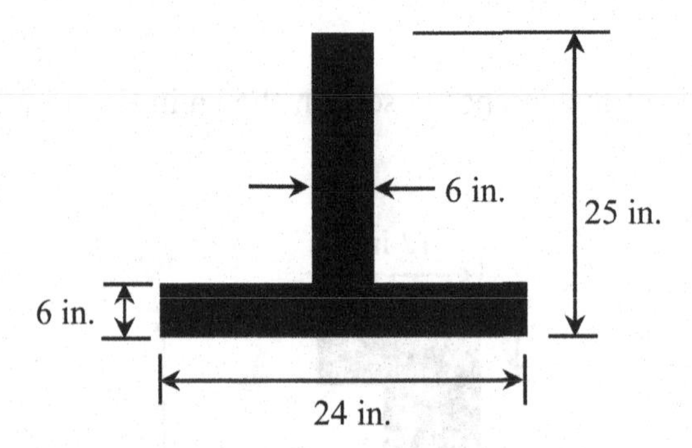

FIGURE 9.26 An invert T-section for Problem 9.8.

PROBLEM 9.9

Determine the centroid of the composite section shown in Figure 9.27 with respect to the bottom-left corner.

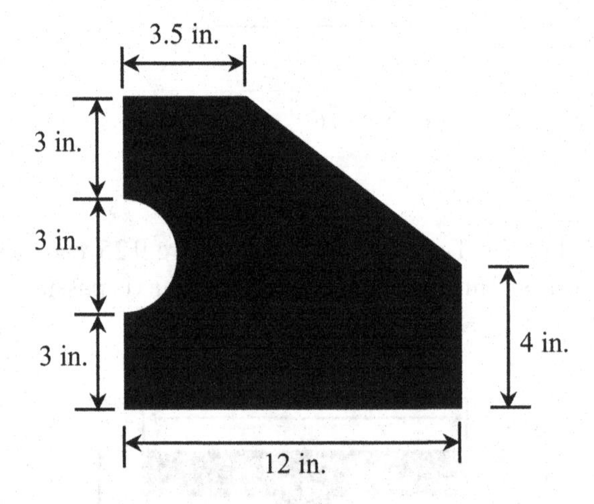

FIGURE 9.27 A composite section for Problem 9.9.

PROBLEM 9.10

Determine the centroid of the culvert section shown in Figure 9.28 with respect to the bottom-left corner. The section is symmetrical about the centroidal vertical axis. The semi-circular opening has a diameter of 4 ft.

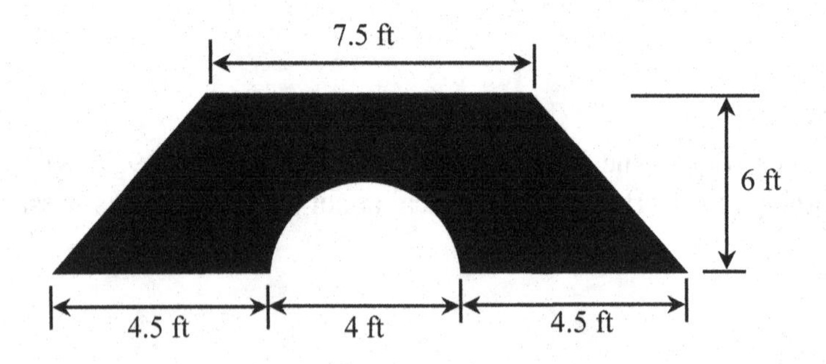

FIGURE 9.28 A culvert section for Problem 9.10.

PROBLEM 9.11

Determine the centroid of the section shown in Figure 9.29 with respect to the axis system shown.

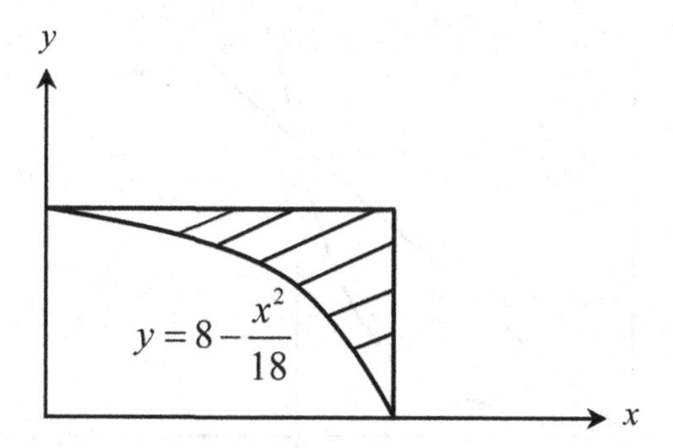

FIGURE 9.29 An arbitrary section for Problem 9.11.

PROBLEM 9.12

Determine the centroid of the section shown in Figure 9.30 with respect to the axis system shown.

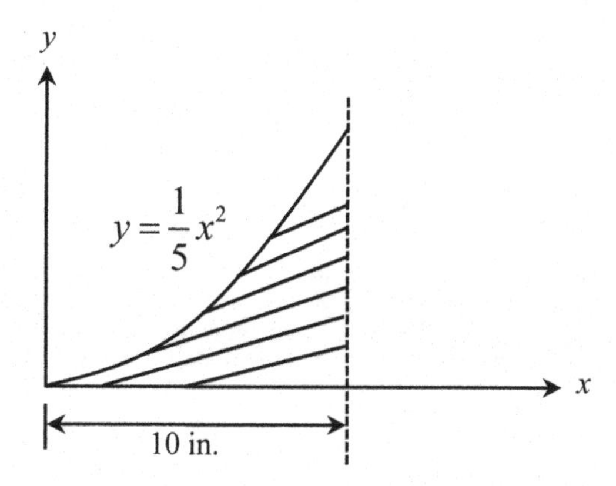

FIGURE 9.30 An arbitrary section for Problem 9.12.

PROBLEM 9.13

Determine the centroid of the bounded section by the two lines shown in Figure 9.31 with respect to the axis system shown.

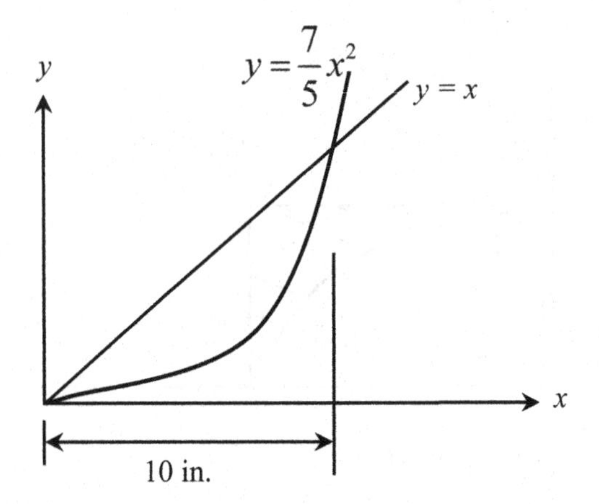

FIGURE 9.31 An arbitrary section for Problem 9.13.

10 Moment of Inertia of Area

10.1 MOMENT OF INERTIA

The moment of inertia (I) of area is a geometrical property of an area that reflects how its points are distributed with regard to an arbitrary axis. It is also known as the second moment of area, moment of inertia of plane area, area moment of inertia, or second area moment. The second moment of area is typically denoted with either an I for an axis that lies in the plane or with a J for an axis perpendicular to the plane. In both cases, it is calculated with a multiple integral over the object in question. Its unit of dimension is m^4, or in.^{4}, and so on.

In civil engineering, the moment of inertia of a beam is a very important property used in the calculation of the beam's deflection caused by a moment applied to the beam. The moment of inertia is an indication of the stiffness of a beam, which is the resistance to deflection of the beam when carrying loads that tend to cause it to bend. It provides the beam's resistance to bending due to an applied moment, force, or distributed load perpendicular to its neutral axis, as a function of its shape. The polar second moment of area provides insight into a beam's resistance to torsional deflection, due to an applied moment parallel to its cross-section, as a function of its shape. In summary, moment of inertia is a geometric property of a section which controls the deflection (Figure 10.1) and the stress-carrying capacity of a section. The larger the moment of inertia, the lower the possible deflection and stress level in the beam. It also helps a column to resist buckling.

The moment of inertia of an area with respect to an axis is defined as the sum of the products obtained by multiplying each infinitesimally small element of the area by the square of its distance from the axis. Mathematically, it can be expressed as follows:

$$\textit{Moment of inertia about x axis}: I_x = \sum y(y\Delta A) = \sum y^2\Delta A$$

$$\textit{Moment of inertia about y axis}: I_y = \sum x(x\Delta A) = \sum x^2\Delta A$$

From the above equation, it can be seen that the I-value is mathematically the second moment of an area. The above equations can be used for the I-value for regular sections (square, rectangle, circle, and so on) or the summation of several regular sections. For irregular section, calculus is the only method to do it. However, in civil engineering related structure, most of the sections are either regular or a summation of several regular sections.

For a rotating member, such as a shaft of motor, the polar moment of inertia is required to analyze the adequacy of the member. The polar moment of inertia (J) is the summation

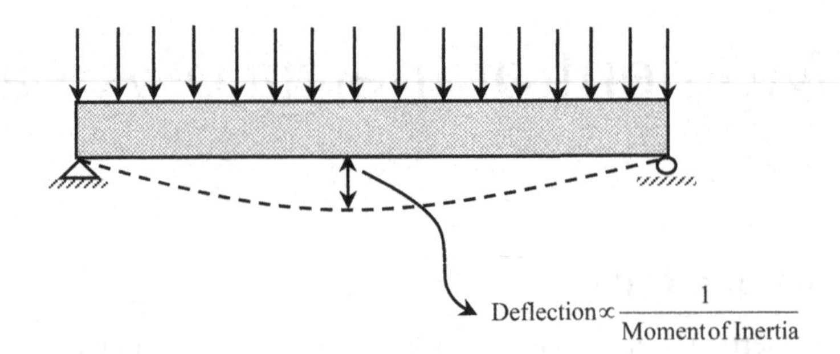

FIGURE 10.1 Deflection of a beam.

of the moment of inertia about two perpendicular axes passing through the centroid. It can be written as:

$$J = \bar{I}_x + \bar{I}_y$$

The bars over I_x and I_y represent the axes passing through the centroid.

10.2 RADIUS OF GYRATION AND SECTION MODULUS

The radius of gyration expresses the relationship between the area of a cross-section and a centroidal moment of inertia. It is a shape factor that measures a column's resistance to buckling about an axis. The radius of gyration of a cross-section (area) is defined as the distance from its moment of inertia axis at which the entire area could be considered as being concentrated without changing its moment of inertia. The radius of gyration (r) is calculated as follows:

$$\text{Radius of gyration about the } x \text{ axis, } r_x = \sqrt{\frac{\bar{I}_x}{A}}$$

$$\text{Radius of gyration about the } y \text{ axis, } r_y = \sqrt{\frac{\bar{I}_y}{A}}$$

$$\text{Polar radius of gyration about the } x \text{ axis, } r_j = \sqrt{\frac{J}{A}} = \sqrt{\frac{\bar{I}_x + \bar{I}_y}{A}}$$

Another geometric parameter called the section modulus (S) is very often used in analyzing beam. Mathematically, it is expressed as:

$$S = \frac{I}{c}$$

The symbol c is taken as the distance from the centroidal axis of the cross-section to the outermost fiber of the beam. Both I and c are properties of the cross-section of a member and are used for analysis and design. The polar section modulus (Z_p) is expressed as:

$$Z_p = \frac{J}{c}$$

10.3 MOMENT OF INERTIA OF COMMON SECTIONS

The moment of inertia of a few common sections about their centroidal axes are given in Figure 10.2. Unless said otherwise, the moment of inertia is always determined about the centroid axes.

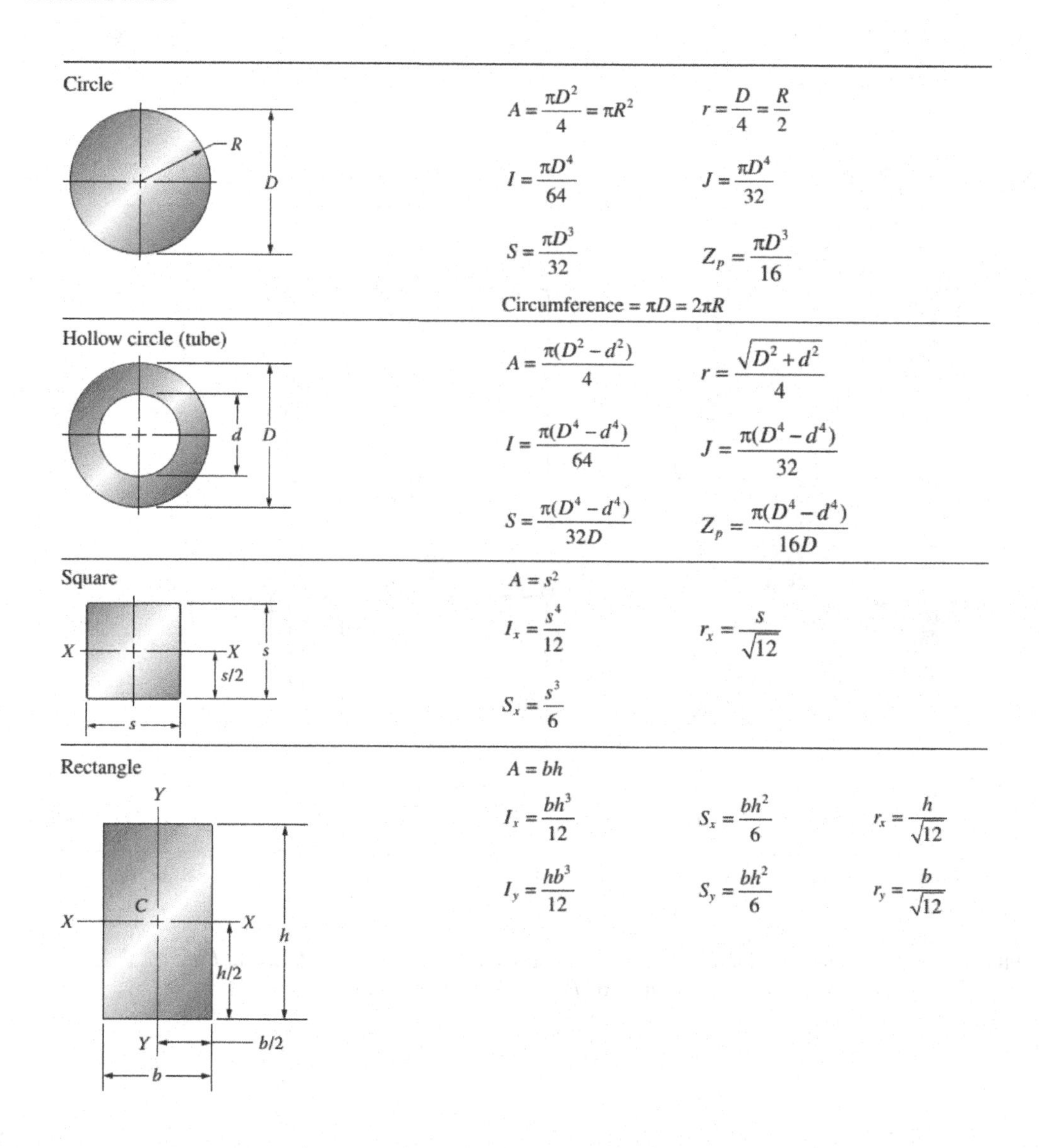

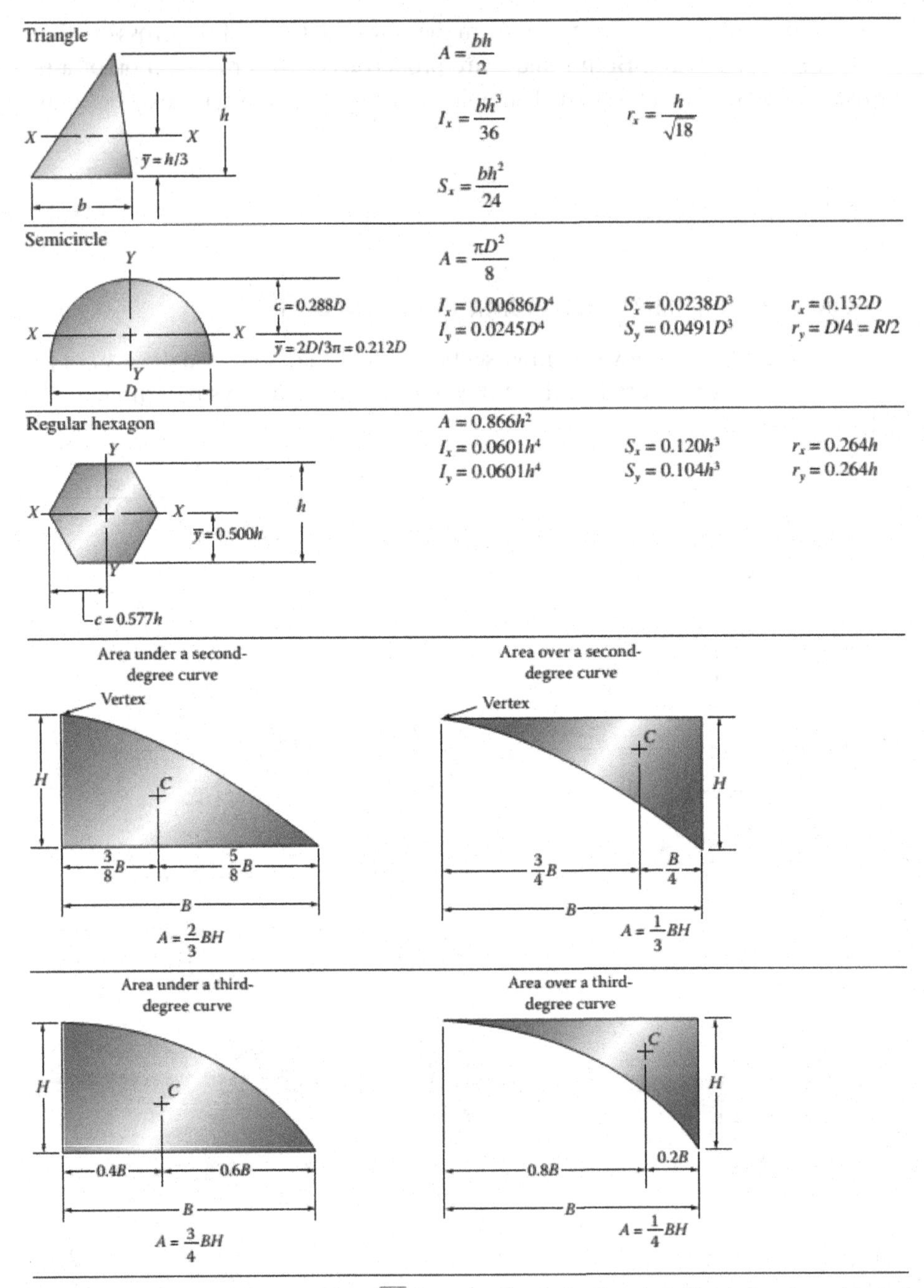

[a]Symbols used are A, area; r, radius of gyration $= \sqrt{I/A}$; I, moment of inertia; J, polar moment of inertia; S, section modulus; and Z_p, polar section modulus.

FIGURE 10.2 Moments of inertia of some common sections. *(From Mott, R. L. and Untener, J. A. (2017). Applied Strength of Materials, 6th edition, CRC Press, Boca Raton, FL.)*

10.4 MOMENT OF INERTIA OF COMPOSITE SECTIONS

Composite sections are composed of several regular component sections for which formulas are available to calculate the moment of inertia. A special case is when all component sections of a composite section have the same centroidal axis. If the component parts of a composite section all have the same centroidal axis, the combined moment of inertia can be found by adding or subtracting the moments of inertia of the component sections with respect to the centroidal axis. Example 10.1 will clarify this rule.

Example 10.1:

Determine the moment of the inertia of the section in Figure 10.3 about the centroidal axis. There is a 20 x10 in. rectangular hollow inside the 44 in. x 24 in. rectangular section.

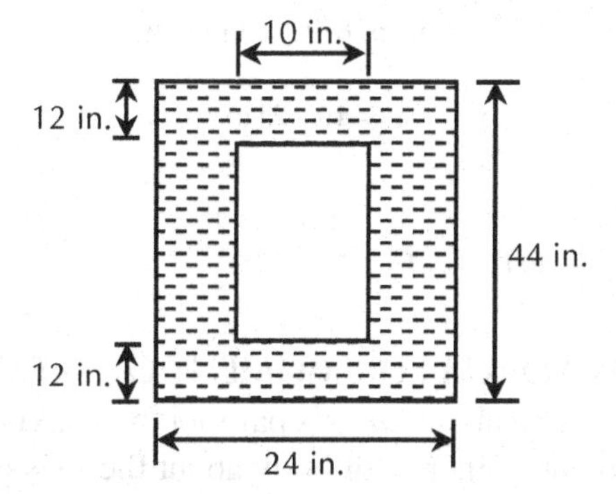

FIGURE 10.3 A rectangular hollow section for Example 10.1.

SOLUTION

The moment of inertia of the combined section can be determined by determining the moment of inertia of the whole section first and then deducting the moment of inertia of the inner rectangular section shown in Figure 10.4. Both sections have the same centroidal axes. Here the *x*- or *y*-coordinates of the centroid of the whole section and the hollow inner section coincide with the *x*- or *y*-coordinates of the combined section.

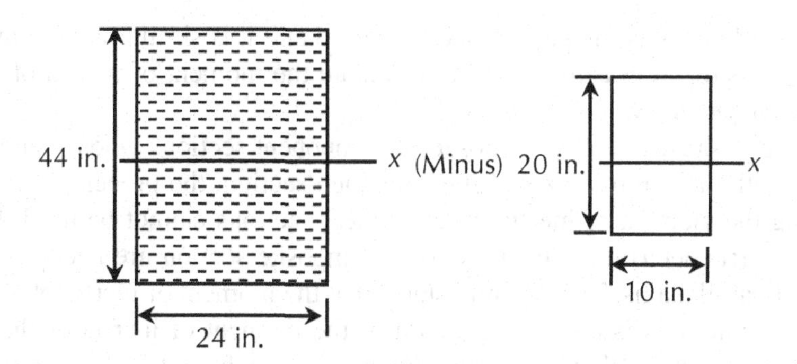

FIGURE 10.4 Analysis of the rectangular hollow section for Example 10.1.

The moment of inertia of the section about the centroidal horizontal axis is $\bar{I}_x = \frac{bh^3}{12}$

Thus, the combined moment of inertia about the centroidal horizontal axis is

$$\bar{I}_x = \frac{24\text{ in.}(44\text{ in.})^3}{12} - \frac{10\text{ in.}(20\text{ in.})^3}{12}$$

$$\bar{I}_x = 170{,}368\text{ in.}^4 - 6{,}666\text{ in.}^4$$

$$\bar{I}_x = 163{,}702\text{ in.}^4$$

Similarly, the moment of inertia of the section about the centroidal vertical axis is $I_y = \frac{hb^3}{12}$

The combined moment of inertia about the centroidal vertical axis is

$$\bar{I}_y = \frac{44\text{ in.}(24\text{ in.})^3}{12} - \frac{20\text{in.}(10\text{ in.})^3}{12}$$

$$\bar{I}_y = 50{,}688\text{ in.}^4 - 1{,}666\text{ in.}^4$$

$$\bar{I}_y = 49{,}021\text{ in.}^4$$

Answer:

$\bar{I}_x = 163{,}702$ in.^{4} *$\bar{I}_y = 49{,}021$ in.^{4}*

10.5 PARALLEL AXIS THEOREM FOR MOMENT OF INERTIA

The moment of inertia of an area about an axis parallel to the axis passing through its center is equal to the sum of moment of inertia of body about the axis passing through the center and the product of area times the square of distance between the two axes. For the x axis, it is expressed as follows:

$$I_x = \sum \bar{I}_x + \sum A d_y^2$$

where:

$\bar{I}_x$ =the moment of inertia of an area about the centroidal x axis (in.^{4} or mm^4)
I_x = the moment of inertia of the area about other parallel x axis (in.^{4} or mm^4)
A = the area of the area (in.^{2} or mm^2)
d_y = the perpendicular distance between the centroidal x axis and the parallel x axis (in. or mm)

This theorem is known as the parallel axis theorem and it is similar about any other axes (say, y axis). This theorem can be used to calculate the moment of inertia of a composite section as discussed below.

When a composite section is composed of component sections whose centroid axes do not coincide with the centroid axis of the combined section, the process of simply adding or subtracting the moment of inertia of component sections cannot be used. In this case, moments of inertia determined for parts of a composite section are needed to transfer it the centroidal parallel axis. The theorem states that the moment of inertia of a section with respect to a certain axis is equal to the sum of the moment of inertia of the shape with respect to its own centroidal axis plus an amount of Ad^2, where A is the area of the section and d is the distance from the centroid of the section to the axis of concern.

For the x axis, it is expressed as follows:

$$\bar{I}_x = \sum I_x + \sum Ad_y^2$$

where:

$\bar{I}_x$ = the moment of inertia of the total cross-section about the centroidal x axis (in.^{4} or mm^4)

Ix = the moment of inertia of the component section about its own centroidal x axis (in.^{4} or mm^4)

A = the area of the component (in.^{2} or mm^2)

dy = the perpendicular distance between the centroidal x axis and the parallel axis that passes through the centroid of the component (in. or mm)

Similarly, for the y axis, $\bar{I}_y = \sum I_y + \sum Ad_x^2$

Alternatively, from the parallel axis theorem can be written as follows:

$$\bar{I}_x = \left(I_{x1} + A_1(\bar{y} - y_1)^2\right) + \left(I_{x2} + A_2(\bar{y} - y_2)^2\right) + \ldots\ldots$$

$$\bar{I}_y = \left(I_{y1} + A_1(\bar{x} - x_1)^2\right) + \left(I_{y2} + A_2(\bar{x} - x_2)^2\right) + \ldots\ldots$$

The steps of determining the moment of inertia of a composite section are listed below:

Step 1. Divide the composite section into several component sections whose moment of inertia formulas are known.

Step 2. Locate the distance (such as y_1, y_2, etc.) from the centroid of each component to the reference axis.

Step 3. Determine the centroid of the combined section ($\bar{x}, \bar{y}$).

Step 4. Determine the moment of inertia of the individual component section about its own centroidal axis (such as I_{x1}, I_{x2} etc.).

Step 5. Determine the distances from the centroid of the combined section to the centroid of each component section (such as $d_{y1} = \bar{y} - y_1$, etc.)

Step 6. Transfer the moment of inertia to the centroid of the composite using the parallel axis theorem.

Step 7. Sum the moment of inertia to determine the moment of inertia of the composite section.

Example 10.2:

Determine the moment of inertia of the section in Figure 10.5 with respect to its centroid x axis.

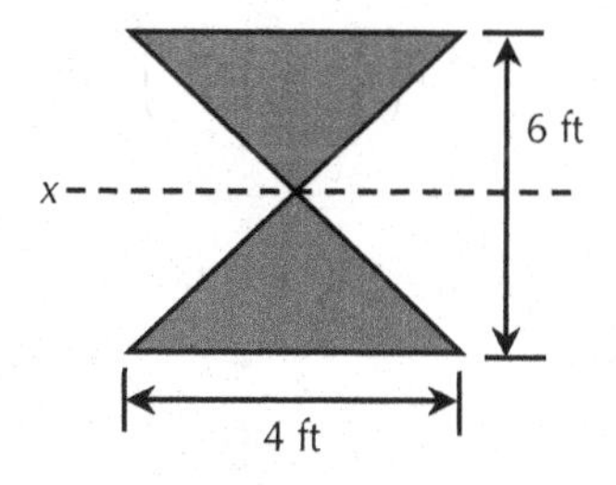

FIGURE 10.5 A composite section for Example 10.2.

SOLUTION

Step 1. The composite section is divided into two component sections: the upper triangle and the lower triangle

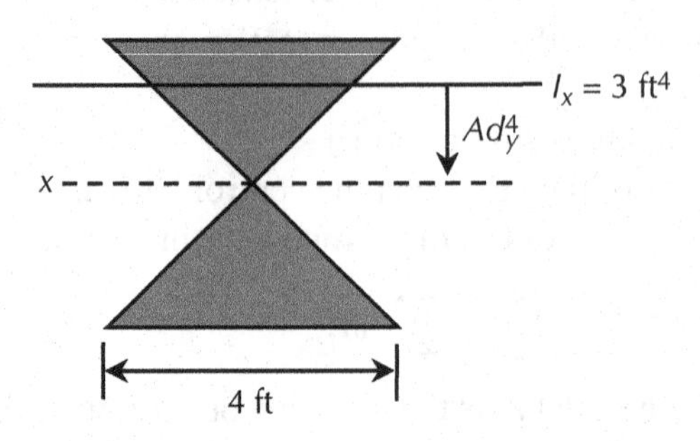

FIGURE 10.6 Analysis of the composite section for Example 10.2.

Step 2. The problem asks you to determine the moment of inertia with respect to the centroidal x axis. Therefore, the y-coordinate of the centroid of the composite section is required. By visual inspection, it can be found that the y-coordinate of the centroid of the composite section is located at the dashed line. The distance from the centroid of upper triangle to the reference axis is $y_I = 2$ ft.

Step 3. Let the centroid of the combined section be $\bar{y} = 0$ (reference line).

Step 4. The moment of inertia of the upper triangle is $I_x = \dfrac{bh^3}{36} = \dfrac{4\,\text{ft}(3\,\text{ft})^3}{36} = 3\ \text{ft}^4$ about its own centroidal horizontal axis.

Step 5. The distances from the centroid of the combined section to the centroid of the upper triangle is $d_y = \bar{y} - y_1 = 0 - 2\ \text{ft} = -\,2\ \text{ft}$.

Step 6. Transfer the moment of inertia to the centroid of the composite using the parallel axis theorem.

$$Ad_y^2 = \left(\frac{1}{2} \text{x } 4 \text{ x } 3\text{ft}^2\right)(2\,\text{ft})^2 = 24\ \text{ft}^4$$

$$\bar{I}_x = I_x + Ad_y^2$$

$$\bar{I}_x = 3\ \text{ft}^4 + \left(\frac{1}{2} \text{x } 4 \text{ x } 3\text{ft}^2\right)(-2\,\text{ft})^2$$

$$\bar{I}_x = 3\ \text{ft}^4 + 24\,\text{ft}^4$$

$$\bar{I}_x = 27\ \text{ft}^4$$

Step 7. The lower triangle is the same as the upper triangle. So, the composite moment of inertia can be determined as:

$$\bar{I}_x = 2(27\ \text{ft}^4) = 54\ \text{ft}^4$$

Answer: the moment of inertia with respect to its centroid x *axis* is *54 ft⁴*

Example 10.3:

For the following symmetric W-section in Figure 10.7, determine the following:

a. moment of inertia with respect to the centroidal horizontal axis.
b. radius of gyration with respect to the centroidal horizontal axis.
c. moment of inertia with respect to the centroidal vertical axis.
d. radius of gyration with respect to the centroidal vertical axis.
e. polar moment of inertia.
f. polar radius of gyration.

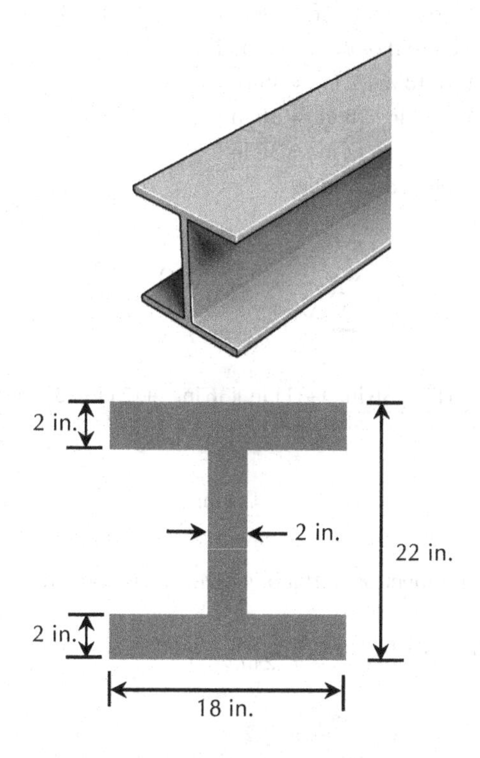

FIGURE 10.7 An I-section for Example 10.3.

SOLUTION

Step 1. The section is divided into smaller regular (rectangular) sections 1, 2 and 3 as shown in Figure 10.8.

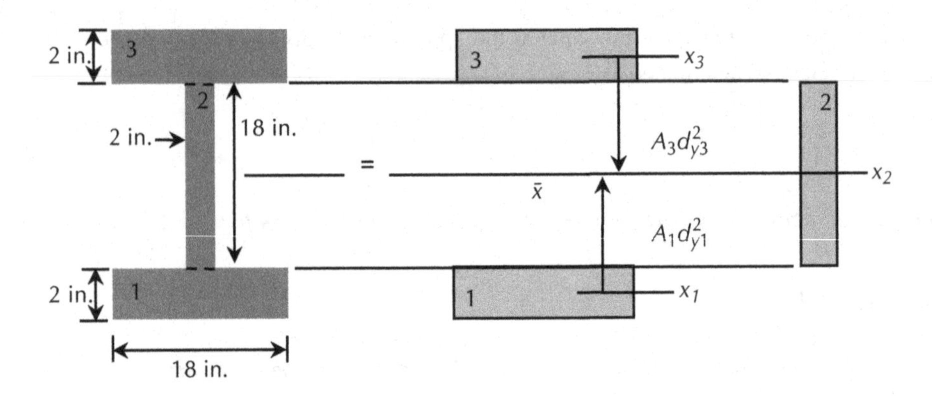

FIGURE 10.8 Division of the I-section for Example 10.3.

Step 2. Locate the distance from the centroid of each component to the reference axis.

The y-coordinate of the centroid of section 1, $y_1 = 1$ in.

The y-coordinate of the centroid of section 2, $y_2 = 2 + 9 = 11$ in.

The y-coordinate of the centroid of section 3, $y_3 = 20 + 1 = 21$ in.

Step 3. Determine the centroid of the combined section.

Area of section 1, $A_1 = (18\text{ in.})(2\text{ in.}) = 36\text{ in.}^2$

Area of section 2, $A_2 = (2\text{ in.})(18\text{ in.}) = 36\text{ in.}^2$

Area of section 3, $A_3 = (18\text{ in.})(2\text{ in.}) = 36\text{ in.}^2$

The y-coordinate of the combined centroid,

$$\bar{y} = \frac{\sum y\Delta A}{\sum \Delta A} = \frac{y_1A_1 + y_2A_2 + y_3A_3}{A_1 + A_2 + A_3}$$

$$\bar{y} = \frac{(1\text{ in.})(36\text{ in.}^2) + (11\text{ in.})(36\text{ in.}^2) + (21\text{ in.})(36\text{ in.}^2)}{36\text{ in.}^2 + 36\text{ in.}^2 + 36\text{ in.}^2}$$

$$\bar{y} = 11\text{ in.}$$

Step 4. Determine the moment of inertia of individual component section about their own centroidal axes

Section 1: $I_{x1} = \dfrac{b_1h_1^3}{12} = \dfrac{18\text{ in.}(2\text{ in.})^3}{12} = 12\text{ in.}^4$

Section 2: $I_{x2} = \dfrac{b_2h_2^3}{12} = \dfrac{2\text{ in.}(18\text{ in.})^3}{12} = 972\text{ in.}^4$

Section 3: Same as Section 1.

Step 5. Determine the distances from the centroid of the combined section to the centroid of each component section.

Section 1: $d_{y1} = \bar{y} - y_1 = 11\text{ in.} - 1\text{ in.} = 10.0\text{ in.}$

Section 2: $d_{y2} = \bar{y} - y_2 = 11\text{ in.} - 11\text{ in.} = 0\text{ in.}$

Section 3: $d_{y3} = \bar{y} - y_3 = 11\text{ in.} - 21\text{ in.} = -10.0\text{ in.}$

Step 6. Transfer the moment of inertia to the centroid of the composite using the parallel axis theorem.

Section 1: $I_{x1} + A_1 d_{y1}^2$

$$= 12\text{ in.}^4 + (18\text{ in. } x\, 2\text{ in.})(10\text{ in.})^2$$
$$= 12\text{ in.}^4 + 3{,}600\text{ in.}^4$$
$$= 3{,}612\text{ in.}^4$$

Section 2: $I_{x2} + A_2 d_{y2}^2$

$$= 972\text{ in.}^4 + A_2(0\text{ in.})^2$$
$$= 972\text{ in.}^4 + 0$$
$$= 972\text{ in.}^4$$

Section 3: $I_{x3} + A_3 d_{y3}^2$

$$= 12\text{ in.}^4 + (18\text{ in. } x\, 2\text{ in.})(-10\text{ in.})^2$$
$$= 12\text{ in.}^4 + 3{,}600\text{ in.}^4$$
$$= 3{,}612\text{ in.}^4$$

Step 7. Sum the moment of inertia to determine the moment of inertia of the composite section.

a) The combined *I*-value, $\bar{I}_x = 3{,}612\text{ in.}^4 + 972\text{ in.}^4 + 3{,}612\text{ in.}^4$

$$\bar{I}_x = 8{,}196\text{ in.}^4$$

b) Radius of gyration about the x axis

$$r_x = \sqrt{\frac{\bar{I}_x}{A}}$$
$$r_x = \sqrt{\frac{8{,}196\text{ in.}^4}{(36\text{ in.}^2 + 36\text{ in.}^2 + 36\text{ in.}^2)}}$$
$$r_x = 8.71\text{ in.}$$

c) Moment of inertia with respect to the centroidal vertical axis.

The section is divided into smaller regular (rectangular) sections 1, 2 and 3 as shown in Figure 10.8. This section is symmetrical about the vertical axis. So, the *x*-coordinate of the centroid = 9 in. Fortunately, the *x*-coordinates of the centroids of sections 1, 2 and 3 are coinciding with the *x*-coordinates of the composite section.

The moment of inertia of the combined section can be determined by determining the moment of inertia of each of the components about their centroidal vertical axes and then adding them.

Section 1: $I_{y1} = \dfrac{h_1 b_1^3}{12} = \dfrac{2\text{ in.}(18\text{ in.})^3}{12} = 972\text{ in.}^4$

Section 2: $I_{y2} = \dfrac{h_2 b_2^3}{12} = \dfrac{18\text{ in.}(2\text{ in.})^3}{12} = 12\text{ in.}^4$

Section 3: $I_{y3} = \dfrac{h_3 b_3^3}{12} = \dfrac{2\text{ in.}(18\text{ in.})^3}{12} = 972\text{ in.}^4$

The moment of inertia of the composite section about the centroidal vertical axis is

$$\bar{I}_y = I_{y1} + I_{y2} + I_{y3}$$
$$\bar{I}_y = 972\text{ in.}^4 + 12\text{ in.}^4 + 972\text{ in.}^4$$
$$\bar{I}_y = 1{,}956\text{ in.}^4$$

d) Radius of gyration with respect to the centroidal vertical axis

$$r_y = \sqrt{\frac{\bar{I}_y}{A}}$$
$$r_y = \sqrt{\frac{1{,}956\text{ in.}^4}{(36\text{ in.}^2 + 36\text{ in.}^2 + 36\text{ in.}^2)}}$$
$$r_y = 4.26\text{ in.}$$

e) Polar moment of inertia

$$J = \bar{I}_x + \bar{I}_y$$
$$J = 8{,}196\text{ in.}^4 + 1{,}956\text{ in.}^4$$
$$J = 10{,}152\text{ in.}^4$$

f) Polar radius of gyration

$$r_j = \sqrt{\frac{J}{A}}$$
$$r_j = \sqrt{\frac{\bar{I}_x + \bar{I}_y}{A}}$$
$$r_j = \sqrt{\frac{10{,}152\text{ in.}^4}{(36\text{ in.}^2 + 36\text{ in.}^2 + 36\text{ in.}^2)}}$$
$$r_j = 9.7\text{ in.}$$

Note that $r_j \neq r_x + r_y$

Answers:

a. $\bar{I}_x = 8{,}196\text{ in.}^4$

b. $r_x = 8.71\text{ in.}$

c. $\bar{I}_y = 1{,}956\text{ in.}^4$

d. $r_y = 4.26\text{ in.}$

e. $J = 10{,}152\text{ in.}^4$

f. $r_j = 9.7\text{ in.}$

Example 10.4:

For the following symmetrical T-section, determine the following in Figure 10.9:

a. moment of inertia with respect to the centroidal horizontal axis.
b. radius of gyration with respect to the centroidal horizontal axis.
c. moment of inertia with respect to the centroidal vertical axis.
d. radius of gyration with respect to the centroidal vertical axis.
e. polar moment of inertia.
f. polar radius of gyration.

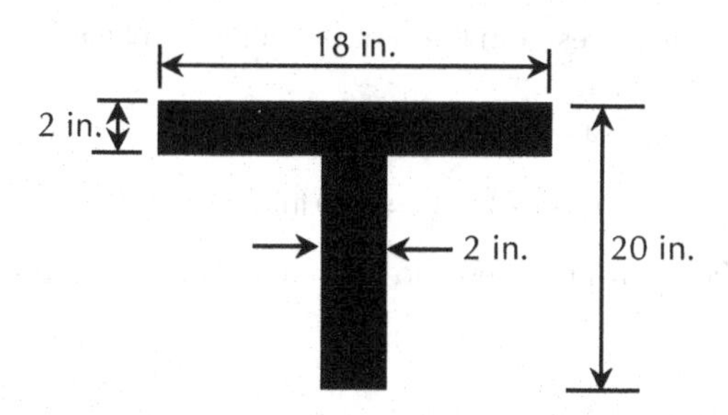

FIGURE 10.9 A T-section for Example 10.4.

SOLUTION

a. *Moment of inertia with respect to the centroidal horizontal axis.*

Step 1. The section is divided into smaller regular (rectangular) sections 1 and 2 as shown in Figure 10.10.

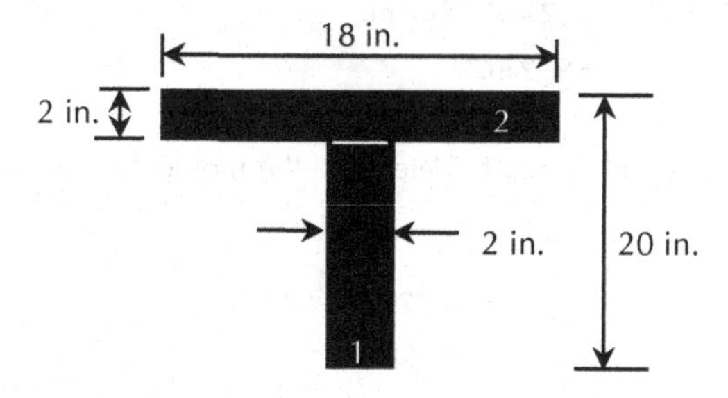

FIGURE 10.10 Division of the T-section for Example 10.4.

Step 2. Locate the distance from the centroid of each component to the reference axis.
y-coordinate of the centroid of section 1, $y_1 = 9.0$ in.
y-coordinate of the centroid of section 2, $y_2 = 19.0$ in.

Step 3. Determine the centroid of the combined section.
Area of section 1, $A_1 = (2 \text{ in.})(18 \text{ in.}) = 36 \text{ in.}^2$
Area of section 2, $A_2 = (18 \text{ in.})(2 \text{ in.}) = 36 \text{ in.}^2$

y-coordinate of the combined centroid,

$$\bar{y} = \frac{y_1 A_1 + y_2 A_2}{A_1 + A_2} = \frac{(9.0\text{ in.})(36\text{ in.}^2) + (19.0\text{ in.})(36\text{ in.}^2)}{(36\text{ in.}^2) + (36\text{ in.}^2)} = 14\text{ in.}$$

Step 4. Determine the moment of inertia of the individual component section about their own centroidal axes

Section 1: $I_{x1} = \frac{b_1 h_1^3}{12} = \frac{2\text{ in.}(18\text{ in.})^3}{12} = 972\text{ in.}^4$

Section 2: $I_{x2} = \frac{b_2 h_2^3}{12} = \frac{18\text{ in.}(2\text{ in.})^3}{12} = 12\text{ in.}^4$

Step 5. Determine the distances from the centroid of the combined section to the centroid of each component section

Section 1: $d_{y1} = \bar{y} - y_1 = 14\text{ in.} - 9\text{ in.} = 5.0\text{ in.}$

Section 2: $d_{y2} = \bar{y} - y_2 = 14\text{ in.} - 19\text{ in.} = -5.0\text{ in.}$

Step 6. Transfer the moment of inertia to the centroid of the composite using the parallel axis theorem.

Section 1: $I_{x1} + A_1 d_{y1}^2$

$$\begin{aligned} &= 972 + A_1 d_{y1}^2 \\ &= (18 \text{ x } 2\text{ in.}^2)(5\text{ in.})^2 \\ &= 972\text{ in.}^4 + 900\text{ in.}^4 \\ &= 1{,}672\text{ in.}^4 \end{aligned}$$

Section 2: $I_{x2} + A_2 d_{y2}^2$

$$\begin{aligned} &= 12\text{ in.}^4 + (18 \text{ x } 2\text{ in.}^2)(-5.0\text{ in.})^2 \\ &= 12\text{ in.}^4 + 900\text{ in.}^4 \\ &= 912\text{ in.}^4 \end{aligned}$$

Step 7. Sum the moment of inertia to determine the moment of inertia of the composite section.

$$\begin{aligned} \bar{I}_x &= 1{,}672\text{ in.}^4 + 912\text{ in.}^4 \\ \bar{I}_x &= 2{,}784\text{ in.}^4 \end{aligned}$$

b. *Radius of gyration with respect to the centroidal horizontal axis.*

$$\begin{aligned} r_x &= \sqrt{\frac{\bar{I}_x}{A}} \\ r_x &= \sqrt{\frac{2{,}784\text{ in.}^4}{(36+36)\text{ in.}^2}} \\ r_x &= 6.21\text{ in.} \end{aligned}$$

c. *Moment of inertia with respect to the centroidal vertical axis.*

The section is divided into smaller regular (rectangular) sections, 1 and 2, as shown in Figure 10.7. This section is symmetrical about the vertical axis. So, the x-coordinate of the centroid = 9 in. Fortunately, the x-coordinates of the centroids of sections 1 and 2 are coinciding with the x-coordinate of the composite section.

The moment of inertia of the combined section can be determined by determining the moment of inertia of both components about their centroidal vertical axes and then adding them.

Section 1: $I_{y1} = \frac{h_1 b_1^3}{12} = \frac{18\,\text{in.}(2\,\text{in.})^3}{12} = 12\,\text{in.}^4$

Section 2: $I_{y2} = \frac{h_2 b_2^3}{12} = \frac{2\,\text{in.}(18\,\text{in.})^3}{12} = 972\,\text{in.}^4$

The moment of inertia of the composite section about the centroidal vertical axis is

$$\bar{I}_y = I_{y1} + I_{y2}$$
$$\bar{I}_y = 12\,\text{in.}^4 + 972\,\text{in.}^4$$
$$\bar{I}_y = 984\,\text{in.}^4$$

d. *Radius of gyration with respect to the centroidal vertical axis.*

$$r_y = \sqrt{\frac{\bar{I}_y}{A}}$$
$$r_y = \sqrt{\frac{984\,\text{in.}^4}{(36\,\text{in.}^2 + 36\,\text{in.}^2)}}$$
$$r_y = 3.7\,\text{in.}$$

e. *Polar moment of inertia.*

$$J = \bar{I}_x + \bar{I}_y$$
$$J = 2{,}784\,\text{in.}^4 + 984\,\text{in.}^4$$
$$J = 3{,}768\,\text{in.}^4$$

f. *Polar radius of gyration.*

$$r_j = \sqrt{\frac{J}{A}}$$
$$r_j = \sqrt{\frac{\bar{I}_x + \bar{I}_y}{A}}$$
$$r_j = \sqrt{\frac{3{,}768\,\text{in.}^4}{(36\,\text{in.}^2 + 36\,\text{in.}^2)}}$$
$$r_j = 7.23\,\text{in.}$$

Note that $r_j \neq r_x + r_y$

Answers:

a. $\bar{I}_x = 2{,}784 \text{ in.}^4$

b. $r_x = 6.21 \text{ in.}$

c. $\bar{I}_y = 984 \text{ in.}^4$

d. $r_y = 3.7 \text{ in.}$

e. $J = 3{,}768 \text{ in.}^4$

f. $r_j = 7.2 \text{ in.}$

Example 10.5:

Two ½ x 13-inch plates are added to the top and bottom of a W 18 x 86 steel section shown in Figure 10.11. The moment of inertia of the W 18 x 86 section is 1,530 in.^{4} about its centroidal horizontal axis. Determine the moment of inertia of the newly built section about the centroidal horizontal axis.

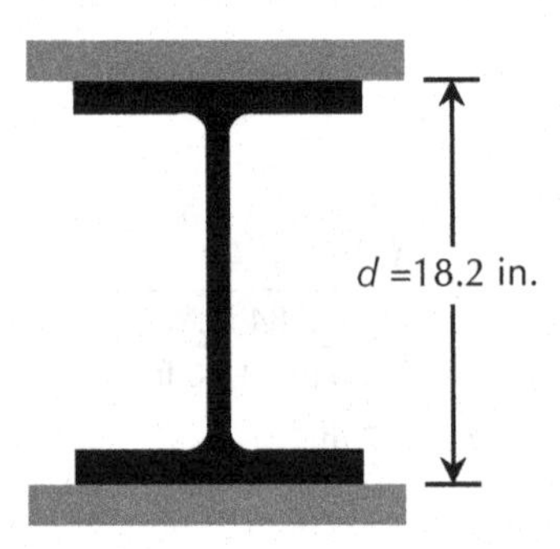

FIGURE 10.11 A W-section for Example 10.5.

SOLUTION

The moment of inertia of the W 18 x 86 section is given as *Ioriginal* = 1,530 in.^{4}. We need to add the moment of inertia of the top and the bottom plates and transfer them to the combined centroid as shown in Figure 10.12.

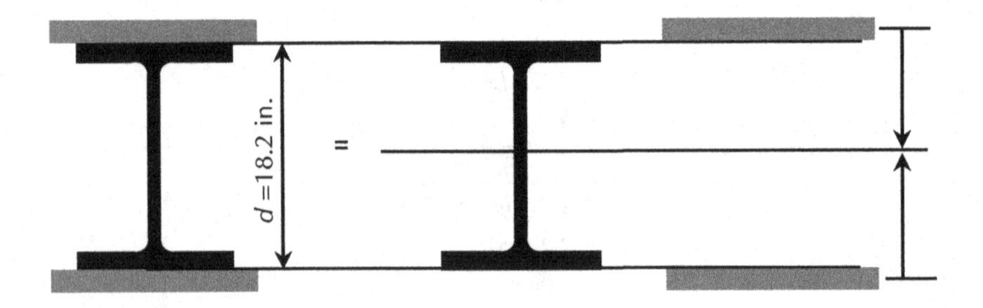

FIGURE 10.12 Analysis of the W-section for Example 10.5.

$$\overline{I}_x = I_{original} + I_{added}$$

$$I_{added} = (2\text{ plates})x\left(\frac{bh^3}{12} + Ad_y^2\right)$$

$$b = 13\text{ in.}$$

$$h = 0.5\text{ in.}$$

$$A = 13\text{ in. x }0.5\text{ in.} = 6.5\text{ in.}^2$$

$$dy = \frac{18.2\text{ in.}}{2} + \frac{0.5\text{ in.}}{2}$$

$$dy = 9.35\text{ in.}$$

$$\overline{I}_x = 1{,}530\text{ in.}^4 + 2\left(\frac{13\text{ in.} \times (0.5\text{ in.})^3}{12} + (6.5\text{ in.}^2)(9.35\text{ in.})^2\right)$$

$$\overline{I}_x = 2{,}667\text{ in.}^4$$

Answer:

$$\overline{I}_x = 2{,}667\ in.^4$$

Example 10.6:

Determine the moment of inertia of the rectangle shown in Figure 10.13 about the x axis, 2 m below the base of the rectangular section.

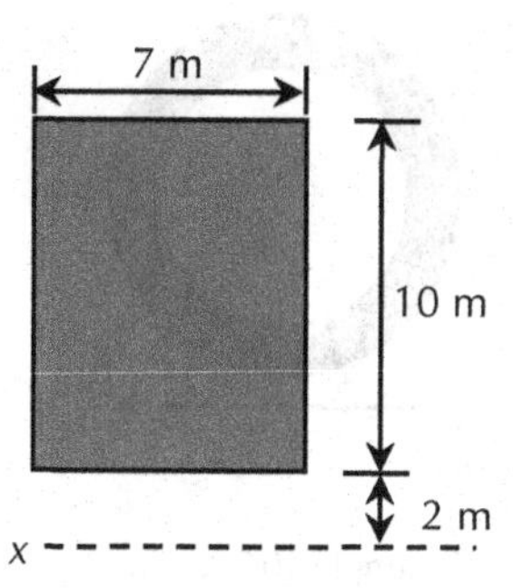

FIGURE 10.13 A rectangular section for Example 10.6.

SOLUTION

The I-value of the rectangular section about its own centroidal x axis is

$$I = \frac{bh^3}{12} = \frac{7\text{m}(10\text{m})^3}{12} = 583\text{ m}^4$$

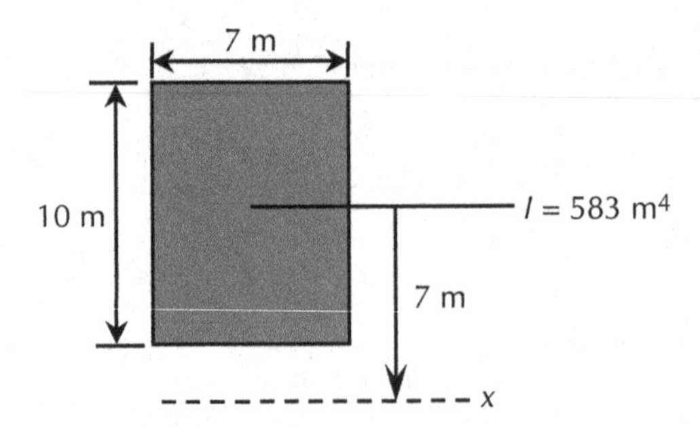

FIGURE 10.14 Analysis of the rectangular section for Example 10.6.

This I-value is to be transferred to the x axis which is 2 m below the base of the rectangular section. Therefore, Ad^2 must be added to it as follows

$$I_x = I + Ad^2$$
$$I_x = 583\ \text{m}^4 + (7x10\ \text{m}^2)(7\ \text{m})^2$$
$$I_x = 4{,}013\ \text{m}^4$$

Answer: $I_x = 4{,}013\ m^4$

Example 10.7:

Determine the radius of gyration of the hollow section in Figure 10.15 about horizontal centroid axis. The outer diameter of the section is 7 ft and wall thickness is 1 ft.

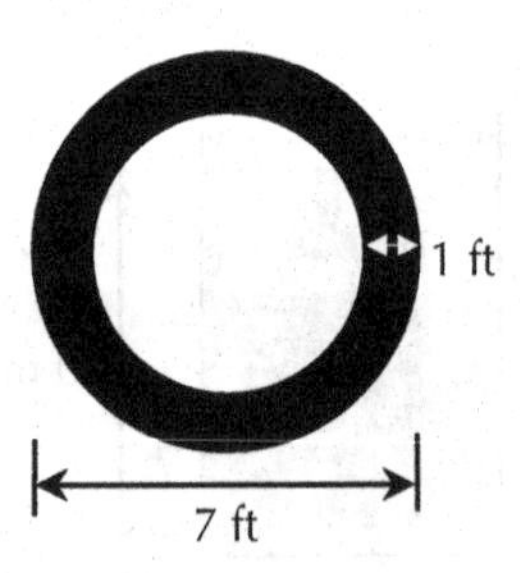

FIGURE 10.15 A hollow section for Example 10.7.

SOLUTION

Moment of inertia about the x axis, $\bar{I}_x = \dfrac{\pi r_o^4}{4} - \dfrac{\pi r_i^4}{4}$

$$\bar{I}_x = \frac{\pi}{4}((3.5\ \text{ft})^4 - (2.5\ \text{ft})^4)$$
$$\bar{I}_x = 87.2\ \text{ft}^4$$

Radius of gyration, $r_x = \sqrt{\frac{\bar{I}_x}{A}}$

$$r_x = \sqrt{\frac{87.2\ \text{ft}^4}{\pi((3.5\text{ft})^2 - (2.5\ \text{ft})^2)}}$$

$$r_x = 2.15\ \text{ft}$$

Answer:

$$r_x = 2.2\ ft$$

Example 10.8:

Determine the following for the culvert section shown in Figure 10.16. The section is symmetrical about the centroidal vertical axis. The semi-circular opening has a diameter of 4 ft.

a. moment of inertia with respect to the centroidal horizontal axis
b. radius of gyration with respect to the centroidal horizontal axis
c. moment of inertia with respect to the centroidal vertical axis
d. radius of gyration with respect to the centroidal vertical axis
e. polar moment of inertia
f. polar radius of gyration

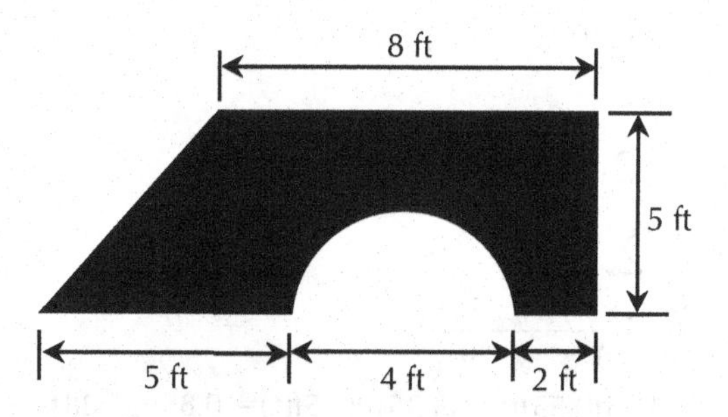

FIGURE 10.16 A composite section for Example 10.8.

SOLUTION

Consider, the whole rectangle as '1', the triangle as '2', and the semi-circle as '3' as shown in Figure 10.17.

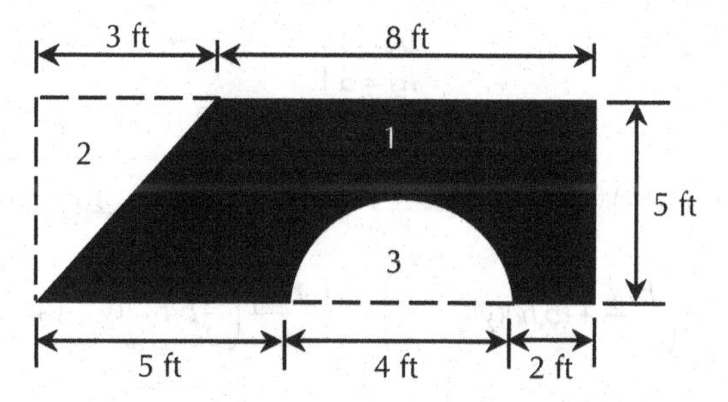

FIGURE 10.17 Division of the composite section for Example 10.8.

At the beginning, we need to determine the centroid. Let us find it with respect to the bottom-left corner:

Area of section 1, $A_1 = (11\ \text{ft})(5\ \text{ft}) = 55\ \text{ft}^2$
Area of section 2, $A_2 = \frac{1}{2}\,(3\ \text{ft})(5\ \text{ft}) = 7.5\ \text{ft}^2$
Area of section 3, $A_3 = \frac{1}{2}\,(\pi)(2\ \text{ft})^2 = 6.28\ \text{ft}^2$
Combined Area, $A = A_1 - A_2 - A_3 = 55\,\text{ft}^2 - 7.5\,\text{ft}^2 - 6.28\,\text{ft}^2 = 41.22\,\text{ft}^2$

The x-coordinate of the centroid of section 1, $x_1 = 5.5$ ft
The x-coordinate of the centroid of section 2, $x_2 = 1.0$ ft
The x-coordinate of the centroid of section 3, $x_3 = 7.0$ ft
The y-coordinate of the centroid of section 1, $y_1 = 2.5$ ft
The y-coordinate of the centroid of section 2, $y_2 = 2/3\ (5.0) = 3.35$ ft
The y-coordinate of the centroid of section 3, $y_3 = \frac{4a}{3\pi} = \frac{4(2.0\,\text{ft})}{3\pi} = 0.85\,\text{ft}$

$$\bar{x} = \frac{\sum x\Delta A}{\sum \Delta A}$$
$$= \frac{x_1A_1 - x_2A_2 - x_3A_3}{A_1 - A_2 - A_3}$$
$$= \frac{(5.5\,\text{ft})(55\,\text{ft}^2) - (1.0\,\text{ft})(7.5\,\text{ft}^2) - (7.0\,\text{ft})(6.28\,\text{ft}^2)}{41.22\,\text{ft}^2}$$
$$= 6.1\ \text{ft}$$

$$\bar{y} = \frac{\sum y\Delta A}{\sum \Delta A}$$
$$= \frac{y_1A_1 - y_2A_2 - y_3A_3}{A_1 - A_2 - A_3}$$
$$= \frac{(2.5\,\text{ft})(55\,\text{ft}^2) - (3.35\,\text{ft})(7.5\,\text{ft}^2) - (0.85\,\text{ft})(6.28\,\text{ft}^2)}{41.22\,\text{ft}^2}$$
$$= 2.6\ \text{ft}$$

a. *Moment of inertia with respect to the centroidal horizontal axis.*

$$b_1 = 5\ \text{ft} + 4\ \text{ft} + 2\ \text{ft} = 11\ \text{ft}$$
$$h_1 = 5\ \text{ft}$$
$$b_2 = 3\ \text{ft}$$
$$h_2 = 5\ \text{ft}$$
$$a = \text{radius of semi-circle} = 2\ \text{ft}$$

$$\bar{I}_x = \left(I_{x1} + A_1(\bar{y} - y_1)^2\right) - \left(I_{x2} + A_2(\bar{y} - y_2)^2\right) - \left(I_{x3} + A_3(\bar{y} - y_3)^2\right)$$

$$\bar{I}_x = \left(\frac{b_1h_1^3}{12} + (b_1h_1)(\bar{y} - y_1)^2\right) - \left(\frac{b_2h_2^3}{36} + \left(\frac{1}{2}b_2h_2\right)(\bar{y} - y_2)^2\right) - \left(0.1098a^4 + \left(\frac{1}{2}\pi a^2\right)(\bar{y} - y_3)^2\right)$$

$$\bar{I}_x = \left(\frac{(11\,\text{ft})(5\,\text{ft})^3}{12} + (55\,\text{ft}^2)(2.6\,\text{ft} - 2.5\,\text{ft})^2\right) - \left(\frac{(3\,\text{ft})(5\,\text{ft})^3}{36} + (7.5\,\text{ft}^2)(2.6\,\text{ft} - 3.35\,\text{ft})^2\right)$$
$$- (0.1098(2\,\text{ft})^4 + (6.28\,\text{ft}^2)(2.6\,\text{ft} - 0.85\,\text{ft})^2)$$

$$\bar{I}_x = 115.13\,\text{ft}^4 - 14.644\,\text{ft}^4 - 20.99\,\text{ft}^4$$
$$= 79.5\ \text{ft}^4$$

b. *Radius of gyration with respect to the centroidal horizontal axis.*

$$r_x = \sqrt{\frac{\bar{I}_x}{A}}$$
$$r_x = \sqrt{\frac{79.5\,\text{ft}^4}{41.22\,\text{ft}^2}}$$
$$r_x = 1.39\ \text{ft}$$

c. *Moment of inertia with respect to the centroidal vertical axis.*

$$\bar{I}_y = \left(I_{y1} + A_1(\bar{x} - x_1)^2\right) - \left(I_{y2} + A_2(\bar{x} - x_2)^2\right) - \left(I_{y3} + A_3(\bar{x} - x_3)^2\right)$$

$$\bar{I}_y = \left(\frac{h_1 b_1^3}{12} + (b_1 h_1)(\bar{x} - x_1)^2\right) - \left(\frac{h_2 b_2^3}{36} + \left(\frac{1}{2} b_2 h_2\right)(\bar{x} - x_2)^2\right) -$$
$$\left(\frac{\pi a^4}{8} + \left(\frac{1}{2}\pi a^2\right)(\bar{x} - x_3)^2\right)$$

$$\bar{I}_y = \left(\frac{(5\,\text{ft})(11\,\text{ft})^3}{12} + (55\,\text{ft}^2)(6.1\,\text{ft} - 5.5\,\text{ft})^2\right) - \left(\frac{(5\,\text{ft})(3\,\text{ft})^3}{36} + (7.5\,\text{ft}^2)(6.1\,\text{ft} - 1.0\,\text{ft})^2\right)$$
$$- \left(\frac{\pi(2)^4}{8}\,\text{ft}^4 + (6.28\,\text{ft}^2)(6.1\,\text{ft} - 7.0\,\text{ft})^2\right)$$

$$\bar{I}_y = 574.4\,\text{ft}^4 - 198.8\,\text{ft}^4 - 11.4\,\text{ft}^4$$
$$= 364.2\ \text{ft}^4$$

d. *Radius of gyration with respect to the centroidal vertical axis.*

$$r_y = \sqrt{\frac{\bar{I}_y}{A}}$$
$$r_y = \sqrt{\frac{364.2\ \text{ft}^4}{41.22\ \text{ft}^2}}$$
$$r_y = 2.97\ \text{ft}$$

e. *Polar moment of inertia.*

$$J = \overline{I}_x + \overline{I}_y$$
$$J = 79.5\ \text{ft}^4 + 364.2\,\text{ft}^4$$
$$J = 443.7\,\text{ft}^4$$

f. *Polar radius of gyration.*

$$r_j = \sqrt{\frac{J}{A}}$$
$$r_j = \sqrt{\frac{\overline{I}_x + \overline{I}_y}{A}}$$
$$r_j = \sqrt{\frac{443.7\ \text{ft}^4}{41.22\ \text{ft}^2}}$$
$$r_j = 3.28\ \text{ft}$$

Answers:

a. $\overline{I}_x = 79.5\ \text{ft}^4$

b. $r_x = 1.39\,\text{ft}$

c. $\overline{I}_y = 364.2\ \text{ft}^4$

d. $r_y = 2.97\ \text{ft}$

e. $J = 443.7\ \text{ft}^4$

f. $r_j = 3.3\,\text{ft}$

10.6 MOMENT OF INERTIA OF IRREGULAR AREAS

The moment of inertia of an area is defined as the second moment of an area with respect to a reference line. Figure 10.18 shows an arbitrary area with an infinitesimal area of dA.

The moment of inertia of the entire area about the x axis can be written as, $I_x = \int y^2 dA$

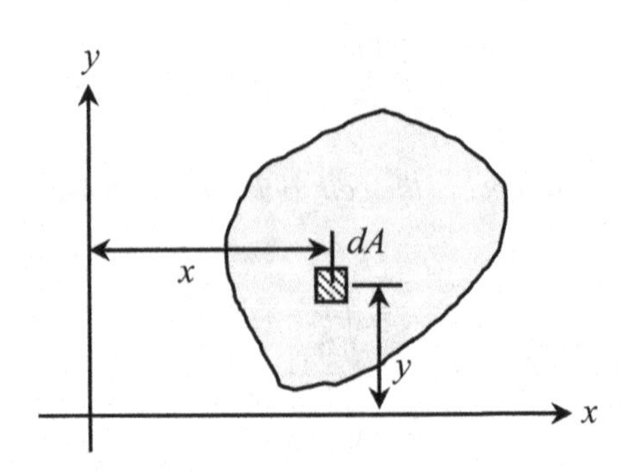

FIGURE 10.18 An arbitrary area for determining the moment of inertia.

The moment of inertia of the entire area about the y axis can be written as, $I_y = \int x^2 dA$

The polar moment of inertia can be written as, $J = \int r^2 dA = \int (x^2 + y^2) dA$

10.6.1 The Moment of Inertia of a Rectangle

Let us assume a rectangle with base b and height h on an xy plane as shown in Figure 10.19. Also, assume that the xy-coordinate system is originates at the centroid of the rectangle. This means we will derive the moment of inertia about the centroidal axis. An infinitesimal segment dx is assumed at x-distance from the origin; the ordinate is $h/2$ at that infinitesimal segment.

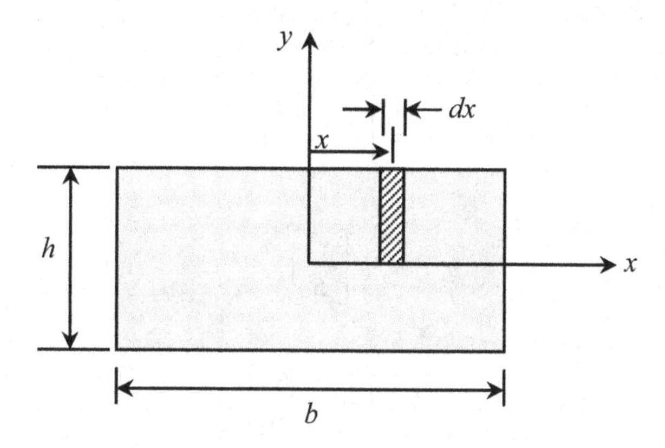

FIGURE 10.19 Moment of inertia of a rectangle.

The infinitesimal segment can be a assumed to be a rectangle and its area can be written as $(h/2)dx$. Therefore, the x-coordinate of the centroid can be found as

$$\bar{I}_y = \int x^2\, dA = \int x^2 \left(\frac{h}{2} dx\right) = \frac{h}{2} \int x^2\, dx = \frac{h}{2}\left[\frac{x^3}{3}\right]_{x=0}^{\frac{b}{2}} = \frac{h}{2}\left[\frac{1}{3}\left(\left(\frac{b}{2}\right)^3 - 0\right)\right] = \frac{hb^3}{48}$$

Considering the whole area, $I_y = 4\left(\frac{hb^3}{48}\right) = \frac{hb^3}{12}$

Similarly, $\bar{I}_x = \frac{bh^3}{12}$ considering the whole area.

10.6.2 The Moment of Inertia of an Irregular Section

Let us assume that a shape can be defined as $y = x^2$ as shown in Figure 10.20. We are required to determine the moment of inertia about the y axis. Let us assume an infinitesimal segment at x-distance from the y axis. The y-coordinate of this segment is y. The area of this infinitesimal segment is ydx.

$$I_y = \int x^2 dA$$

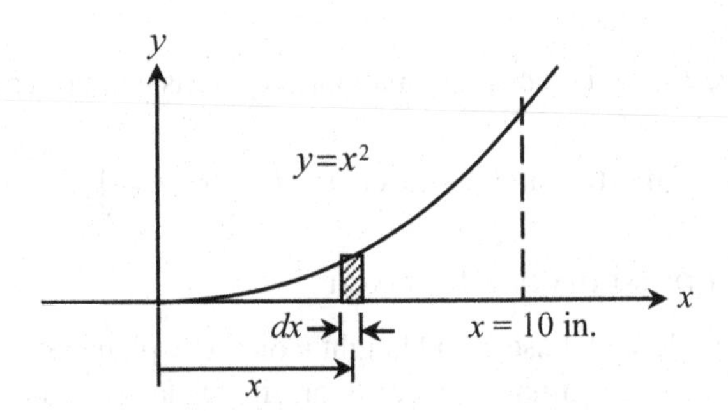

FIGURE 10.20 Determining the moment of inertia about the y axis.

$$I_y = \int (x^2)(ydx)$$

$$I_y = \int (x^2)(x^2)dx$$

$$I_y = \left[\frac{x^5}{5}\right]_0^{10}$$

$$I_y = \left[\frac{10^5}{5} - \frac{0^5}{5}\right]$$

$$I_y = 20{,}000 \text{ in.}^4$$

From Figure 10.21:

$$I_x = \int y^2 dA = \int y^2((10 - x)dy)$$

$$I_x = \int (10y^2 - y^{2.5})dy$$

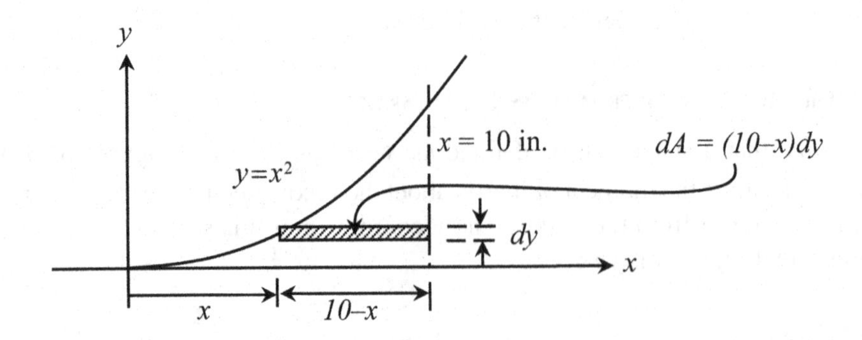

FIGURE 10.21 Determining the moment of inertia about the x axis.

$$I_x = \left[\frac{10y^3}{3} - \frac{y^{3.5}}{3.5}\right]_0^{100}$$

$$I_x = \left[\left(\frac{10(100)^3}{3} - \frac{(100)^{3.5}}{3.5}\right) - \left(\frac{10(0)^3}{3} - \frac{(0)^{3.5}}{3.5}\right)\right]_0^{100}$$

$$I_x = 476{,}187 \text{ in.}^4$$

PRACTICE PROBLEMS

PROBLEM 10.1

Determine the following properties of the section shown in Figure 10.22:

a. moment of inertia with respect to the centroidal horizontal axis
b. radius of gyration with respect to the centroidal horizontal axis
c. moment of inertia with respect to the centroidal vertical axis
d. radius of gyration with respect to the centroidal vertical axis
e. polar moment of inertia
f. polar radius of gyration

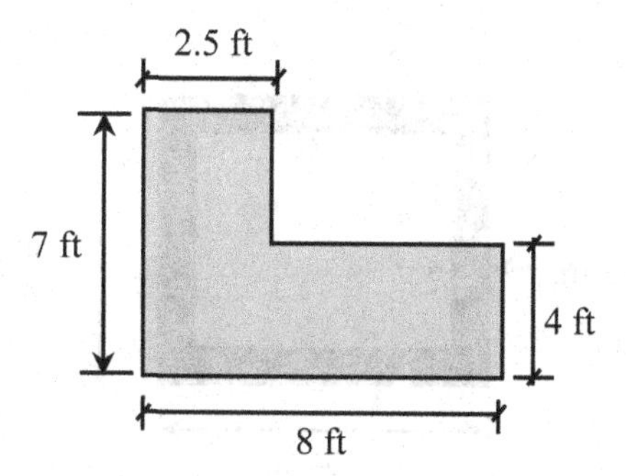

FIGURE 10.22 A L-section for Problem 10.1.

PROBLEM 10.2

Determine the following properties of the section shown in Figure 10.23:

a. moment of inertia with respect to the centroidal horizontal axis
b. radius of gyration with respect to the centroidal horizontal axis
c. moment of inertia with respect to the centroidal vertical axis
d. radius of gyration with respect to the centroidal vertical axis
e. polar moment of inertia
f. polar radius of gyration

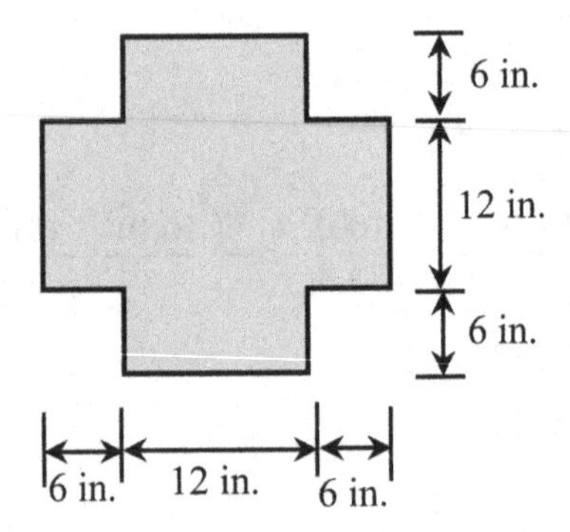

FIGURE 10.23 A composite section for Problem 10.2.

PROBLEM 10.3

Determine the following properties of the section shown in Figure 10.24:

a. moment of inertia with respect to the centroidal horizontal axis
b. radius of gyration with respect to the centroidal horizontal axis
c. moment of inertia with respect to the centroidal vertical axis
d. radius of gyration with respect to the centroidal vertical axis
e. polar moment of inertia
f. polar radius of gyration

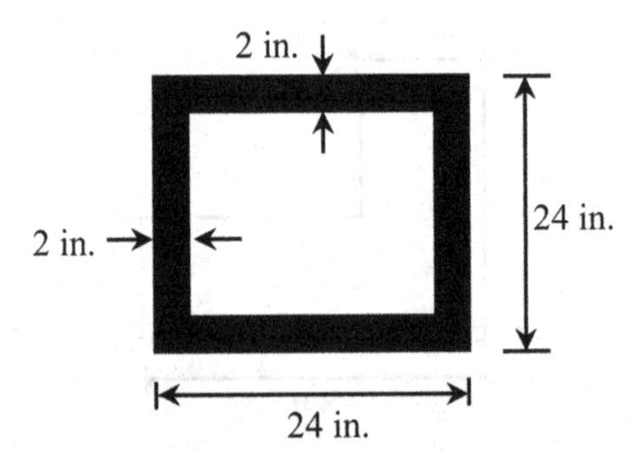

FIGURE 10.24 A square hollow section for Problem 10.3.

PROBLEM 10.4

Determine the following properties of the section shown in Figure 10.25:

a. moment of inertia with respect to the centroidal horizontal axis
b. radius of gyration with respect to the centroidal horizontal axis
c. moment of inertia with respect to the centroidal vertical axis
d. radius of gyration with respect to the centroidal vertical axis
e. polar moment of inertia
f. polar radius of gyration

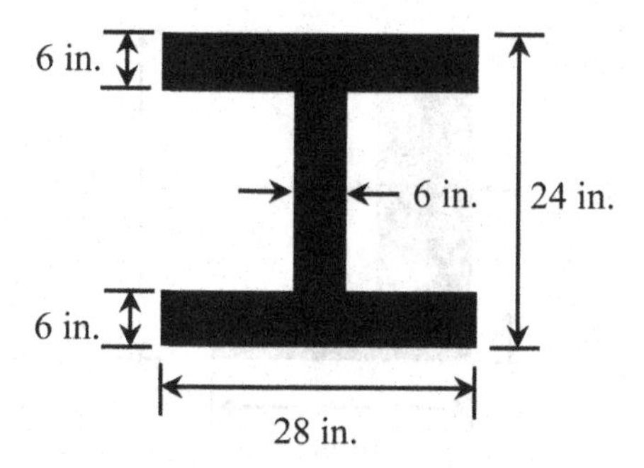

FIGURE 10.25 An I-section for Problem 10.4.

PROBLEM 10.5

Determine the following properties of the rectangular section with a circular hollow as shown in Figure 10.26. The circular hollow with 6-in. diameter is located at the center of the rectangle.

a. moment of inertia with respect to the centroidal horizontal axis
b. radius of gyration with respect to the centroidal horizontal axis
c. moment of inertia with respect to the centroidal vertical axis
d. radius of gyration with respect to the centroidal vertical axis
e. polar moment of inertia
f. polar radius of gyration

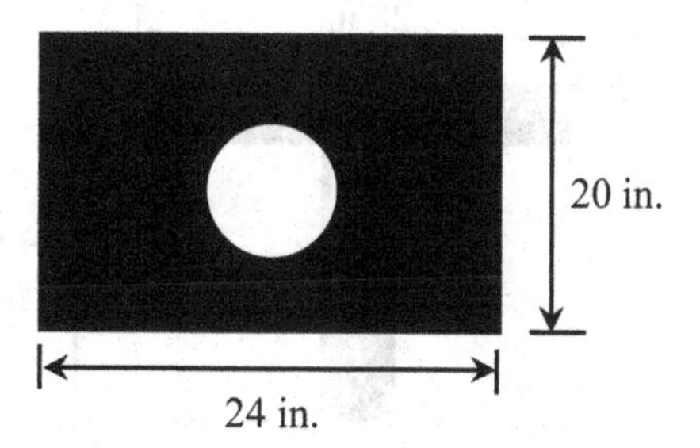

FIGURE 10.26 A hollow rectangular section for Problem 10.5.

PROBLEM 10.6

Determine the following properties of the channel section shown in Figure 10.27.

a. moment of inertia with respect to the centroidal horizontal axis
b. radius of gyration with respect to the centroidal horizontal axis
c. moment of inertia with respect to the centroidal vertical axis
d. radius of gyration with respect to the centroidal vertical axis
e. polar moment of inertia
f. polar radius of gyration

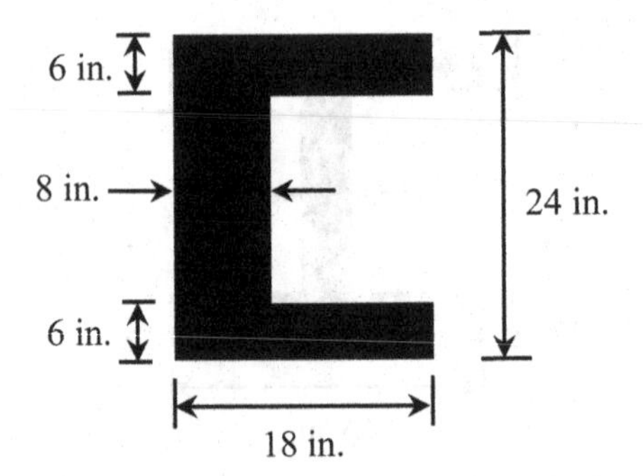

FIGURE 10.27 A channel section for Problem 10.6.

PROBLEM 10.7

Determine the following properties of the T-section shown in Figure 10.28.

a. moment of inertia with respect to the centroidal horizontal axis
b. radius of gyration with respect to the centroidal horizontal axis
c. moment of inertia with respect to the centroidal vertical axis
d. radius of gyration with respect to the centroidal vertical axis
e. polar moment of inertia
f. polar radius of gyration

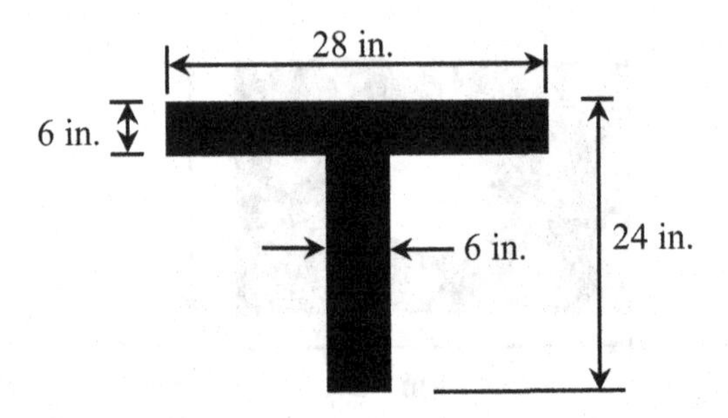

FIGURE 10.28 A T-section for Problem 10.7.

PROBLEM 10.8

Determine the following properties of the inverse T-section shown in Figure 10.29.

a. moment of inertia with respect to the centroidal horizontal axis
b. radius of gyration with respect to the centroidal horizontal axis
c. moment of inertia with respect to the centroidal vertical axis
d. radius of gyration with respect to the centroidal vertical axis
e. polar moment of inertia
f. polar radius of gyration

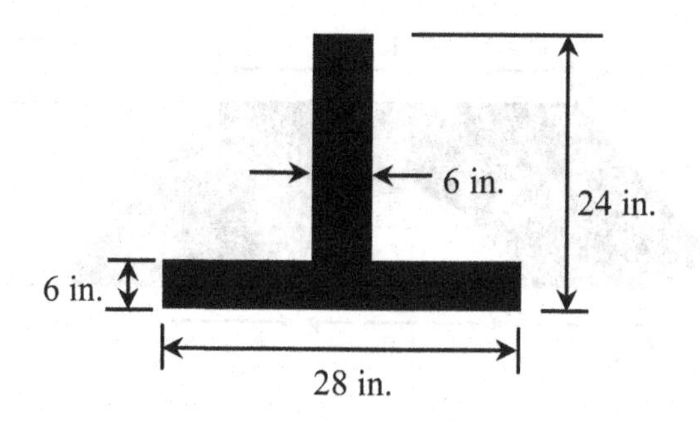

FIGURE 10.29 An inverted T-section for Problem 10.8.

PROBLEM 10.9

Determine the following properties of the composite section with a semi-circular and a triangular cut as shown in Figure 10.30.

a. moment of inertia with respect to the centroidal horizontal axis
b. radius of gyration with respect to the centroidal horizontal axis
c. moment of inertia with respect to the centroidal vertical axis
d. radius of gyration with respect to the centroidal vertical axis
e. polar moment of inertia
f. polar radius of gyration

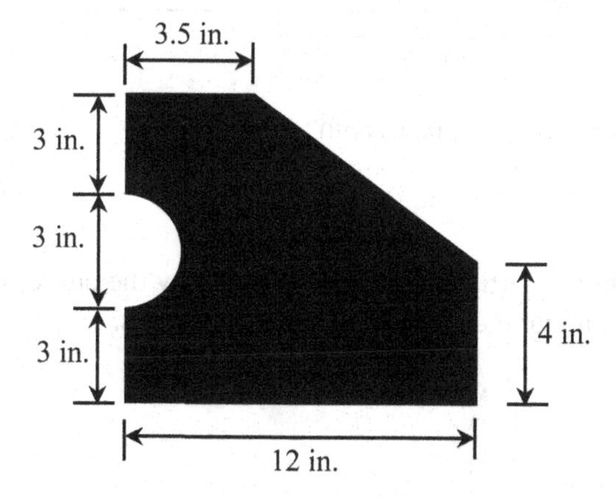

FIGURE 10.30 A composite section for Problem 10.9.

PROBLEM 10.10

Determine the following for the culvert section shown in Figure 10.31. The section is symmetric about the centroidal vertical axis. The semi-circular opening has a diameter of 4 ft.

a. moment of inertia with respect to the centroidal horizontal axis
b. radius of gyration with respect to the centroidal horizontal axis
c. moment of inertia with respect to the centroidal vertical axis
d. radius of gyration with respect to the centroidal vertical axis
e. polar moment of inertia
f. polar radius of gyration

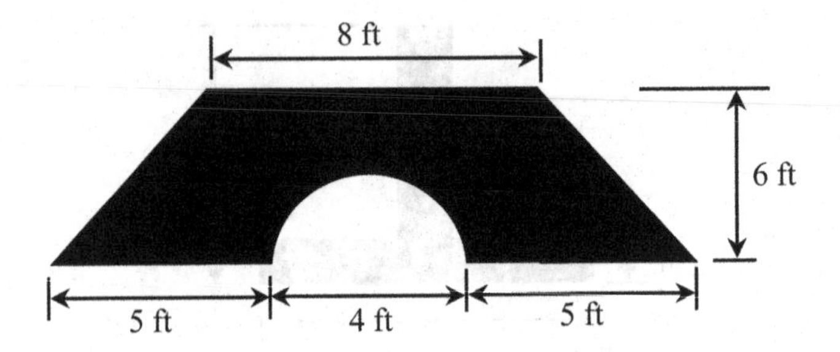

FIGURE 10.31 A composite section for Problem 10.10.

PROBLEM 10.11

Determine the moment of inertia of the area bounded by the curve, x axis and $x = 10$ in. as shown in Figure 10.32, about the x axis and about the y axis.

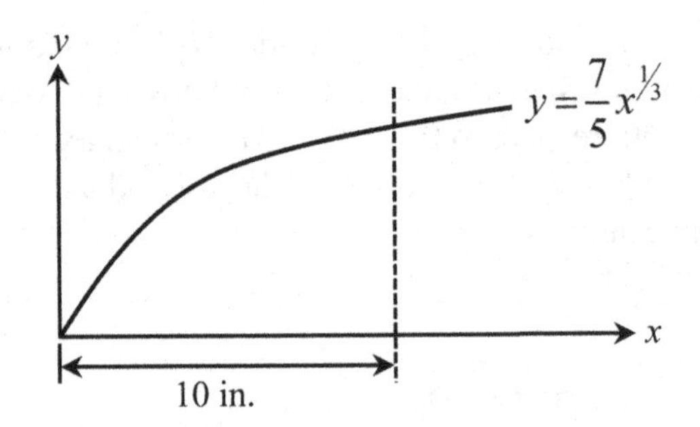

FIGURE 10.32 An irregular section for Problem 10.11.

PROBLEM 10.12

Determine the moment of inertia of the area bounded by the curve, y axis and $y = 6$ in. as shown in Figure 10.33, about the x axis and about the y axis.

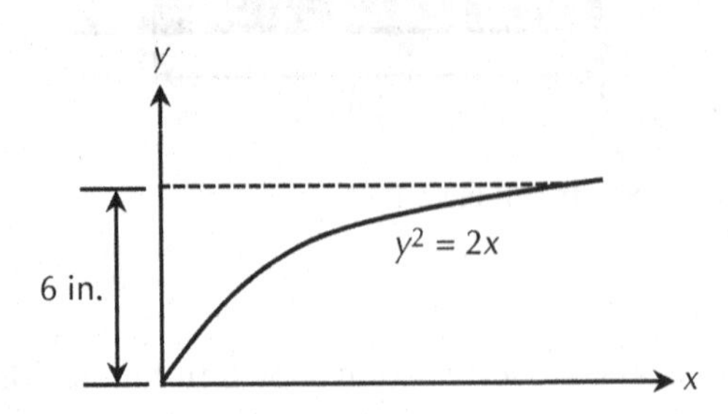

FIGURE 10.33 An irregular section for Problem 10.12.

11 Friction

11.1 STATIC AND KINETIC FRICTIONS

Surface friction or simply friction is the resistance between two surfaces when they are in contact. If the surfaces are at rest, the friction between them is called static friction. It helps an object to stay at rest. If one of the surfaces wants to move, the surface must overcome static friction. Once the object is in motion, it also has some sort of friction called kinetic friction. Consider driving over ice. Some friction is required to drive safely. When tires cannot get enough friction with roads due to the slippery ice, cars skid.

Friction is not a fundamental force. It is generated by inter-surface attraction and surface roughness. The complexity of these interactions makes the calculation of friction force very inaccurate. In engineering, an approximation is made to calculate the friction. Logical determination of friction force is essential in engineering especially in machinery design, slope analysis, and roads and highway design, etc.

If a small amount of force is applied to an object, the static friction has an equal magnitude in the opposite direction. If the force is increased, at some point the value of the maximum static friction will be reached, and the object will move. The largest frictional force is called the limiting friction. Any further increase in applied forces will cause motion. This means:

$$F_f \leq \mu_s R$$

where:

F_f = frictional force (limiting friction, $F_{limit} = \mu_s R$)
μ_s = coefficient of static friction, and
R = normal force between surfaces in contact

Once the applied force (F) is greater than the limiting friction (F_{limit}), the object will move, and the friction force in motion can be determined as:

$$F_k \leq \mu_k R$$

where:

F_k = frictional force in motion
μ_k = coefficient of kinetic friction

The applied force must overcome the limiting friction ($F_{limit} = \mu_s R$) to have impending motion or motion. Once the object is in motion, the frictional force suddenly decreases (shown in Figure 11.1) as the coefficient of kinetic friction is commonly less than the coefficient of static friction.

The following example will clarify the above-mentioned mechanics of friction.

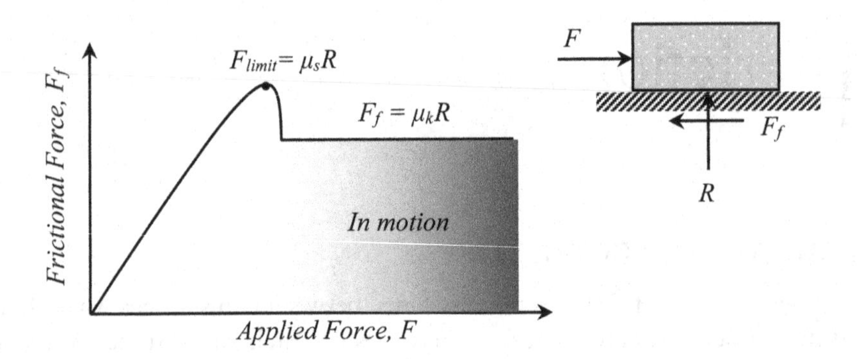

FIGURE 11.1 Mechanics of friction.

Example 11.1:

A 1 m x 1 m x 1 m block is resting on a horizontal surface as shown in Figure 11.2. The density of the block is 2,000 kg/m³. The static and kinetic frictional coefficients (μ_s and μ_k) are also mentioned. The gravitational constant = 9.81 m/sec². Determine the following:

a. The frictional force; if a force of $F = 1.886$ kN is applied, will the block move?
b. The frictional force; if a force of $F = 5.886$ kN is applied, will the block move?
c. The frictional force; if a force of $F = 8.886$ kN is applied, will the block move?

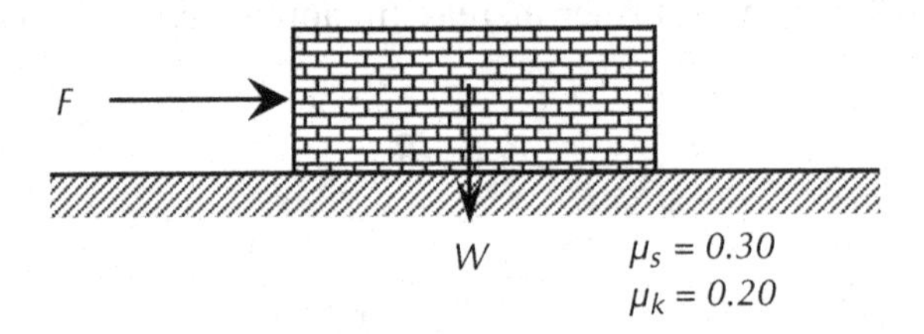

FIGURE 11.2 A block to be pulled over for Example 11.1.

SOLUTION

Step 1. Draw the free-body diagram as shown in Figure 11.3

$$
\begin{aligned}
\text{Weight, } W &= \rho g V \\
&= (2{,}000 \text{ kg/m}^3)(9.81 \text{ m/sec}^2)(1 \text{ m}^3) \\
&= 19{,}620 \text{ N} \\
&= 19.62 \text{ kN}
\end{aligned}
$$

5 kN
W
$F_f = \mu R$
R

FIGURE 11.3 The free-body diagram of the block for Example 11.1.

Step 2. Apply equilibrium equations

From the free-body diagram of the system as shown in Figure 11.3:

$$\Sigma F_y = 0 \ (\uparrow +ve)$$
$$R - W = 0$$
$$R = W$$
$$R = 19.62 \text{ kN}$$

Maximum static frictional force possible, $F_f = \mu_s R = 0.3$ (19.62 kN) = 5.886 kN

a. Driving force, $F = 1.886$ kN $< F_f$

   The block will not move.
   The frictional force = 1.886 kN

Answer:

   The block will not move.
   The frictional force = 1.886 kN

The value is given to three decimal places. Therefore, the answer is given to three decimal places.

b. Driving force, $F = 5.886$ kN $= F_f$

   The block will intend to move, and the static coefficient will apply.
   The frictional force = $\mu_s R = 0.3$ (19.62 kN) = 5.886 kN

Answer:

   The block will move.
   The frictional force = 5.886 kN

c. Driving force, $F = 8.886$ kN $< F_f$

   The block will move, and the kinetic coefficient will apply.
   The frictional force = $\mu_k R = 0.2$ (19.62 kN) = 3.924 kN

Answers:

   The block will move.
   The frictional force = 3.924 kN.

Example 11.2:

A 1 m x 1 m x 1 m block is resting on an inclined surface as shown in Figure 11.4. Most nearly at what inclination (θ), will the block intend to slide?

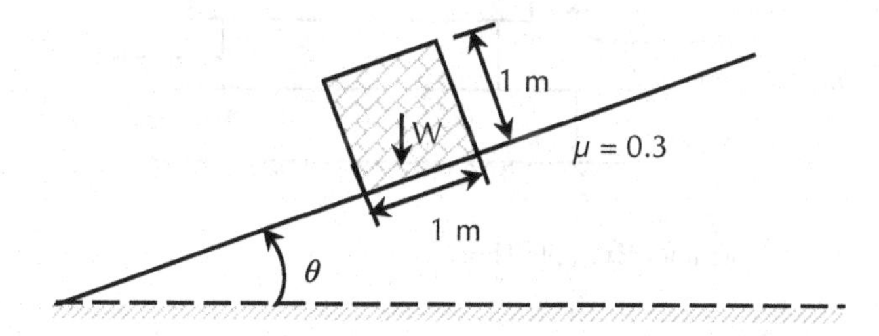

FIGURE 11.4 A body on an inclined plane for Example 11.2.

SOLUTION

Step 1. Draw the free-body diagram as shown in Figure 11.5

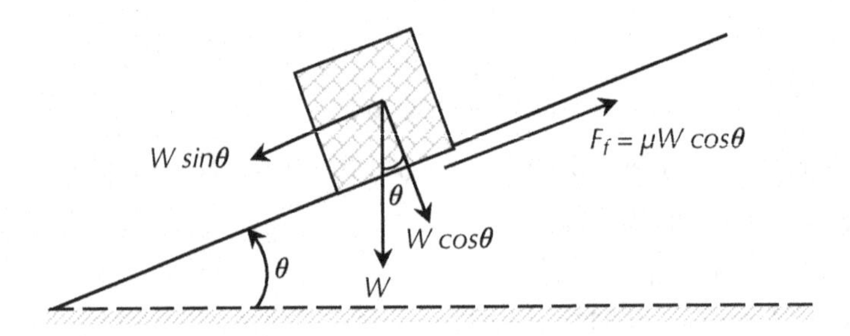

FIGURE 11.5 The free-body diagram of the body for Example 11.2.

Step 2. Apply the equilibrium equations

From the free-body diagram in Figure 11.5, normal reaction to the inclined surface, $R = W\cos\theta$

Friction force, $F_f = \mu R = \mu W\cos\theta = 0.3\ W\cos\theta$

Driving force, $F = W\sin\theta$

At the moment of sliding, Friction Force = Driving Force

$$0.3\ W\cos\theta = W\sin\theta$$
$$\tan\theta = 0.3$$
$$\theta = 16.7°$$

Answer: the block will intend to slide at 16.7°.

Example 11.3:

Three blocks are resting, as shown in Figure 11.6. The weights of block A, block B, and block C are 200 lb, 160 lb, and 100 lb, respectively. The coefficient of friction in any pair of surfaces is 0.30.

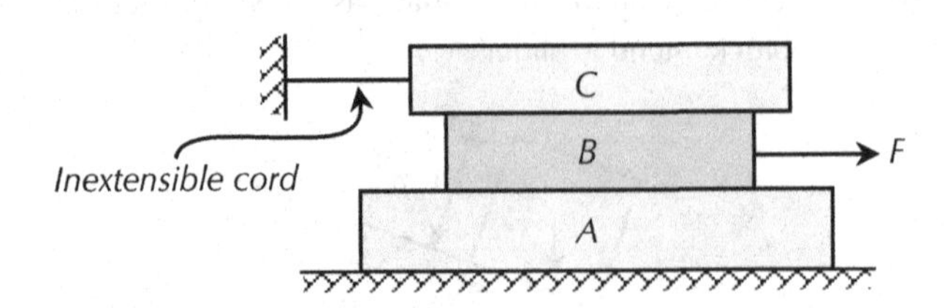

FIGURE 11.6 A block system for Example 11.3.

Determine the maximum force, F (lb) that can be applied.

SOLUTION

Step 1. Draw the free-body diagram.

Assume, blocks A and C remain stationary. Then from the free-body diagram shown in Figure 11.7:

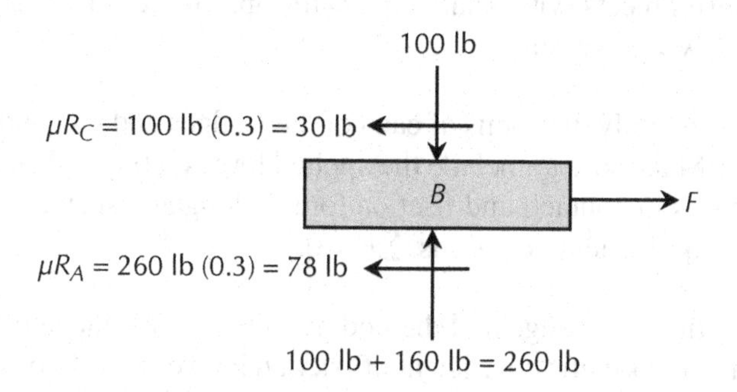

FIGURE 11.7 The free-body diagram of Block B for Example 11.3.

Step 2. Apply the equilibrium equations

$$\sum F_x = 0 \ (\rightarrow +ve)$$

$$F - 30 \text{ lb} - 78 \text{ lb} = 0$$

$$F = 108 \text{ lb}$$

Assume, blocks A and B move at constant speed. Then from the free-body diagram shown in Figure 11.8:

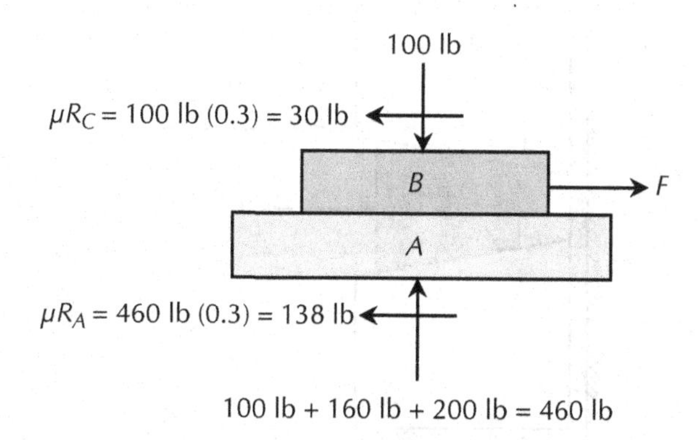

FIGURE 11.8 The free-body diagram of Blocks A and B for Example 11.3.

$$\sum F_x = 0 \ (\rightarrow +ve)$$

$$F - 30 \text{ lb} - 138 \text{ lb} = 0$$

$$F = 168 \text{ lb}$$

The lowest force (108 lb) controls the equilibrium.

Answer: 108 lb.

11.2 FRICTION IN WEDGES

A wedge-shaped object that is used to force two objects apart or to force one object away from a close surface is commonly referred to as a wedge. Wedges help to create very large normal forces to move objects with relatively small input forces. The following steps are followed to analyze a wedge system:

Step 1. Draw the free-body diagrams of each of the wedges and any bodies the wedge will be moving. Make sure to include the applied force on the wedge, normal forces along any surfaces in contact, and friction forces along any surfaces in contact.
Step 2. Apply the equilibrium equations, $\sum F = 0$.

It is usually assumed that the wedge and the bodies slide against one another, so each frictional force is equal to the kinetic coefficient of friction between the two surfaces times the associated normal force between the two forces, i.e., frictional force = $\mu_k R$. This reduces the number of unknowns and allows us to solve for any unknown problems. Work on the following solved examples to improve your understanding of friction.

Example 11.4:

As shown in Figure 11.9, the box, *A*, has a weight of 800 N, and the wedge, *B*, has a weight of 400 N. The coefficient of friction of any two surfaces is 0.20. What would the value of force *P* need to be to move block *A* upward? Blocks *A* and *C* are symmetric.

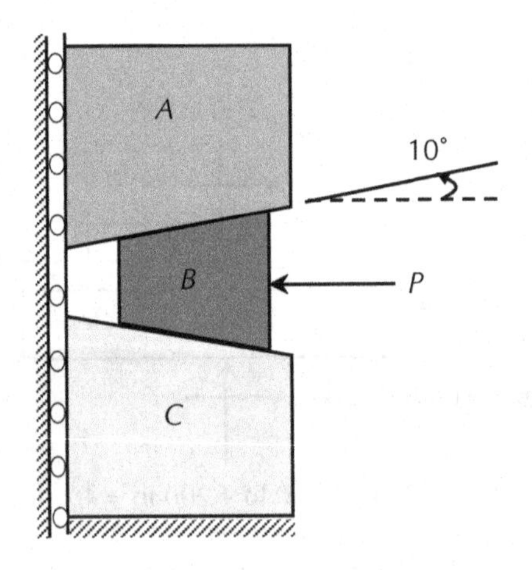

FIGURE 11.9 A wedge system for Example 11.4.

SOLUTION

Step 1. Draw the free-body diagram as shown in Figure 11.10:
Step 2. Apply the equilibrium equations

$$\sum F_y = 0 \ (\uparrow +ve)$$

$$R_A \cos 10 - 800 \text{ N} - \mu R_A \sin 10 = 0$$

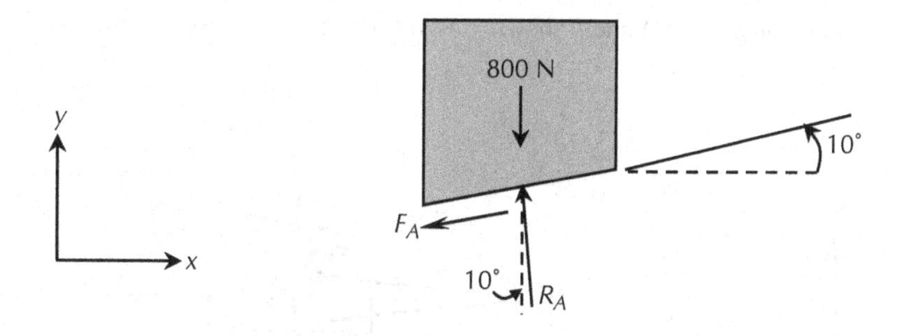

FIGURE 11.10 The free-body diagram of Wedge *A* for Example 11.4.

$$R_A \cos 10 - 800\text{ N} - 0.2\ R_A \sin 10 = 0$$
$$R_A \cos 10 - 0.2\ R_A \sin 10 = 800\text{ N}$$
$$0.95\ R_A = 800\text{ N}$$
$$R_A = 842\text{ N}$$

Friction force, $F_A = \mu R_A = 0.2\ (842\text{ N}) = 168.4\text{ N}$

From the free-body diagram shown in Figure 11.11:

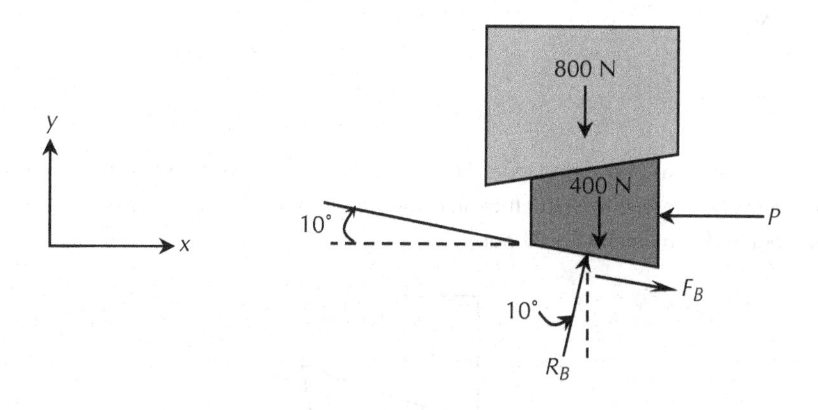

FIGURE 11.11 The free-body diagram of Wedges *A* and *B* for Example 11.4.

$$\sum F_y = 0\ (\uparrow +ve)$$

$$R_B \cos 10 - 800\text{ N} - 400\text{ N} - \mu R_B \sin 10 = 0$$
$$R_B \cos 10 - 800\text{ N} - 400\text{ N} - 0.2\ R_B \sin 10 = 0$$
$$R_B \cos 10 - 0.2\ R_B \sin 10 = 1{,}200\text{ N}$$
$$R_B = 1{,}263.1\text{ N}$$

Friction force, $F_B = \mu R_B = 0.2\ (1{,}263.1\text{ N}) = 252.6\text{ N}$

From the free-body diagram of A and B shown in Figure 11.12:

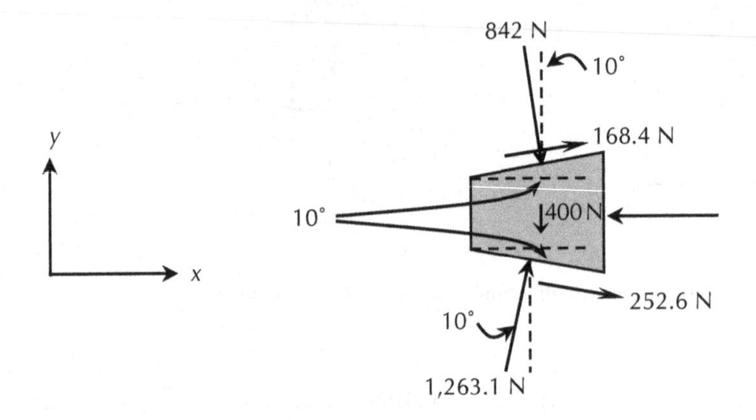

FIGURE 11.12 The free-body diagram of Wedge B for Example 11.4.

$$\sum F_x = 0\ (\leftarrow +ve)$$

$$-168.4\text{ N}\cos 10 - 252.6\text{ N}\cos 10 - 842\text{ N}\sin 10 - 1{,}263.1\text{ N}\sin 10 + P = 0$$

$$P = 780.2\text{ N}$$

Answer:

$P = 780.2$ N.

Example 11.5:

Two blocks are resting as shown in Figure 11.13. The weights of Blocks A and B are 400 N and 600 N, respectively. The coefficient of friction of any two surfaces is 0.20. Determine the amount of force (P) required to maintain equilibrium.

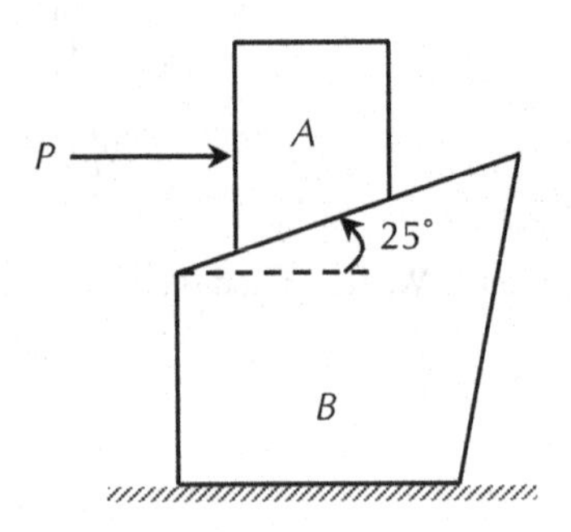

FIGURE 11.13 A wedge system for Example 11.5.

SOLUTION

Step 1. Draw the free-body diagram as shown in Figure 11.14: Block A will slide down. Force, P and the friction will resist the sliding.

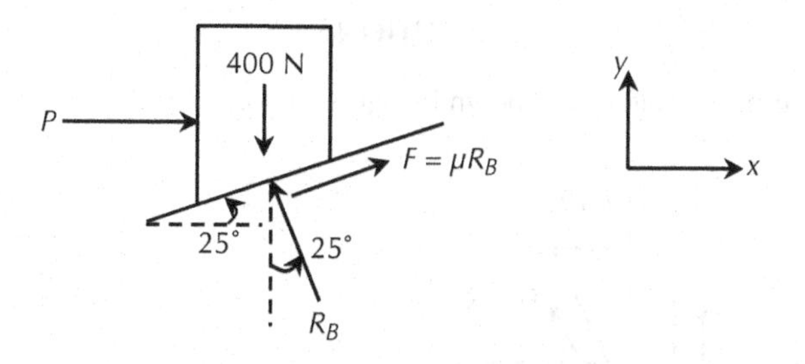

FIGURE 11.14 The free-body diagram of Wedge A for Example 11.5.

Step 2. Apply the equilibrium equations

$$\sum F_y = 0 \; (\uparrow +ve)$$

$$R_B \cos 25 - 400 \text{ N} + \mu R_B \sin 25 = 0$$
$$R_B \cos 25 - 400 \text{ N} + 0.2\, R_B \sin 25 = 0$$
$$R_B = 403.7 \text{ N}$$

Friction force, $F = =\mu R_B = 0.2\,(403.7 \text{ N}) = 80.7 \text{ N}$

$$\sum F_x = 0 \; (\leftarrow +ve)$$

$$R_B \sin 25 - P - F \cos 25 = 0$$
$$P = 403.7 \sin 25 - 80.7 \cos 25$$
$$P = 97.5 \text{ N}$$

Answer:

P = 97.5 N.

Example 11.6:

A 5-kN block is being pushed away from a rigid wall using a 20°-wedge as shown in Figure 11.15. The frictional coefficients (μ) for different surfaces are shown in the figure. Determine the pushing force required to create an impending motion.

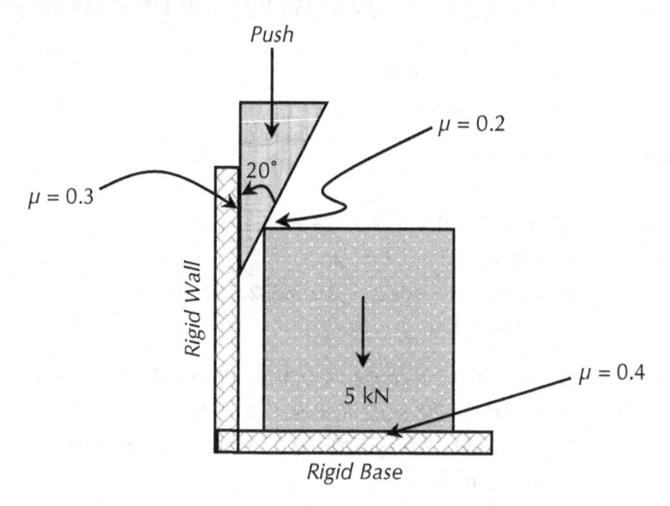

FIGURE 11.15 A wedge system for Example 11.6.

SOLUTION

Step 1. Draw the free-body diagram as shown in Figure 11.16:

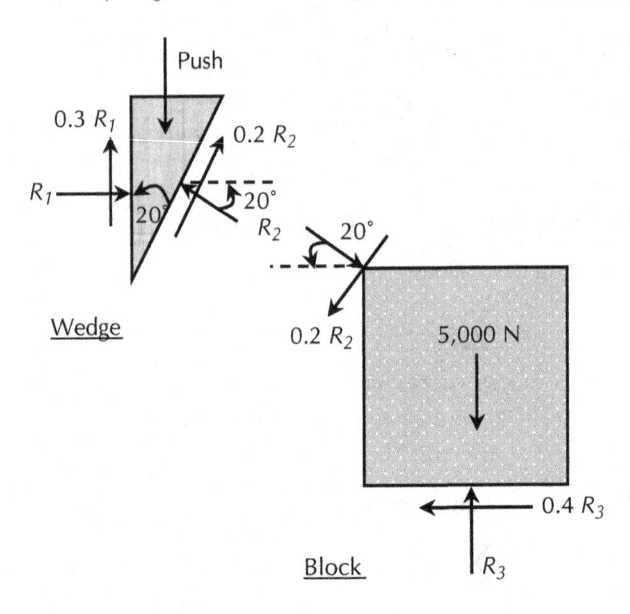

FIGURE 11.16 The free-body diagram of wedges for Example 11.6.

Step 2. Apply the equilibrium equations

Whatever we apply ($\Sigma F_y = 0$ or $\Sigma F_x = 0$) and wherever (block or wedge), there are more than one unknowns. Therefore, we need to develop and solve simultaneous equations. Let us start with the block:

$\Sigma F_x = 0$ ($\rightarrow$ +ve): $R_2 \cos 20 - 0.2\, R_2 \sin 20 - 0.4\, R_3 = 0$

$R_2 = 0.459\, R_3$

$\Sigma F_y = 0$ ($\uparrow$ +ve): $-R_2 \sin 20 - 0.2\, R_2 \cos 20 + R_3 - 5{,}000\text{ N} = 0$

$-(0.459\, R_3) \sin 20 - 0.2\,(0.459\, R_3) \cos 20 + R_3 - 5{,}000\text{ N} = 0$

$R_3 = 6{,}607.2\text{ N}$

Therefore, $R_2 = 0.459\, R_3$

$= 0.459\,(6{,}607.2\text{ N}) = 3{,}032.7\text{ N}$

Now, let us start with the wedge:

$\Sigma F_x = 0$ ($\rightarrow$ +ve): $-R_2 \cos 20 + 0.2\, R_2 \sin 20 + R_1 = 0$

$R_1 = R_2 \cos 20 - 0.2\, R_2 \sin 20$

$R_1 = 3{,}032.7\text{ N} \cos 20 - 0.2(3{,}032.7\text{ N}) \sin 20$

$R_1 = 2{,}642.4\text{ N}$

$\Sigma F_y = 0$ ($\rightarrow$ +ve): $R_2 \sin 20 + 0.2\, R_2 \cos 20 + 0.3 R_1 - \text{Push} = 0$

$\text{Push} = R_2 \sin 20 + 0.2\, R_2 \cos 20 + 0.3 R_1$

$\text{Push} = (3{,}032.7\text{ N}) \sin 20 + 0.2\,(3{,}032.7\text{ N}) \cos 20 + 0.3(2{,}642.4\text{ N})$

$\text{Push} = 2{,}400\text{ N}$

Answer:
Push = 2,400 N.

Example 11.7:

A wedge, *P*, is being used to lift the foundation of a brick column with the calculated load of 4,000 lb as shown in Figure 11.17. The portion *Q* of the foundation will remain stationary. The frictional coefficient of all the surfaces is 0.40. Determine the push force required to have impending motion of the column.

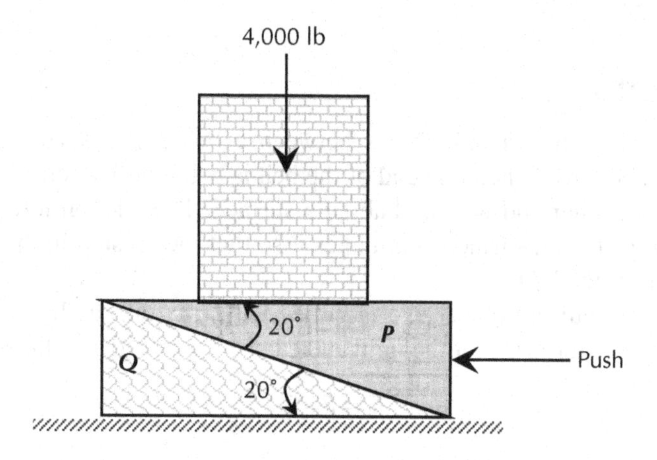

FIGURE 11.17 A wedge system for Example 11.7.

SOLUTION

Step 1. Draw the free-body diagram as shown in Figure 11.18:

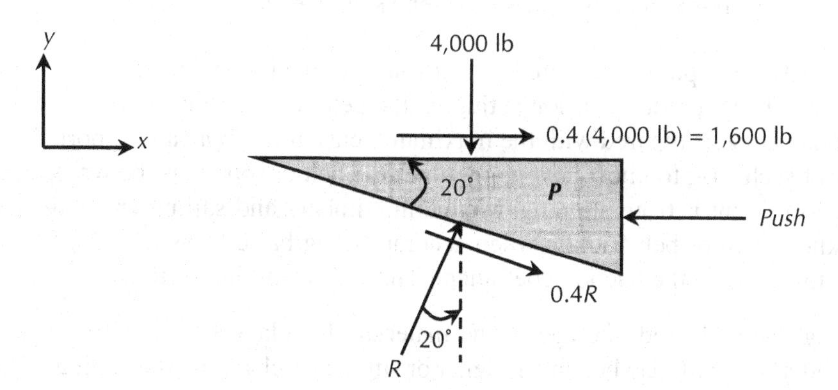

FIGURE 11.18 The free-body diagram of Wedge *P* for Example 11.7.

Step 2. Apply the equilibrium equations

$\Sigma F_y = 0$ ($\uparrow$ +ve)
$R \cos 20 - 0.4\, R \sin 20 - 4{,}000 \text{ lb} = 0$
$R \cos 20 - 0.4\, R \sin 20 = 4{,}000 \text{ lb}$
$R = 4{,}982 \text{ lb}$

$\Sigma Fx = 0$ ($\rightarrow$ +ve)

$R \sin 20 + 0.4\,R \cos 20 + 1{,}600$ lb – Push = 0

(4,982 lb) sin20 + 0.4(4,982 lb) cos20 + 1,600 lb – Push = 0

Push = (4,982 lb) sin20 + 0.4(4,982 lb) cos20 + 1,600 lb

Push = 5,176.6 lb

Answer:

Push = 5,176.6 lb.

11.3 BELT FRICTION

Belt friction describes the friction forces between a belt and a surface, such as a belt wrapped around a bollard. When one end of the belt is being pulled, only part of this force is transmitted to the other end wrapped about a surface. The friction force increases with the amount of wrap about a surface and makes it so that the tension in the belt can be different at both ends of the belt.

If a belt is used to pull an object over a pulley with friction, the belt forces at the two ends are different. The force (F_1) at the tight end can be expressed as follows (Figure 11.19):

$$F_1 = F_2 e^{\mu\theta}$$

where:

F_1 = the force being applied in the direction of the impending motion
F_2 = the force applied to resist the impending motion
μ = the coefficient of static friction, and
θ = the total angle of contact between the surfaces expressed in radian

As $F_1 > F_2$, the above equation is also written as: $T_{high} = T_{low} e^{\mu\theta}$

For the frictionless pulley, the forces at both sides of the pulley are equal, $F_1 = F_2$.

In practice, the theoretical tension acting on the belt or rope calculated by the belt friction equation can be compared with the maximum tension the belt can support. This helps a designer of such a rig to know how many times the belt or rope must be wrapped around the pulley to prevent it from slipping. Mountain climbers and sailing crews demonstrate standard knowledge of belt friction when accomplishing basic tasks. Certain factors help determine the value of the friction coefficient. These determining factors are:

- Belting material used: the age of the material also plays a part, where worn-out and older material may become rougher or smoother, changing the sliding friction.

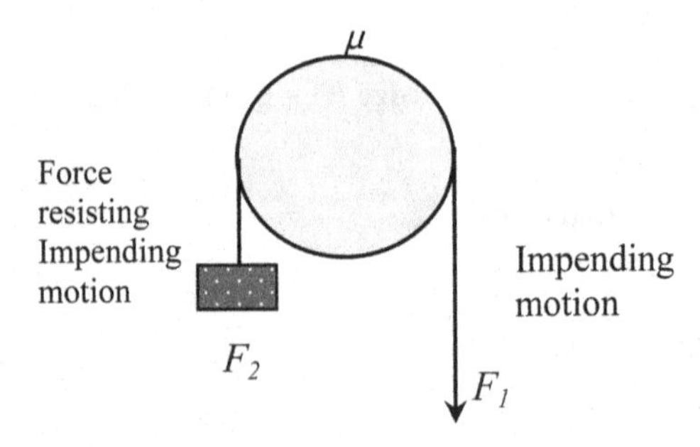

FIGURE 11.19 A pulley system.

- Construction of the drive-pulley system: this involves the strength and stability of the material used, as the pulley, and how greatly it will oppose the motion of the belt or rope.
- Conditions under which the belt and pulleys are operating: the friction between the belt and pulley may decrease substantially if the belt happens to be muddy or wet, as it may act as a lubricant between the surfaces. This also applies to extremely dry or warm conditions, which will evaporate any water naturally found in the belt, making friction greater.
- Overall design of the setup: the setup involves the initial conditions of the construction, such as the angle at which the belt is wrapped and the geometry of the belt and pulley system.

An understanding of belt friction is essential for sailing crews and mountain climbers. Their professions require being able to understand the amount of weight a rope with a specific tension capacity can hold versus the number of wraps around a pulley. Too many revolutions around a pulley make it inefficient to retract or release the rope, and too few may cause the rope to slip. Misjudging the ability of a rope and capstan system to maintain the proper frictional forces may lead to failure and injury.

Example 11.8:

Determine the minimum force, T, to lift the 150-lb man shown in Figure 11.20.

SOLUTION

To lift the man, the 'T' will need to be higher than the man's weight.

$T_{low} = 150$ lb

$$T_{high} = T_{low}e^{\mu\theta}$$
$$T_{high} = (150\text{ lb})e^{(0.2)(\pi)}$$
$$T_{high} = (150\text{ lb})(1.87)$$
$$T_{high} = 281\text{ lb}$$

Answer: 281 lb.

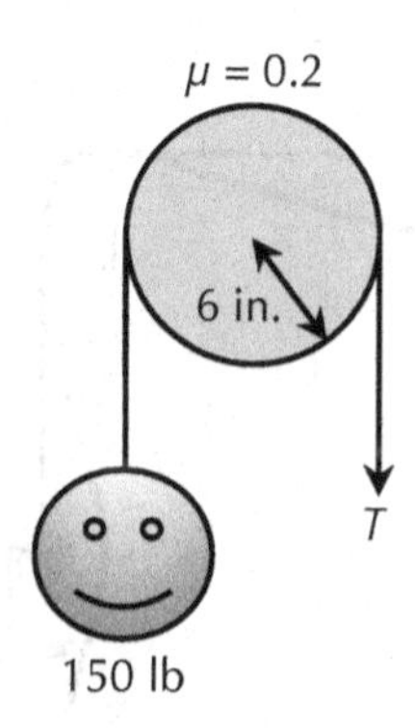

FIGURE 11.20 A pulley system for Example 11.8.

Example 11.9:

Determine the minimum force, T, to hold the 150-lb man in position shown in Figure 11.21.

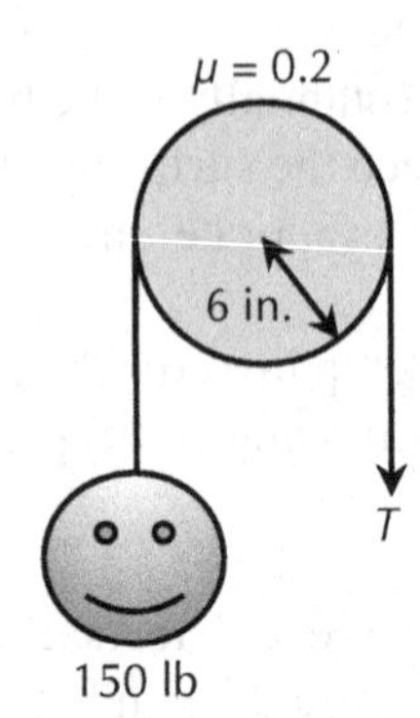

FIGURE 11.21 A pulley system for Example 11.9.

SOLUTION

To hold the man, the 'T' will need to be lower than the man's weight.

$T_{high} = 150$ lb

$T_{high} = T_{low} e^{\mu\theta}$

$T_{high} = T_{low} e^{(0.2)(\pi)}$

$150 \text{ lb} = T_{low}(1.87)$

$T_{low} = 150 \text{ lb}/1.87$

$T_{low} = 80$ lb

Answer: 80 lb.

Example 11.10:

A weight of 100 kg is to be lowered over the quarter-circled rock cliff with a constant velocity as shown in Figure 11.22. Determine the force required to do this operation.

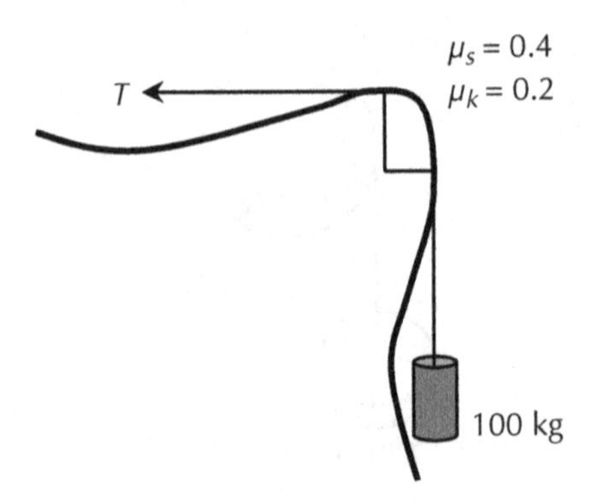

FIGURE 11.22 A pulley system for Example 11.10.

SOLUTION

$\mu = \mu_k = 0.2$ (as the weight will be in motion)
$\theta = 90° = \pi/2$
$T = T_{low}$ as the weight is to go down

$$T_{high} = T_{low}e^{\mu\theta}$$

$$(100\ \text{kg})\left(9.81\frac{\text{m}}{\text{sec}^2}\right) = T_{low}e^{(0.2)\left(\frac{\pi}{2}\right)}$$

$981\ \text{N} = T_{low}\,(1.37)$
$T_{low} = 981\ \text{N}/1.37$
$T_{low} = 716\ \text{N}$

Answer: 716 N.

Example 11.11:

Determine the minimum force, *T*, to lift the 100-kg man shown in Figure 11.23.

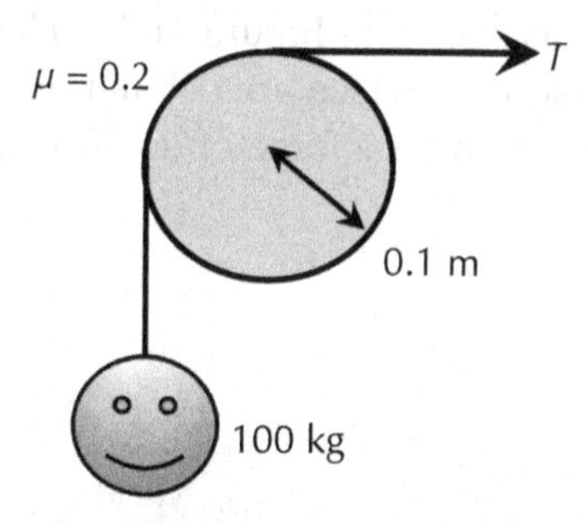

FIGURE 11.23 A pulley system for Example 11.11.

SOLUTION

To lift the man, the 'T' will need to be higher than the man's weight. T_{low} = 100 kg = 100 kg (9.81 m/sec²) = 981 N

$$T_{high} = T_{low}e^{\mu\theta}$$

$$T_{high} = (981\ \text{N})e^{(0.2)\left(\frac{\pi}{2}\right)}$$

$$T_{high} = 981\ \text{N}(1.369)$$

$$T_{high} = 1{,}343\ \text{N}$$

Answer: 1,343 N.

PRACTICE PROBLEMS

PROBLEM 11.1

The following recreational trailer has a mass of 2,000 kg and acts along the centerline of the tire as shown in Figure 11.24. How much pull force must be applied to start the motion? When it runs, how much applied force does it lose due to the friction?

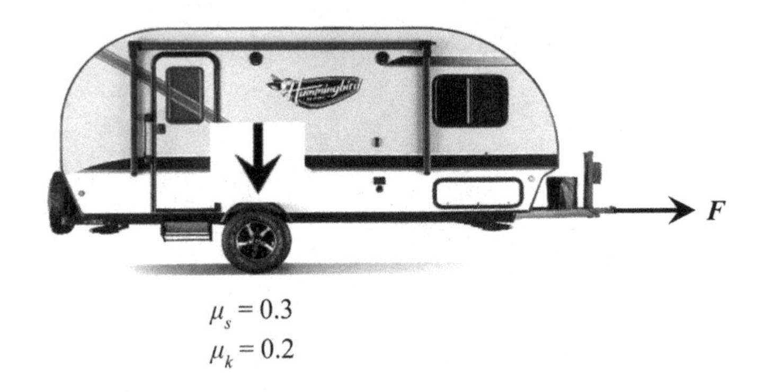

FIGURE 11.24 A recreational trailer for Problem 11.1.

PROBLEM 11.2

The mass of the crate is 100 kg as shown in Figure 11.25. The static and kinetic frictional coefficients between the crate and the surface are 0.4 and 0.3, respectively. Determine the frictional force and the state of the box if a force of (a) 200 N, and (b) 700 N is applied to the crate as shown below.

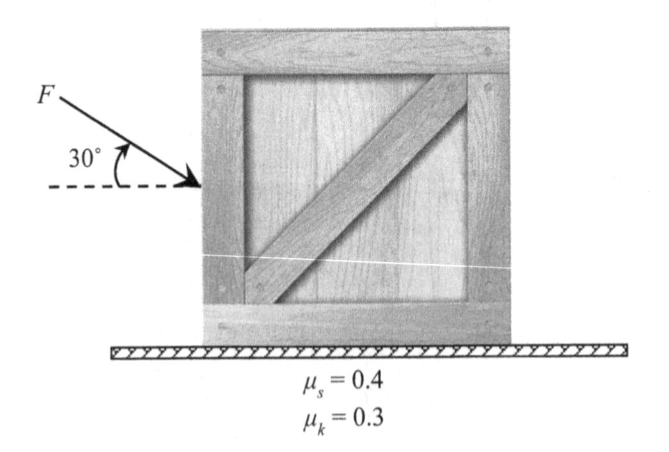

FIGURE 11.25 The crate to be pulled for Problem 11.2.

PROBLEM 11.3

Determine the maximum force T the connection can support so that no slipping occurs between the plates as shown in Figure 11.26. There is one bolt used for the connection and

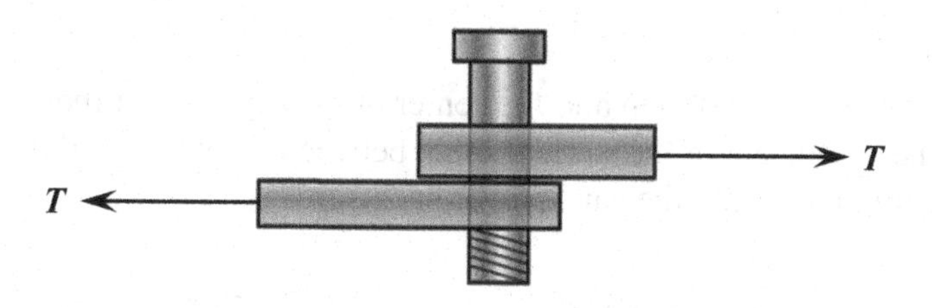

FIGURE 11.26 A single shear system Problem 11.3.

it is tightened so that it is subjected to a tension of 20 kip. The coefficient of the static friction between the plates is 0.3.

PROBLEM 11.4

Determine the maximum force, T, the connection can support so that no slipping occurs between the plates. There is one bolt used for the connection and each is tightened so that it is subjected to a tension of 20 kip as shown in Figure 11.27. The coefficient of the static friction between the plates is 0.3.

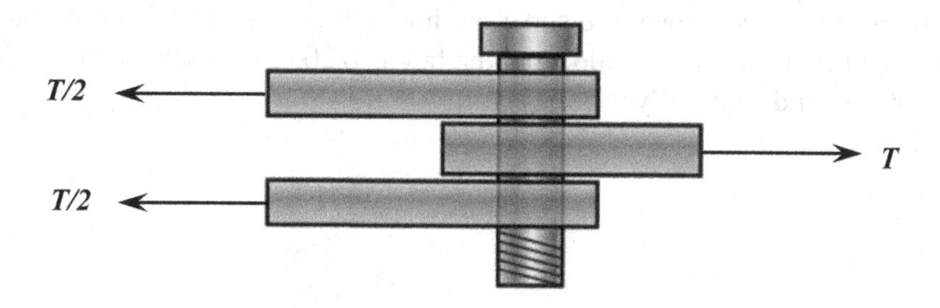

FIGURE 11.27 A double shear system Problem 11.4.

PROBLEM 11.5

A car has a total weight of 6,000 lb as shown in Figure 11.28. The coefficient of static friction between the wheel and the road surface is 0.7. Determine the maximum amount of towing force it can apply. Can this car tow another similar car of 8,000 lb with a similar frictional coefficient?

FIGURE 11.28 A car to be pulled for Problem 11.5.

PROBLEM 11.6

A car has a total weight of 7,000 lb and the center of gravity (c.g.) of the car is as shown in Figure 11.29. The coefficient of the static friction between the wheel and the road surface is 0.5. At what inclination would the car tip over backward?

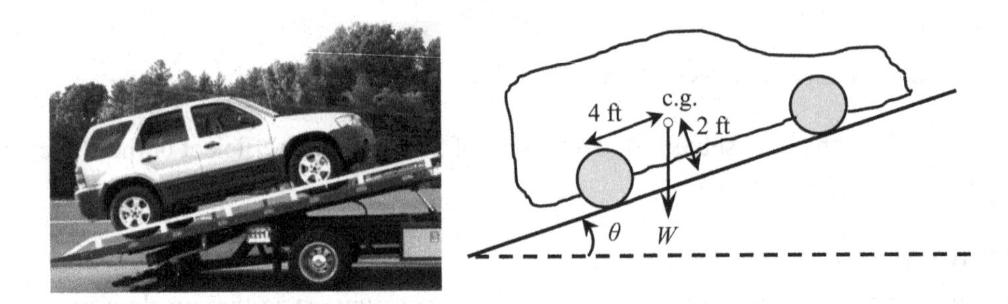

FIGURE 11.29 A car in an inclined plane for Problem 11.6.

PROBLEM 11.7

A 65-lb boy wants to slide down the slider as shown in Figure 11.30. The frictional coefficient between the surface of the slide and the boy is 0.20. Below which angle, θ, will the boy not be able to slide naturally?

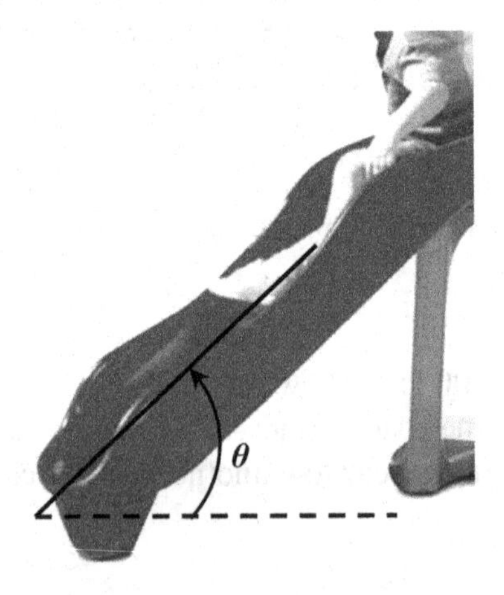

FIGURE 11.30 A sliding boy for Problem 11.7.

PROBLEM 11.8

A car has a total weight of 4,000 lb, and a dark dot in Figure 11.31 shows the center of gravity (c.g.) of the car. The coefficient of the static friction between the wheel and the road surface is 0.5. What are the factors of safety against the car tipping over and sliding?

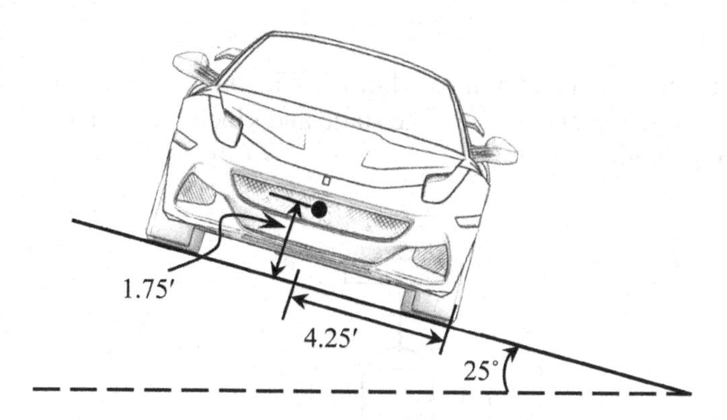

FIGURE 11.31 A car on an inclined plane for Problem 11.8.

PROBLEM 11.9

A car has a total weight of 4,000 lb, and a dark dot in Figure 11.32 shows the center of gravity (c.g.) of the car. The coefficient of static friction between the wheel and the road surface is 0.5. If the inclination continues to increase, which one will occur first, tipping over or sliding?

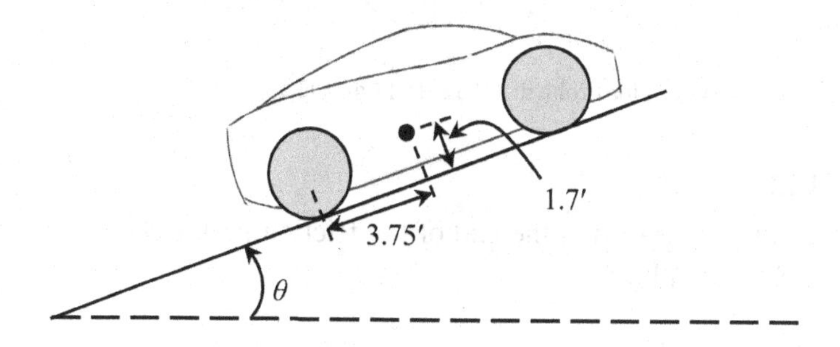

FIGURE 11.32 A car on an inclined plane for Problem 11.9.

PROBLEM 11.10

A car has a total weight of 4,500 lb, and a dark dot in Figure 11.33 shows the center of gravity (c.g.) of the car. The coefficient of static friction between the wheel and the road surface is 0.5. Determine the factors of safety against the car tipping over and sliding.

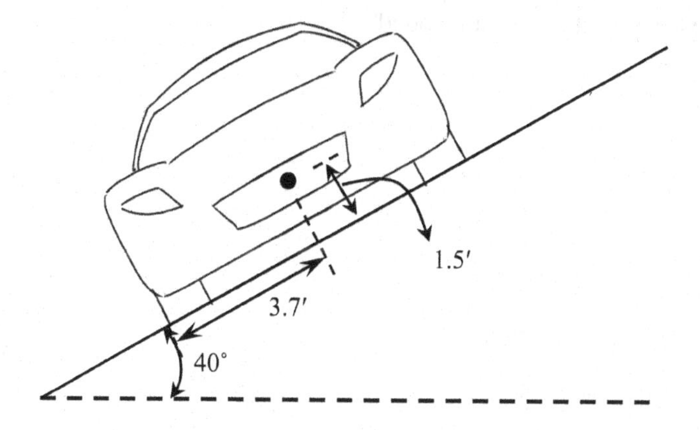

FIGURE 11.33 A car on an inclined plane for Problem 11.10.

PROBLEM 11.11

How much load is to be applied at the end of the friction pulley shown in Figure 11.34 to make the 100-kg object intend to rise? The static and kinetic frictional coefficients (μ_s and μ_k) are noted in the figure.

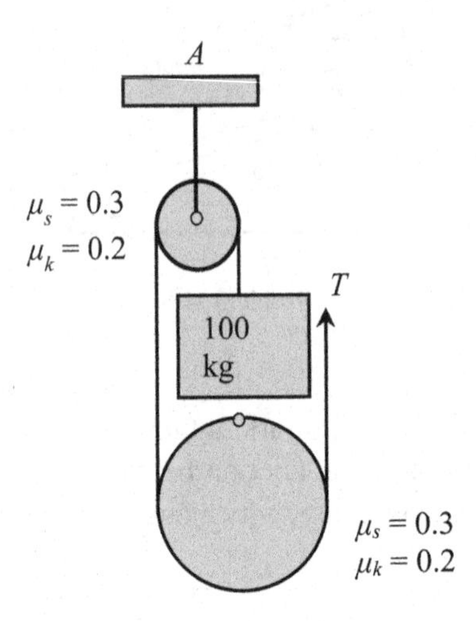

FIGURE 11.34 A pulley system for Problems 11.11, 11.12 and 11.13.

PROBLEM 11.12

How much load is to be applied at the end of the friction pulley shown in Figure 11.34 to hold the 100-kg object in place?

PROBLEM 11.13

How much load is to be applied at the end of the friction pulley shown in Figure 11.34 to lift the 100-kg object at a constant speed?

PROBLEM 11.14

A 30-lb girl is hanging using a pulley system, as shown in Figure 11.35. The static and kinetic frictional coefficients (μ_s and μ_k) are noted in the figure. How much force, T, is to be applied to raise the girl at a constant speed?

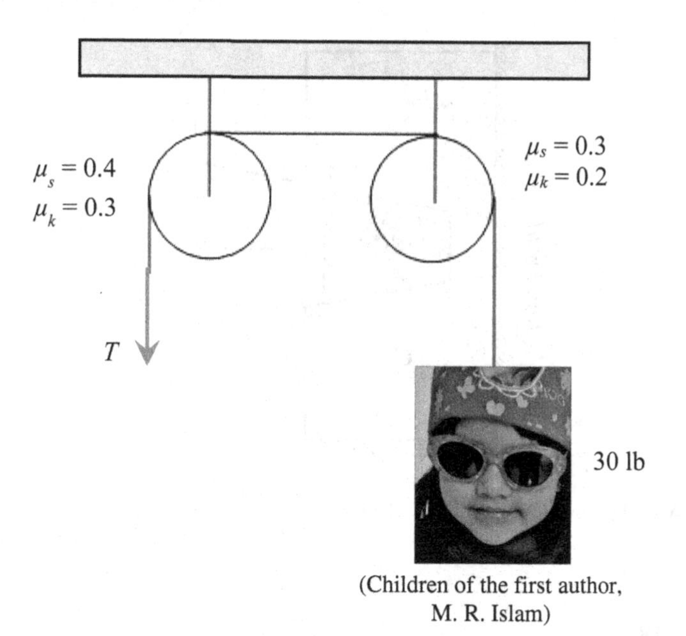

(Children of the first author, M. R. Islam)

FIGURE 11.35 A pulley system for Problem 11.14.

PROBLEM 11.15

How much force is to be applied to hold the girl in Problem 11.14 in place without letting her down?

PROBLEM 11.16

A clamp can apply the maximum of 300 lb force on a piece of timber as shown in Figure 11.36. If the static frictional coefficient between the jaw of the clamp and the wood piece is 0.4, how much tension force (F) can be applied to the timber?

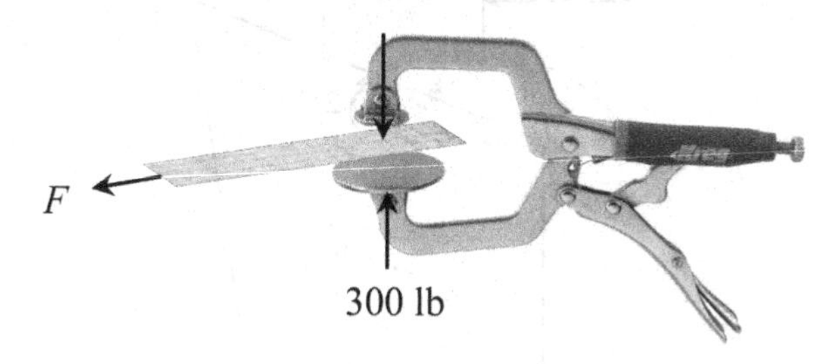

FIGURE 11.36 A clamping system for Problem 11.16.

PROBLEM 11.17

A box, A, weighs 800 N, and the wedge, B, weighs 400 N, as shown in Figure 11.37. The coefficient of friction of any two surfaces is 0.20. Determine the value of force, P, required to cause the impending motion of block A upward; blocks A and C are symmetric.

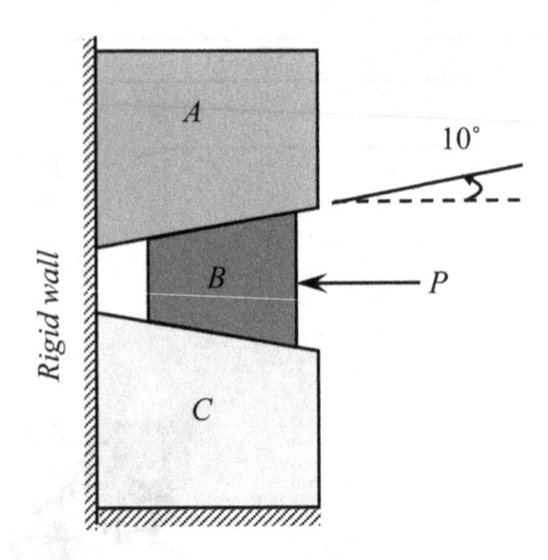

FIGURE 11.37 A wedge system for Problem 11.17.

PROBLEM 11.18

Three blocks, *A*, *B*, and *C*, are resting, as shown in Figure 11.38. The weights of blocks *A*, *B*, and *C* are 400 N, 300 N, and 600 N, respectively. The coefficient of friction (μ) is shown in the figure. Block *A* is restrained from horizontal movement. Determine the amount of force (P) required to maintain the equilibrium.

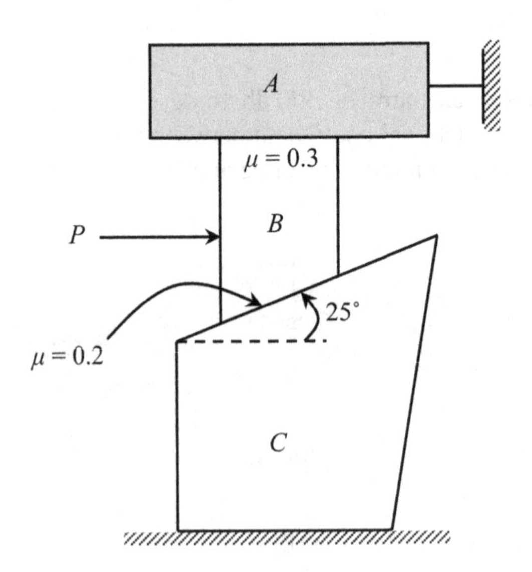

FIGURE 11.38 A wedge system for Problem 11.18.

PROBLEM 11.19

A 5-kN block is being raised against a rigid wall using a 30°-wedge as shown in Figure 11.39. The frictional coefficients (μ) for different surfaces are shown in the figure. Determine the pushing force required to create an impending motion.

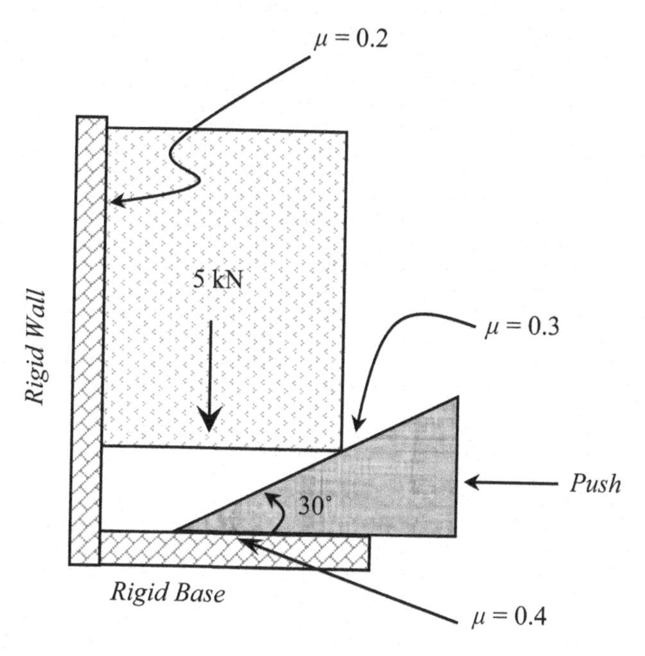

FIGURE 11.39 A wedge system for Problem 11.19.

Index

Q

R

S

T

U

V

W

X

Y

Z

Printed in the United States
By Bookmasters